Eugen Netto

The Theory of Substitution and its Applications to Algebra

Rev. by the author and translated with his permission by F.N. Cole

Eugen Netto

The Theory of Substitution and its Applications to Algebra

Rev. by the author and translated with his permission by F.N. Cole

ISBN/EAN: 9783337189266

Printed in Europe, USA, Canada, Australia, Japan

Cover: Foto ©Andreas Hilbeck / pixelio.de

More available books at **www.hansebooks.com**

THEORY OF SUBSTITUTIONS

AND ITS

APPLICATIONS TO ALGEBRA.

By DR. EUGEN NETTO,
PROFESSOR OF MATHEMATICS IN THE UNIVERSITY OF GIESSEN.

Revised by the Author and Translated with his Permission
BY F. N. COLE, PH. D.,
Assistant Professor of Mathematics in the
University of Michigan.

ANN ARBOR, MICH.:
THE REGISTER PUBLISHING COMPANY.
The Inland Press.
1892.

PREFACE.

The presentation of the Theory of Substitutions here given differs in several essential features from that which has heretofore been customary. It will accordingly be proper in this place to state in brief the guiding principles adopted in the present work.

It is unquestionable that the sphere of application of an Algorithm is extended by eliminating from its fundamental principles and its general structure all matters and suppositions not absolutely essential to its nature, and that through the general character of the objects with which it deals, the possibility of its employment in the most varied directions is secured. That the theory of the construction of groups admits of such a treatment is a guarantee for its far-reaching importance and for its future.

If, on the other hand, it is a question of the application of an auxiliary method to a definitely prescribed and limited problem, the elaboration of the method will also have to take into account only this one purpose. The exclusion of all superfluous elements and the increased usefulness of the method is a sufficient compensation for the lacking, but not defective, generality. A greater efficiency is attained in a smaller sphere of action.

The following treatment is calculated solely to introduce in an elementary manner an important auxiliary method for algebraic investigations. By the employment of integral functions from the outset, it is not only possible to give to the Theory of Substitutions, this operating with operations, a concrete and readily comprehended foundation, but also in many cases to simplify the demonstrations, to give the various conceptions which arise a precise form, to define sharply the principal question, and—what does not appear to be least important—to limit the extent of the work.

The two comprehensive treatises on the Theory of Substitutions which have thus far appeared are those of J. A. Serret and of C. Jordan.

The fourth section of the "Algèbre Supérieure" of Serret is devoted to this subject. The radical difference of the methods involved here and there hardly permitted an employment of this highly deserving work for our purposes. Otherwise with the more extensive work of Jordan, the "Traité des substitutions et des équations algébriques." Not only the new fundamental ideas were taken from this book, but it is proper to mention expressly here that many of its proofs and pro-

PREFACE.

The presentation of the Theory of Substitutions here given differs in several essential features from that which has heretofore been customary. It will accordingly be proper in this place to state in brief the guiding principles adopted in the present work.

It is unquestionable that the sphere of application of an Algorithm is extended by eliminating from its fundamental principles and its general structure all matters and suppositions not absolutely essential to its nature, and that through the general character of the objects with which it deals, the possibility of its employment in the most varied directions is secured. That the theory of the construction of groups admits of such a treatment is a guarantee for its far-reaching importance and for its future.

If, on the other hand, it is a question of the application of an auxiliary method to a definitely prescribed and limited problem, the elaboration of the method will also have to take into account only this one purpose. The exclusion of all superfluous elements and the increased usefulness of the method is a sufficient compensation for the lacking, but not defective, generality. A greater efficiency is attained in a smaller sphere of action.

The following treatment is calculated solely to introduce in an elementary manner an important auxiliary method for algebraic investigations. By the employment of integral functions from the outset, it is not only possible to give to the Theory of Substitutions, this operating with operations, a concrete and readily comprehended foundation, but also in many cases to simplify the demonstrations, to give the various conceptions which arise a precise form, to define sharply the principal question, and—what does not appear to be least important—to limit the extent of the work.

The two comprehensive treatises on the Theory of Substitutions which have thus far appeared are those of J. A. Serret and of C. Jordan.

The fourth section of the "Algèbre Supérieure" of Serret is devoted to this subject. The radical difference of the methods involved here and there hardly permitted an employment of this highly deserving work for our purposes. Otherwise with the more extensive work of Jordan, the "Traité des substitutions et des équations algébriques." Not only the new fundamental ideas were taken from this book, but it is proper to mention expressly here that many of its proofs and processes of thought also permitted of being satisfactorily employed in the present work in spite of the essential difference of the general treatment. The investigations of Jordan not contained in the "Traité" which have been consulted are cited in the appropriate places.

But while many single particulars are traceable to this "Traité" and to these investigations, nevertheless, the author is indebted to his honored teacher, L. Kronecker, for the ideas which lie at the foundation of his entire work. He has striven to employ to best advantage the benefit which he has derived from the lectures and from the study of the works of this scholarly man, and from the inspiring personal intercourse with him; and he hopes that traces of this influence may appear in many places in his work. One thing he regrets: that the recent important publication of Kronecker, "Grundzüge einer arithmetischen Theorie der algebraischen Grössen," appeared too late for him to derive from it the benefit which he would have wished for himself and his readers.

The plan of the present book is as follows:

In the first part the leading principles of the theory of substitutions are deduced with constant regard to the theory of the integral functions; the analytical treatment retires almost wholly to the background, being employed only at a late stage in reference to the groups of solvable equations.

In the second part, after the establishment of a few fundamental principles, the equations of the second, third and fourth degrees, the Abelian and the Galois equations are discussed as examples. After this follows a chapter devoted to an arithmetical discussion the necessity of which is there explained. Finally the more general, but still elementary questions with regard to solvable equations are examined.

STRASSBURG, 1880.

To the preceding I have now to add that the present translation differs from the German edition in many important particulars. Many new investigations have been added. Others, formerly included, which have shown themselves to be of inferior importance, have been omitted. Entire chapters have been rearranged and demonstrations simplified. In short, the whole material which has accumulated in the course of time since the first appearance of the book is now turned to account.

In conclusion the author desires to express his warmest thanks to Mr. F. N. Cole who has disinterestedly assumed the task of translation and performed it with care and skill.

GIESSEN, 1892.

EUGEN NETTO.

TRANSLATOR'S NOTE.

The translator has confined himself almost exclusively to the function of rendering the German into respectable English. My thanks are especially due to The Register Publishing Company for their generous assumption of the expense of publication and to Mr. C. N. Jones, of Milwaukee, for valuable assistance while the book was passing through the press.

F. N. COLE.

Ann Arbor, February 27, 1892.

TABLE OF CONTENTS.

PART I.

Theory of Substitutions and of Integral Functions.

CHAPTER I.

Symmetric or Single-Valued Functions—Alternating and Two-Valued Functions.

1-3.	Symmetric and single-valued functions.
4.	Elementary symmetric functions.
5-10.	Treatment of the symmetric functions.
11.	Discriminants.
12.	Euler's formula.
13.	Two-valued functions; substitutions.
14.	Decomposition of substitutions into transpositions.
15.	Alternating functions.
16-20.	Treatment and group of the two-valued functions.

CHAPTER II.

Multiple-Valued Functions and Groups of Substitutions.

22.	Notation for substitutions.
24.	Their number.
25.	Their applications to functions.
26-27.	Products of substitutions.
28.	Groups of substitutions.
29-32.	Correlation of function and group.
34.	Symmetric group.
35.	Alternating group.
36-38.	Construction of simple groups.
39-40.	Group of order p^f.

CHAPTER III.

The Different Values of a Multiple-Valued Function and Their Algebraic Relation to One Another.

41-44.	Relation of the order of a group to the number of values of the corresponding function.

45.	Groups belonging to the different values of a function.
46–47.	Transformation.
48–50.	The Cauchy-Sylow Theorem.
51.	Distribution of the elements in the cycles of a group.
52.	Substitutions which belong to all values of a function.
53.	Equation for a p-valued function.
55.	Discriminants of the functions of a group.
56–59.	Multiple-valued functions, powers of which are single-valued.

CHAPTER IV.

TRANSITIVITY AND PRIMITIVITY.—SIMPLE AND COMPOUND GROUPS.—ISOMORPHISM.

60–61.	Simple transitivity.
62–63.	Multiple transitivity.
64.	Primitivity and non-primitivity.
65–67.	Non-primitive groups.
68.	Transitive properties of groups.
69–71.	Commutative substitutions; self-conjugate subgroups.
72–73.	Isomorphism.
74–76.	Substitutions which affect all the elements.
77–80.	Limits of transitivity.
81–85.	Transitivity of primitive groups.
86.	Quotient groups.
87.	Series of composition.
88–89.	Constant character of the factors of composition.
91.	Construction of compound groups.
92.	The alternating group is simple.
93.	Groups of order p^a.
94.	Principal series of composition.
95.	The factors of composition equal prime numbers.
96.	Isomorphism.
97–98.	The degree and order equal.
99–101.	Construction of isomorphic groups.

CHAPTER V.

ALGEBRAIC RELATIONS BETWEEN FUNCTIONS BELONGING TO THE SAME GROUP.—FAMILIES OF MULTIPLE-VALUED FUNCTIONS.

103–105.	Functions belonging to the same group can be rationally expressed one in terms of another.
106.	Families; conjugate families.
107.	Subordinate families.

108–109.	Expression of the principal functions in terms of the subordinate.
110.	The resulting equation binomial.
111.	Functions of the family with non-vanishing discriminant.

CHAPTER VI.

THE NUMBER OF THE VALUES OF INTEGRAL FUNCTIONS.

112.	Special cases.
113.	Change in the form of the question.
114–115.	Functions whose number of values is less than their degree.
116.	Intransitive and non-primitive groups.
117–121.	Groups with substitutions of four elements.
122–127.	General theorem of C. Jordan.

CHAPTER VII.

CERTAIN SPECIAL CLASSES OF GROUPS.

128.	Preliminary theorem.
129.	Groups Ω with $r = n = p$. Cyclical groups.
130.	Groups Ω with $r = n = p \cdot q$.
131.	Groups Ω with $r = n = p^2$.
132–135.	Groups which leave, at the most, one element unchanged.— Metacyclic and semi-metacyclic groups.
136.	Linear fractional substitutions. Group of the modular equations.
137–139.	Groups of commutative substitutions.

CHAPTER VIII.

ANALYTICAL REPRESENTATION OF SUBSTITUTIONS.—THE LINEAR GROUP.

140.	The analytical representation.
141.	Condition for the defining function.
143.	Arithmetic substitutions.
144.	Geometric substitutions.
145.	Condition among the constants of a geometric substitution.
146–147.	Order of the linear group.

PART II.

APPLICATION OF THE THEORY OF SUBSTITUTIONS TO THE ALGEBRAIC EQUATIONS.

CHAPTER IX.

THE EQUATIONS OF THE SECOND, THIRD AND FOURTH DEGREES.—GROUP OF AN EQUATION.—RESOLVENTS.

148.	The equations of the second degree.

	CONTENTS.
149.	The equations of the third degree.
150.	The equations of the fourth degree.
152.	The general problem formulated. Galois resolvents.
153-154.	Affect equations. Group of an equation.
156.	Fundamental theorems on the group of an equation.
157.	Group of the Galois resolvent equation.
158-159.	General resolvents.

CHAPTER X.

THE CYCLOTOMIC EQUATIONS.

161.	Definition and irreducibility.
162.	Solution of cyclic equations.
163.	Investigation of the operations involved.
164-165.	Special resolvents.
166.	Construction of regular polygons by ruler and compass.
167.	The regular pentagon.
168.	The regular heptadecagon.
169-170.	Decomposition of the cyclic polynomial.

CHAPTER XI.

THE ABELIAN EQUATIONS.

171-172.	One root of an equation a rational function of another.
173.	Construction of a resolvent.
174-175.	Solution of the simplest Abelian equations.
176.	Employment of special resolvents for the solution.
177.	Second method of solution.
178-180.	Examples.
181.	Abelian equations. Their solvability.
182.	Their group.
183.	Solution of the Abelian equations; first method.
184-186.	Second method.
187.	Analytical representation of the groups of primitive Abelian equations.
188-189.	Examples.

CHAPTER XII.

EQUATIONS WITH RATIONAL RELATIONS BETWEEN THREE ROOTS.

190-193.	Groups analogous to the Abelian groups.
194.	Equations all the roots of which are rational functions of two among them.
196.	Their group in the case $n = p$.
197.	The binomial equations.

199.	Triad equations.
200–201.	Constructions of compound triad equations.
202.	Group of the triad equation for $n=7$.
203–205.	Group of the triad equation for $n=9$.
206.	Hessian equation of the ninth degree.

CHAPTER XIII.

THE ALGEBRAIC SOLUTION OF EQUATIONS.

207–209.	Rational domain. Algebraic functions.
210–211.	Preliminary theorem.
212–216.	Roots of solvable equations.
217.	Impossibility of the solution of general equations of higher degrees.
218.	Representation of the roots of a solvable equation.
219.	The equation which is satisfied by any algebraic expression.
220–221.	Changes of the roots of unity which occur in the expressions for the roots.
222–224.	Solvable equations of prime degree.

CHAPTER XIV.

THE GROUP OF AN ALGEBRAIC EQUATION.

226.	Definition of the group.
227.	Its transitivity.
228.	Its primitivity.
229.	Galois resolvents of general and special equations.
230.	Composition of the group.
231.	Resolvents.
232–234.	Reduction of the solution of a compound equation.
235.	Decomposition of the equation into rational factors.
236–238.	Adjunction of the roots of a second equation.

CHAPTER XV.

ALGEBRAICALLY SOLVABLE EQUATIONS.

239–241.	Criteria for solvability.
242.	Applications.
243.	Abel's theorem on the decomposition of solvable equations.
244.	Equations of degree p^k; their group.
245.	Solvable equations of degree p.
246.	Solvable equations of degree p^2.
248–249.	Expression of all the roots in terms of a certain number of them.

ERRATA.

p. 7, footnote, for transformatione read transmutatione.
p. 15, line 10, read $\psi = S\sqrt{\Delta}$.
p. 16, line 5, read $\varphi_1 - \varphi_2 = 2S_2\sqrt{\Delta}$.
p. 28, line 9, for φ read Φ.
p. 29, line 12, for Ψ read ψ.
p. 29, line 9, from bottom, for ψ read Ψ.
p. 31, line 8, read $G = [1, (x_1x_2)(x_3x_4), (x_1x_3)(x_2x_4), (x_1x_4)(x_2x_3)]$.
p. 41, line 7, for p read p'.
p. 52, line 13, read $a\varphi_\sigma + b\psi_\tau$.
p. 52, line 5, from bottom, for $\dfrac{d}{d_a}$ read $\dfrac{h_1}{d_a}$.
p. 89, line 2, for *not more* read *less*.
p. 93, line 9, for a group H read a primitive group H.
p. 94, line 2, for $n-q+2$ read $n-q+k$.
p. 98, line 19, for $\mathrm{S}_a\mathrm{S}'_b\mathrm{S}_c$ read $\mathrm{S}_a\mathrm{S}'_b\sigma_c$.
p. 101, line 3, for Ω_2 read Ω'_1.
p. 103, line 14, read $[1, (z_1z_2)]$.
p. 125, THEOREM XI, read: *In order that there may be a $p\rho$-valued function χ a prime power χ^p of which shall have ρ values*, etc.
p. 159, lines 10, 11 from bottom, read: Since σ_1 belongs to r_1, at least one of the exponents $\mu, \nu, \ldots$, the s of which belongs to r_1, must be prime to r_1.
p. 166, line 4, read $\tau_2, t_2\tau_2, t_3\tau_2, \ldots$
p. 174, line 2, for c_1 read $2c_1$.
p. 210, foot note, for No. XI, etc., read 478–507, edition of Sylow and Lie.
p. 219, line 10, for $_1(cos\ a)$ read $\theta_1(cos\ a)$.
p. 224, line 2 from bottom, read: which leaves two elements with successive indices unchanged.
p. 248, line 3 from bottom, read $V_{a+1}^{pa+1} = F_{a+1}(V_{a+2}, \ldots)$.

PART I.

THEORY OF SUBSTITUTIONS AND OF THE INTEGRAL FUNCTIONS.

CHAPTER I.

SYMMETRIC OR SINGLE-VALUED FUNCTIONS. ALTERNATING AND TWO-VALUED FUNCTIONS.

§ 1. In the present investigations we have to deal with n elements $x_1, x_2, \ldots x_n$, which are to be regarded throughout as entirely independent quantities, unless the contrary is expressly stated. It is easy to construct integral functions of these elements which are unchanged in *form* when the x_λ's are permuted or interchanged in any way. For example the following functions are of this kind:

$$x_1^a + x_2^a + x_3^a + \ldots + x_n^a,$$
$$x_1^a x_2^\beta + x_1^a x_3^\beta + \ldots + x_2^a x_1^\beta + x_2^a x_3^\beta + \ldots$$
$$+ x_n^a x_1^\beta \ldots + x_n^a x_{n-1}^\beta,$$
$$(x_1 - x_2)^2 (x_1 - x_3)^2 (x_2 - x_3)^2 \ldots (x_{n-1} - x_n)^2,$$
etc.

Such functions are called *symmetric* functions. We confine ourselves, unless otherwise noted, to the case of integral functions.

If the x_λ's be put equal to any arbitrary quantities, $a_1, a_2, \ldots a_n$, so that $x_1 = a_1, x_2 = a_2, \ldots x_n = a_n$, it is clear that the symmetric functions of the x_λ's will be unchanged not only in *form*, but also in *value* by any change in the order of assignment of the values a_λ to the x_λ's. Such a reassignment may be denoted by

$$x_1 = a_{i_1}, x_2 = a_{i_2}, \ldots x_n = a_{i_n}$$

where the $a_{i_1}, a_{i_2}, \ldots$ denote the same quantities $a_1, a_2, \ldots$ in any one of the possible $n!$ orders.

Conversely, it can be shown that every integral function, $\varphi(x_1, x_2 \ldots x_n)$, of n independent quantities $x_1, x_2, \ldots x_n$, which is unchanged in value by all the possible permutations of arbitrary values of the x_λ's, is also unchanged in form by these permutations.

Theorem I. *Every single-valued integral function of n independent elements $x_1, x_2, \ldots x_n$ is symmetric in these elements.*

§ 2. The reasoning on which the proof of this theorem is based will be of frequent application in the following treatment. It seems, therefore, desirable to present it here in full detail.

(A). If in the integral function

(1) $$f(x) = \sum_{\lambda=0}^{\lambda=n} a_\lambda x^\lambda$$

all the coefficients a_λ are equal to zero, then $f(x)$ vanishes identically, *i. e.*, $f(x)$ is equal to zero for every value of x. Conversely, if $f(x)$ vanishes for every value of x, then all the coefficients a_λ are equal to zero.

For if $f(x)$ is not identically zero, then there is a value ζ_0 such that for every real x of which the absolute value $|x|$ is greater than ζ_0, the value of the function $f(x)$ is different from zero. For ζ_0 we may take the highest of the absolute values of the several roots of the equation $f(x) = 0$. Without assuming the existence of roots of algebraic equations, we may also obtain a value of ζ_0 as follows:*

Let a_k be the numerically greatest of the n coefficients $a_0, a_1, \ldots a_{n-1}$ in (1), and denote $\dfrac{|a_k| + |a_n|}{|a_n|}$ by r. We have then

$$\left| \frac{a_{n-1} x^{n-1} + a_{n-2} x^{n-2} + \ldots + a_0}{a_n} \right| \leq \left| \frac{a_k}{a_n} \right| (|x|^{n-1} + |x|^{n-2} + \ldots + 1)$$

$$\leq \left| \frac{a_k}{a_n} \right| \cdot \frac{|x|^n - 1}{|x| - 1}$$

$$\leq |(r-1)| \frac{|x|^n - 1}{|x| - 1}$$

Hence, for any value of x not lying between $-r$ and $+r$,

$$\left| \frac{a_{n-1} x^{n-1} + a_{n-2} x^{n-2} + \ldots + a_0}{a_n} \right| < |x|^n$$

$$|a_{n-1} x^{n-1} + a_{n-2} x^{n-2} + \ldots + a_0| < |a_n x^n|$$

so that the sign of $f(x)$ is the same as that of $a_n x^n$. Consequently, we may take $\zeta_0 = r$.

---

*L. Kronecker. Crelle 101. p. 347.

(B). If no two of the integral functions
(2) $$f_1(x), f_2(x), \ldots f_m(x)$$
are identically equal to each other, then there is always a quantity ζ_0 such that for every x the absolute value of which is greater than ζ_0, the values of the functions (2) are different from each other.

For, if we denote by $r_{\alpha\beta}$ the value determined for the function $f_\alpha(x) - f_\beta(x)$, as r was determined in (A), we may take for ζ_0 the greatest of the quantities $r_{\alpha\beta}$.

(C). If in the integral function
$$f(x_1, x_2, \ldots x_n) = \Sigma\, a_{\lambda_1 \lambda_2 \ldots \lambda_m}\, x_1^{\lambda_1} x_2^{\lambda_2} \ldots x_m^{\lambda_m}$$
all the coefficients a are equal to zero, then the function f vanishes identically, i. e., the value of f is equal to zero for every system of values of $x_1, x_2, \ldots x_n$.

To prove the converse proposition we put
(3) $$x_1 = g, \quad x_2 = g^\nu, \quad x_3 = g^{\nu^2}, \quad \ldots x_n = g^{\nu^{m-1}}$$
$x_1^{\lambda_1} x_2^{\lambda_2} \ldots x_m^{\lambda_m}$ then becomes a power of g, the exponent of which is
$$r_{\lambda_1 \lambda_2 \ldots \lambda_m} = \lambda_1 + \lambda_2 \nu + \lambda_3 \nu^2 + \ldots \lambda_m \nu^{m-1}$$

From (B), we can find a value for ν such that for all greater values of ν, the various $r_{\lambda_1 \lambda_2 \ldots \lambda_m}$ are all different from one another. We have then
$$f(x_1, x_2, \ldots x_m) = \Sigma\, a_{\lambda_1 \lambda_2 \ldots \lambda_m}\, g^{r_{\lambda_1 \lambda_2 \ldots \lambda_m}}$$

But, from (A), if all the coefficients a do not disappear, we can take g so large that f is different from 0. The converse proposition is then proved.

(D). If a product of integral functions
(4) $$f_1(x_1, x_2, \ldots x_n)\, f_2(x_1, x_2, \ldots x_n) \ldots f_m(x_1, x_2, \ldots x_n)$$
is equal to zero for all systems of values of the x_λ's, then one of the factors is identically zero.

For, if we employ again the substitution (3), we can, from (C), select such values g_α and ν_α for any factor $f_\alpha(x_1, x_2, \ldots x_n)$ which does not vanish identically, that for every system of values which arises from (3) when $g > g_\alpha$ and $\nu > \nu_\alpha$ the value of f_α is different from zero. If then we take g greater than $g_1, g_2, \ldots g_m$ and at the same time ν greater than $\nu_1, \nu_2, \ldots \nu_m$ we obtain systems of values of the x_λ's for which (4) does not vanish, unless one of the factors vanishes identically.

The proof of Theorem I follows now directly from (C). For if $\varphi(x_1, x_2, \ldots x_n)$ is a single-valued function, and if $\varphi_1(x_1, x_2 \ldots x_n)$ arises from φ by any rearrangement of the x_λ's, then it follows from the fact that φ has only one value, that the difference

$$\varphi(x_1, x_2, \ldots x_n) - \varphi_1(x_1, x_2, \ldots x_n)$$

vanishes identically.

If the elements x_λ are not independent, Theorem I is no longer necessarily true. For instance, if all the x_λ's are equal, then any arbitrary function of the x_λ's is single-valued. Again the function $x_1^3 + 3x_1^2 x_2 - 4x_1 x_2^2 + 3 x_2^3$ is single-valued if $x_1 = 2x_2$, although it is unsymmetric.

§ 3. If, in any symmetric function, we combine all terms which only differ in their coefficients into a single term, and consider any one of these terms, $Cx_1^\alpha x_2^\beta x_3^\gamma, \ldots$, then the symmetric character of the function requires that it should contain every term which can be produced from the one considered by any rearrangement of the x_λ's. If these terms do not exhaust all those present in the function there will still be some term, $C'x_1^{\alpha'} x_2^{\beta'} x_3^{\gamma'} \ldots$ in which the system of exponents is not the same as in the preceding case. This term then gives rise to a new series of terms, and so on. Every symmetric function is therefore reducible to a sum of simpler symmetric functions in each of which all the terms proceed from any single one among them by rearrangement of the x_λ's. The several terms of any one of these simple functions are said to be *of the same type* or *similar*. Since these functions are deducible from a single term, it will suffice to write this one term preceded by an S. Thus $S(x^2)$ denotes in the case of two elements $x_1^2 + x_2^2$, in the case of three $x_1^2 + x_2^2 + x_3^2$, etc.

§ 4. If we regard the elements $x_1, x_2, \ldots x_n$ as the roots of an equation of the nth degree, this equation, apart from a constant factor, has the form

(5) $\qquad f(x) = (x - x_1)(x - x_2) \ldots (x - x_n) = 0$

the left member of which expanded becomes

(6) $\quad x^n - (x_1 + x_2 + \ldots + x_n)\, x^{n-1}$
$\quad\quad + (x_1 x_2 + x_1 x_3 + x_2 x_3 + \ldots + x_{n-1} x_n) x^{n-2}$
$\quad\quad - \ldots + (-1)^n x_1 x_2 \ldots x_n.$
$\quad\quad = x^n - c_1 x^{n-1} + c_2 x^{n-2} - \ldots + (-1)^n c_n$

The coefficients of the powers of x in this equation are therefore simple integral symmetric functions of the x_λ's:

(7) $\quad c_1 = S(x_1), \quad c_2 = S(x_1 x_2), \quad \ldots \quad c_\lambda = S(x_1 x_2 \ldots x_\lambda),$
$\quad \ldots \quad c_n = S(x_1 x_2 \ldots x_n) = x_1 x_2 \ldots x_n$

These combinations c_λ are called the *elementary symmetric functions*. They are of special importance for the reason that every symmetric function of the x_λ's can be expressed as a rational integral function of the c_λ's.

§ 5. Among the many proofs of this proposition we select that of Gauss.*

We call a term $x_1^{m_1} x_2^{m_2} x_3^{m_3} \ldots$ *higher* than $x_1^{\mu_1} x_2^{\mu_2} x_3^{\mu_3} \ldots$ when the first of the differences $m_1 - \mu_1, m_2 - \mu_2, m_3 - \mu_3, \ldots$ which does not vanish is positive. This amounts then to assigning an arbitrary standard order of precedence to the elements x_λ.

In accordance with this convention, $c_1, c_2, c_3, \ldots c_\lambda, \ldots$ have for their highest terms respectively

$$x_1, \quad x_1 x_2, \quad x_1 x_2 x_3, \quad \ldots x_1 x_2 x_3 \ldots x_\lambda, \quad \ldots$$

and the function $c_1^\alpha c_2^\beta c_3^\gamma \ldots$ has for its highest term

$$x_1^{\alpha + \beta + \gamma + \ldots} x_2^{\beta + \gamma + \ldots} x_3^{\gamma + \ldots} \ldots$$

In order, therefore, that the highest terms of the two expressions, $c_1^\alpha c_2^\beta c_3^\gamma \ldots$ and $c_1^{\alpha'} c_2^{\beta'} c_3^{\gamma'} \ldots$ may be equal, we must have

$$\alpha + \beta + \gamma + \ldots = \alpha' + \beta' + \gamma' + \ldots$$
$$\beta + \gamma + \ldots = \beta' + \gamma' + \ldots$$
$$\gamma + \ldots = \gamma' + \ldots$$
$$\cdot \quad \cdot \quad \cdot$$

that is, $\alpha = \alpha', \beta = \beta', \gamma = \gamma', \ldots$

It follows that two different systems of exponents in $c_1^\alpha c_2^\beta c_3^\gamma \ldots$ give two different highest terms in the x_λ's. Again it is clear that

$$x_1^\alpha x_2^\beta x_3^\gamma \ldots \quad\quad (\alpha \geq \beta \geq \gamma \geq \delta \ldots)$$

is the highest term of the expression $c_1^{\alpha-\beta} c_2^{\beta-\gamma} c_3^{\gamma-\delta} \ldots$ and that

---

* Demonstratio nova altera etc. Gesammelte Werke III, § 5, pp. 37-38. *Cf.* Kronecker, Monatsberichte der Berliner Akademie, 1889, p. 943 seq.

all the terms in the expansion of this expression in terms of the x_λ's are of the same degree.

§ 6. If now a symmetric function S be given of which the highest term is
$$A\ x_1^\alpha\ x_2^\beta\ x_3^\gamma\ x_4^\delta \ldots \qquad (\alpha \geq \beta \geq \gamma \geq \delta \ldots)$$
the difference
$$S - A\ c_1^{\alpha-\beta}\ c_2^{\beta-\gamma}\ c_3^{\gamma-\delta} \ldots = S_1$$
will again be a symmetric function; and if, in the subtrahend on the left, the values of the c_λ's given in (7) be substituted, the highest term of S will be removed, and accordingly a reduction will have been effected.

If the highest term of S_1 is now $A_1\ x_1^{\alpha'}\ x_2^{\beta'}\ x_3^{\gamma'}\ x_4^{\delta'} \ldots$, then
$$S_1 - A_1\ c_1^{\alpha'-\beta'}\ c_2^{\beta'-\gamma'}\ c_3^{\gamma'-\delta'} \ldots = S_2$$
is again a symmetric function with a still lower highest term. The degrees of S_2 and S_1 are clearly not greater than that of S, and since there is only a finite number of expressions $x_1^\lambda\ x_2^\mu\ x_3^\nu \ldots$ of a given degree which are lower than $x_1^\alpha\ x_2^\beta\ x_3^\gamma \ldots$, we shall finally arrive by repetition of the same process at the symmetric function 0; that is
$$S_k - A_k\ c_1^{\alpha^{(k)}-\beta^{(k)}}\ c_2^{\beta^{(k)}-\gamma^{(k)}} \ldots = 0;$$
and accordingly we have
$$S = A_1\ c_1^{\alpha-\beta}\ c_2^{\beta-\gamma} \ldots + A_2 c_1^{\alpha'-\beta'}\ c_2^{\beta'-\gamma'} \ldots + \ldots$$
$$+ A_k\ c_1^{\alpha^{(k)}-\beta^{(k)}}\ c_2^{\beta^{(k)}-\gamma^{(k)}} \ldots$$

§ 7. It is also readily shown that the expression of a symmetric function of the x_λ's as a rational function of the c_λ's can be effected in only one way.

For, if an integral symmetric function of $x_1, x_2, \ldots x_n$ could be reduced to two essentially different functions of $c_1, c_2, \ldots c_n$, $\varphi\ (c_1, c_2, \ldots c_n)$ and $\psi\ (c_1, c_2, \ldots c_n)$, then we should have, for all values of the x_λ's, the equation
$$\varphi\ (c_1, c_2, \ldots c_n) = \psi\ (c_1, c_2, \ldots c_n)$$

The difference $\varphi - \psi$, which, as function of the c_λ's, is not identically zero, since otherwise the two functions φ and ψ would coincide, must, as function of the x_λ's, be identically zero.

Suppose, now, that in $\varphi - \psi$ those terms in $c_1, c_2, \ldots c_n$ which cancel each other are removed, and let any remaining term be

$B\, c_1{}^\alpha c_2{}^\beta c_3{}^\gamma \ldots$ This term on being expressed in terms of the x_λ's will give as highest term

$$B\, x_1{}^{\alpha+\beta+\gamma+\cdots}\, x_2{}^{\beta+\gamma+\cdots}\, x_3{}^{\gamma+\cdots}\ldots$$

Now the different remaining terms $B'\, c_1{}^{\alpha'} c_2{}^{\beta'} c_3{}^{\gamma'}\ldots$ give *different* highest terms in the x_λ's (§ 5). Consequently among these highest terms there must be one higher than the others. But the coefficient of this term is not zero; and consequently (§ 2 (C)) the function $\varphi - \varphi'$ cannot be identically zero. We have therefore

Theorem II. *An integral symmetric function of $x_1, x_2, \ldots x_n$ can always be expressed in one and only one way as an integral function of the elementary symmetric functions $c_1, c_2, \ldots c_n$.*

§ 8. If we write $s_\lambda = S\,(x_1{}^\lambda)$ for the sum of the λ powers of the n elements $x_1, x_2, \ldots x_n$, we might attempt to calculate the s_λ's as functions of the c_i's by the above method. It is however simpler to obtain this result by the aid of two recursion formulas first given by Newton[*] and known under his name. These formulas are

A) $s_r - c_1 s_{r-1} + c_2 s_{r-2} - \ldots + (-1)^n c_n s_{r-n} = 0 \quad (r \geq n)$

B) $s_r - c_1 s_{r-1} + c_2 s_{r-2} - \ldots + (-1)^r r\, c_r = 0 \quad (r \leq n)$

These two formulas can be proved in a variety of ways. The formula A) is obtained by multiplying the right member of (6) by x^{r-n}, replacing x by x_λ, and taking the sum over $\lambda = 1, 2, \ldots n$.

The formula B) may be verified with equal ease as follows. If we represent the elementary symmetric functions of $x_2, x_3, \ldots x_n$ by $c_1', c_2', \ldots c'_{n-1}$, we have

$$c_1 = x_1 + c_1', \quad c_2 = x_1 c_1' + c_2', \quad c_3 = x_1 c_2' + c_3', \ldots$$

and accordingly, if $r \leq n$, we have

$$x_1{}^r - c_1 x_1{}^{r-1} + c_2 x_1{}^{r-2} - \ldots (-1)^r c_r$$
$$= x_1{}^r - (x_1 + c_1') x_1{}^{r-1} + (x_2 c_1' + c_2') x_1{}^{r-2} - \ldots$$
$$+ (-1)^r (x_1 c'_{r-1} + c_r') = (-1)^r c_r'$$

and hence, replacing x_1 successively by $x_2, x_3, \ldots x_n$, and, correspondingly, c_r' by $c_r'', c_r''', \ldots c_r^{(n)}$, and taking the sum of the n resulting equations

$$s_r - c_1 s_{r-1} + c_2 s_{r-2} - \ldots (-1)^r c_r\, n$$
$$= (-1)^r (c_r' + c_r'' + c_r''' + \ldots + c_r^{(n)}).$$

---

[*]Newton: Arith. Univ., De Transformatione Aequationum.

The right member is symmetric in $x_1, x_2, \ldots x_n$, and contains all the terms of c_r and no others. Moreover, the term $x_1 x_2 \ldots x_r$, and consequently every term, occurs $n - r$ times. Accordingly we have

$$s_r - c_1 s_{r-1} + c_2 s_{r-2} - \ldots + (-1)^r c_r n = (-1)^r (n-r) c_r,$$
$$\therefore s_r - c_1 s_{r-1} + c_2 s_{r-2} - \ldots (-1)^r c_r \cdot r = 0,$$

and formula B) is proved.* The formula A) can obviously be verified in the same way.

§ 9. The solution of the equations A) and B) for the successive values of the s_λ's gives the expressions for these quantities in terms of the c_λ's. The solution is readily accomplished by the aid of determinants. We add here a few of the results.†

C) $\quad s_0 = n$

$s_1 = c_1$

$s_2 = c_1^2 - 2c_2$

$s_3 = c_1^3 - 3c_1 c_2 + 3c_3$

$s_4 = c_1^4 - 4c_1^2 c_2 + 4c_1 c_3 + 2c_2^2 - 4c_4$

$s_5 = c_1^5 - 5c_1^3 c_2 + 5c_1^2 c_3 + 5c_1 c_2^2 - 5c_1 c_4 - 5c_2 c_3 + 5c_5$

It is to be observed here that all the c_λ's of which the indices are greater than n are to be taken equal to 0. This is obvious if we add to the n elements $x_1, x_2, \ldots x_n$ any number of others with the value 0; for the c_λ's up to c_n will not be affected by this addition, while $c_{n+1}, c_{n+2}, \ldots$ will be 0.

§ 10. The observation of § 5 that $c_1^\alpha c_2^\beta c_3^\gamma \ldots$ gives for its highest term $x_1^{\alpha + \beta + \gamma + \cdots} x_2^{\beta + \gamma + \cdots} x_3^{\gamma + \cdots}$ can be employed to facilitate the calculation of a symmetric function in terms of the c_λ's.

We may suppose that the several terms of the given function are of the same type, that is that they arise from a single term among them by interchanges of the x_λ's. The function is then homogeneous; suppose it to be of degree ν. We can then obtain its literal part at once.

For, if the function contains one element, and consequently all elements, in the mth and no higher power, then every term of the corresponding expression in the c_λ's will be of degree m at the highest. For, in the first place, two different terms $c_1^\alpha c_2^\beta c_3^\gamma \ldots$ and

---

*Another, purely arithmetical, proof is given by Euler; Opuscula Varii Argumenti. Demonstrat. genuina theor. Newtoniani, II, p. 108.

† *Cf.* Faà di Bruno: Formes Binaires.

$c_1{}^{a'} c_2{}^{\beta'} c_3{}^{\gamma'} \ldots$ give different highest terms in the x_λ's, so that two such terms cannot cancel each other; and, in the second place, $c_1{}^a c_2{}^\beta c_3{}^\gamma \ldots$ gives a power $x_\lambda{}^{a+\beta+\gamma+\cdots}$, so that

$$a + \beta + \gamma + \ldots \leq m.$$

Again the degree of $x_1{}^{a+\beta+\gamma+\cdots} x_2{}^{\beta+\gamma+\cdots} x_3{}^{\gamma+\cdots} \ldots$ is

$$a + 2\beta + 3\gamma + \ldots = \nu$$

and since the given expresssion is homogeneous, the sum $a + 2\beta + 3\gamma + \ldots$ must be equal to ν for every term $c_1{}^a c_2{}^\beta c_3{}^\gamma \ldots$

These two limitations imposed on the exponents of the c_λ's that

$$a + \beta + \gamma + \ldots \leq m, \qquad a + 2\beta + 3\gamma \ldots = \nu,$$

exclude a large number of possible terms. The coefficients of those that remain are then calculated from numerical examples. The quantity $a + 2\beta + 3\gamma + \ldots$ is called the *weight* of the term $c_1{}^a c_2{}^\beta c_3{}^\gamma \ldots$ and a function of the c_λ's whose several terms are all of the same weight is called *isobaric*. For example

$$S(x_1{}^2 x_2{}^2 x_3{}^2 x_4) = q_0 c_7 + q_1 c_6 c_1 + q_2 c_5 c_2 + q_3 c_4 c_3 \quad (m=2, \nu=7)$$

$$S[(x_1 - x_2)^2 (x_2 - x_3)^2 (x_3 - x_1)^2] = q_0 c_6 + q_1 c_5 c_1 + q_2 c_4 c_2$$
$$+ q_3 c_4 c_1{}^2 + q_4 c_3{}^2 + q_5 c_3 c_2 c_1 + q_6 c_3 c_1{}^3 + q_7 c_2{}^3 + q_8 c_2{}^2 c_1{}^2$$
$$(m=4, \nu=6)$$

where the q's are as yet undetermined numerical coefficients.

In the second example we will calculate the q's for the case $n=3$, for which therefore $c_4 = c_5 = c_6 = 0$. It is obvious that for different values of n the coefficient q's will be different. Taking

I. $x_1 = 1$, $x_2 = -1$, $x_3 = 0$, we have $c_1 = 0$, $c_2 = -1$, $c_3 = 0$
$\therefore S = 4 = -q_7$; $q_7 = -4$.

II. $x_1 = x_2 = 1$, $x_3 = 0$, $c_1 = 2$, $c_2 = 1$, $c_3 = 0$
$\therefore S = 0 = -4 + 4q_8$; $q_8 = 1$.

III. $x_1 = x_2 = 1$, $x_3 = -1$, $c_1 = 0$, $c_2 = -3$, $c_3 = -2$
$\therefore S = 0 = 4q_4 + 4 \cdot 27$; $q_4 = -27$.

IV. $x_1 = x_2 = 2$, $x_3 = -1$, $c_1 = 3$, $c_2 = 0$, $c_3 = -4$
$\therefore S = 0 = -27 \cdot 16 - 4 \cdot 27 q_6$; $q_6 = -4$.

V. $x_1 = x_2 = x_3 = 1$, $c_1 = 3$, $c_2 = 3$, $c_3 = 1$
$\therefore S = 0 = -27 + 9 q_5 - 135$; $q_5 = 18$.

$\therefore (x_1 - x_2)^2 (x_2 - x_3)^2 (x_3 - x_1)^2 = -27 c_3{}^2$
$\qquad + 18 c_3 c_2 c_1 - 4 c_3 c_1{}^3 - 4 c_2{}^3 + c_2{}^2 c_1{}^2.$

This expression, $(x_1 - x_2)^2 (x_2 - x_3)^2 (x_3 - x_1)^2 = \Delta$, is called the *discriminant* of the quantities x_1, x_2, x_3. The characteristic property of this discriminant is that it is symmetric and that its vanishing is the sufficient and necessary condition that at least two of the x_λ's are equal.

§ 11. In general, we give the name "discriminant of n quantities $x_1, x_2, \ldots x_n$," to the symmetric function of the x_λ's the vanishing of which is the sufficient and necessary condition for the equality of at least two of the x_λ's. If a symmetric function S of the x_λ's is to vanish for $x_1 = x_2$, it must be divisible by $x_1 - x_2$, and consequently by every difference $x_\alpha - x_\beta$. Suppose

$$S = (x_1 - x_2) S_1.$$

Now S, and consequently $(x_1 - x_2) S_1$, is unchanged if x_1 and x_2 be interchanged. But this changes the sign of $x_1 - x_2$ and therefore of S_1. Consequently S_1 vanishes if $x_1 = x_2$, and accordingly S_1 contains $x_1 - x_2$ as a factor.

The symmetric function S is therefore divisible by $(x_1 - x_2)^2$ and consequently by every $(x_\alpha - x_\beta)^2$; that is it is divisible by

$$\Delta = \prod_{\lambda\mu} (x_\lambda - x_\mu)^2 \quad (\lambda < \mu; \lambda = 1, 2, \ldots n-1; \mu = 2, 3, \ldots n)$$

(8)
$$\begin{aligned}
&= (x_1 - x_2)^2 (x_1 - x_3)^2 (x_1 - x_4)^2 \ldots (x_1 - x_n)^2 \\
&\quad (x_2 - x_3)^2 (x_2 - x_4)^2 \ldots (x_2 - x_n)^2 \\
&\quad (x_3 - x_4)^2 \ldots (x_3 - x_n)^2 \\
&\quad \cdot \quad \cdot \quad \cdot \quad \cdot \quad \cdot \\
&\quad (x_{n-1} - x_n)^2.
\end{aligned}$$

This quantity Δ already satisfies the condition as to the equality of the x_λ's, and, being the simplest function with this property, is itself the discriminant. It contains $\frac{1}{2} n(n-1)$ factors of the form $(x_\lambda - x_\mu)^2$; its degree is $n(n-1)$, and the highest power to which any x_λ occurs is the $(n-1)^{\text{th}}$. It is the square of an integral, but, as we shall presently show, unsymmetric function, with which we shall hereafter frequently have to deal.

§ 12. Finally we will consider another symmetric function in which the discriminant occurs as a factor.

Let the equation of which the roots are $x_1, x_2, \ldots x_n$ be, as before, $f(x) = 0$. Then if we write

$$\frac{df(x)}{dx} = f'(x)$$

we have, for all values $\lambda = 1, 2, \ldots n$, the equation

$$f'(x_\lambda) = (x_\lambda - x_1)(x_\lambda - x_2)\ldots(x_\lambda - x_{\lambda-1})(x_\lambda - x_{\lambda+1})\ldots(x_\lambda - x_n).$$

We attempt now to express the integral symmetric function

$$S\left[x_1^a \cdot f'(x_2) \cdot f'(x_3)\ldots f'(x_n)\right]$$

in terms of the coefficients $c_1, c_2, \ldots c_n$ of $f(x)$. Every one of the n terms of S is divisible by $x_1 - x_2$, since either $f'(x_1)$ or $f'(x_2)$ occurs in every term. Consequently, by the same reasoning as in § 11, S is divisible by $(x_1 - x_2)^2$, and therefore being a symmetric function, by every $(x_\alpha - x_\beta)^2$, that is by

$$\mathtt{J} = \prod_{\lambda\mu}(x_\lambda - x_\mu)^2 \quad (\lambda < \mu;\ \lambda = 1, 2, \ldots n-1;\ \mu = 2, 3, \ldots n).$$

S is therefore divisible by the *discriminant of $f(x)$*, i. e., by the discriminant of the n roots of $f(x)$.

Now $f(x_\lambda)$ is of degree of $n-1$ in x_λ and of degree 1 in every other x_μ; and therefore

$x_1^a \cdot f'(x_2) \cdot f'(x_3)\ldots f'(x_n)$ is of degree $a + n - 1$ in x_1

$x_2^a \cdot f'(x_1) \cdot f'(x_3)\ldots f'(x_n)$ is of degree $2n - 3$ in x_1.

Consequently, if $a < n-1$, S is of degree $2n-3$ in x_1, while $\mathtt{J}$ is of degree $2n-2$ in x_1. But since $\mathtt{J}$ is a divisor of S, it follows that S is in this case identically 0.

(9) $\quad S\left[x_1^a \cdot f'(x_2) \cdot f'(x_3)\ldots f'(x_n)\right] = 0, \quad (a < n-1.)$

Again, if $a = n-1$, then S and $\mathtt{J}$ can only differ by a constant factor. To determine this factor we note that the first term of S is of degree $2n-2$ in x_1, while all the other terms are of lower degree in x_1. The coefficient of x_1^{2n-2} is therefore

$$(-1)^{n-1}(x_2 - x_3)\ldots(x_2 - x_n)(x_3 - x_2)\ldots(x_3 - x_n)\ldots$$
$$(x_n - x_2)\ldots(x_n - x_{n-1}) = (-1)^{\frac{n(n-1)}{2}}(x_2 - x_3)^2(x_2 - x_4)^2\ldots$$
$$(x_{n-1} - x_n)^2.$$

In $\mathtt{J}$ the coefficient of x_1^{2n-2} is

$$(x_2 - x_3)^2(x_2 - x_4)^2\ldots(x_{n-1} - x_n)^2.$$

The desired numerical factor is therefore $(-1)^{\frac{n(n-1)}{2}}$ and we have

(10) $\quad S[x_1^{n-1} \cdot f'(x_1) \cdot f'(x_2) \ldots f'(x_n)] = (-1)^{\frac{n(n-1)}{2}} \lrcorner$.

Formulas (9) and (10) evidently still hold if we replace x_1^a or x_1^{n-1} by any integral function $\varphi(x)$ of degree $a \leq n$ respectively. Moreover since

$$(-1)^{\frac{n(n-1)}{2}} \lrcorner = f'(x_1) \cdot f'(x_2) \ldots f'(x_n)$$

we have

(D) $$\sum_{\lambda=1}^{\lambda=n} \frac{\varphi(x_\lambda)}{f'(x_\lambda)} = 0 \text{ or } 1^*,$$

according as the degree of φ is less than or equal to $n-1$.

§ 13. If an integral function of the elements $x_1, x_2, \ldots x_n$ is not symmetric, it will be changed in form, and consequently, if the x_λ's are entirely independent, also in value, by some of the possible interchanges of the x_λ's. The process of effecting such an interchange we shall call a *substitution*. Any order of arrangement of the x_λ's we call a *permutation*. The substitutions are therefore *operations*; the permutations the result. Any substitution whatever leaves a symmetric function unchanged in form; but there are other functions the form of which can be changed by substitutions. For example, the functions

I $\quad x_1^2 - x_2^2 + x_3^2 - x_4^2, \quad x_1 x_2^2 x_3 + x_4 x_5 + x_6, \quad x_1^3 + x_2^2 + x_3$

take new values if certain substitutions be applied to them; thus if x_1 and x_2 be interchanged, these functions become

II $\quad -x_1^2 + x_2^2 + x_3^2 - x_4^2, \quad x_1^2 x_2 x_3 + x_5 x_5 + x_6, \quad x_2^3 + x_1^2 + x_3$.

The first two functions are unchanged if x_1 and x_3 be interchanged, the second also if x_4 and x_5 be interchanged, etc.

Functions are designated as one-, two-, three-, m-valued according to the number of different values they take under the operation of all the $n!$ possible substitutions. The existence of one-valued functions was apparent at the outset. We enquire now as to the possibility of the existence of two-valued functions.

In § 11 we have met with the symmetric function $\lrcorner$, the discriminant of the n quantities $x_1, x_2, \ldots x_n$. The square root of $\lrcorner$ is also a rational integral function of these n quantities:

---

*The formula D is due to Euler; Calc. Int. II § 1169.

SYMMETRIC AND TWO-VALUED FUNCTIONS.

$$\sqrt{J} = (x_1 - x_2)(x_1 - x_3)(x_1 - x_4)\ldots(x_1 - x_n)$$
$$(x_2 - x_3)(x_2 - x_4)\ldots(x_2 - x_n)$$
$$(x_3 - x_4)\ldots(x_3 - x_n)$$
$$\cdot \quad \cdot \quad \cdot \quad \cdot \quad \cdot \quad \cdot \quad \cdot$$
$$(x_{n-1} - x_n).$$

Every difference of two elements $x_a - x_\beta$ occurs once and only once on the right side of this equation. Accordingly if we interchange the x_λ's in any way, every such difference still occurs once and only once, and the only possible change is that in one or more cases an $x_a - x_\beta$ may become $x_\beta - x_a$. The result of any substitution is therefore either $+\sqrt{J}$ or $-\sqrt{J}$, i. e., the function $\sqrt{J}$ is either one-valued or two-valued. But if, in particular, we interchange x_1 and x_2, the first factor of the first row above changes its sign, while the other factors of the first row are converted into the corresponding factors of the second row, and vice versa. No change occurs in the other rows, since these do not contain either x_1 or x_2. Since then, for this substitution, $\sqrt{J}$ becomes $-\sqrt{J}$, it appears that we have in $\sqrt{J}$ a *two-valued function*.

This function is specially characterized by the fact that its two values only differ in algebraic sign. Such two-valued functions we shall call *alternating functions*.

Theorem III. *The square root of the discriminant of the n quantities $x_1, x_2, \ldots x_n$ is an alternating function of these quantities.*

§ 14. Before we can determine all the alternating functions, a short digression will be necessary.

An interchange of two elements we shall call a *transposition*. The transposition of x_a and x_β, we will denote by the symbol $(x_a\ x_\beta)$. We shall now prove the following

Theorem IV. *Every substitution can be replaced by a series of transpositions.*

Thus, if we have to transform the order $x_1, x_2, x_3, \ldots x_n$ into the order $x_{l_1}, x_{l_2}, x_{l_3}, \ldots x_{l_n}$, we apply first the transposition $(x_1\ x_{l_1})$. The order of the x_λ's then becomes $x_{l_1}, x_2, x_3, \ldots x_{l_1-1}$ $x_1, x_{l_1+1}, \ldots x_n$, and we have now only to convert the order $x_2 \ldots x_{l_1-1}, x_1, x_{l_1+1}, \ldots x_n$ into the order $x_{l_2}, x_{l_3}, \ldots x_{l_n}$. By repeating the same process as before, this can be gradually effected, and the theorem is proved.

Since a symmetric function is unaltered by any substitution, we obtain as a direct result

Theorem V. *A function which is unchanged by every transposition is symmetric.*

§ 15. There is therefore at least one transposition which changes the value of any alternating function into the opposite value. We will denote this transposition by $(x_\alpha\ x_\beta)$, and the alternating function by φ, and accordingly we have

$$\varphi(x_1, x_2, \ldots x_\alpha, \ldots x_\beta, \ldots x_n) = -\varphi(x_1, x_2, \ldots x_\beta, \ldots x_\alpha, \ldots x_n)$$

Accordingly, if $x_\alpha = x_\beta$, we must have $\varphi = 0$. Consequently the equation

$$\varphi(x_1, x_2, \ldots z, \ldots x_\beta, \ldots x_n) = 0$$

regarded as an equation in z has a root $z = x_\beta$ and the polynomial φ is therefore divisible by $z - x_\beta$. The function

$$\varphi(x_1, x_2, \ldots x_\alpha \ldots x_\beta \ldots x_n)$$

therefore contains $x_\alpha - x_\beta$ as a factor, and, consequently, φ^2 contains $(x_\alpha - x_\beta)^2$ as a factor.

But since, for all substitutions, φ either remains unchanged or only changes its sign, φ^2 must be a symmetric function; and, accordingly, since φ^2 contains the factor $(x_\alpha - x_\beta)^2$, it must contain all factors of the form $(x_\lambda - x_\mu)^2$, *i. e.*, φ^2 contains Δ as a factor, and consequently φ contains $\sqrt{\Delta}$ as a factor. The remaining factor of φ is determined by aid of the following

Theorem VI. *Every alternating integral function is of the form $S \cdot \sqrt{\Delta}$, where $\sqrt{\Delta}$ is the square root of the discriminant and S is an integral symmetric function.*

That $S \cdot \sqrt{\Delta}$ is an alternating function is obvious. Conversely, if φ is an alternating function, it is, as we have just seen, divisible by $\sqrt{\Delta}$. Let $(\sqrt{\Delta})^m$ be the highest power of $\sqrt{\Delta}$ which occurs as a factor in φ. Then the quotient

$$\frac{\varphi}{(\sqrt{\Delta})^m}$$

is either a one- or a two-valued function, since every substitution either leaves both numerator and denominator unchanged or changes the sign of one or both of them. But this quotient cannot be two-valued, for then it would be again divisible by $\sqrt{J}$, which is contrary to hypothesis. It must therefore be symmetric, and we have accordingly

$$\psi = S_1 . (\sqrt{J})^m$$

Now if m were an even number, the right member of this equation, and consequently the left, would be symmetric. We must therefore have $m = 2n + 1$. And if we write $S_1 . J^n = S$, we have

$$ \mathsf{J} = S . \sqrt{J}$$

Corollary. *From the form of an alternating function it follows that such a function remains unchanged or is changed in sign simultaneously with $\sqrt{J}$ for all substitutions.*

§ 16. Having now shown how to form all the alternating functions, we proceed to the examination of the two-valued functions in general.

Let $\varphi(x_1, x_2, \ldots x_n)$ be any two-valued function, and let the two values of φ be denoted by

$$\varphi_1(x_1, x_2, \ldots x_n) \quad \text{and} \quad \varphi_2(x_1, x_2, \ldots x_n).$$

These two functions must differ in form as well as in value, and since the x_λ's are any arbitrary quantities, if we apply to φ_1 and φ_2 any substitution whatever, the resulting values

(a) $\qquad \varphi_1(x_{i_1}, x_{i_2}, \ldots x_{i_n}) \quad \text{and} \quad \varphi_2(x_{i_1}, x_{i_2}, \ldots x_{i_n})$

will also be different. But whatever substitutions are applied to φ the result is always φ_1 or φ_2. Consequently, of the two expressions (a), one must be identical with φ_1 and the other with φ_2. In other words:

Those substitutions which leave the one value of a two-valued function φ unchanged, leave the other value unchanged also; those substitutions which convert the one value of φ into the other, also convert the second value into the first.

§ 17. From the preceding section it follows at once that $\varphi_1 + \varphi_2$ is a symmetric function

(β) $\qquad\qquad \varphi_1 + \varphi_2 = 2 S_1.$

Again the difference $\varphi_1 - \varphi_2$ is a two-valued function of which the second value is $\varphi_2 - \varphi_1 = -(\varphi_1 - \varphi_2)$. This difference is therefore an alternating function, and accordingly, from Theorem VI, we may write

(γ) $\qquad\qquad S_1 - S_2 = 2S_2 \sqrt{\Delta}.$

From (β) and (γ) we obtain

$$\varphi_1 = S_1 + S_2 \sqrt{\Delta}, \quad \varphi_2 = S_1 - S_2 \sqrt{\Delta}, \quad \varphi = S_1 \pm S_2 \sqrt{\Delta}.$$

Conversely, it is plain that every function of the last form is two-valued.

Theorem VII. *Every two-valued integral function is of the form* $\varphi_1 = S_1 \pm S_2 \sqrt{\Delta}$, *where S_1 and S_2 are integral symmetric functions and $\sqrt{\Delta}$ is the square root of the discriminant. Conversely every function of this form is two-valued.*

Corollary. *Every two-valued integral function is unchanged or changed simultaneously with $\sqrt{\Delta}$ by every substitution.*

§ 18. From the corollaries of the last two theorems we recognize the importance of determining those substitutions which leave the value of $\sqrt{\Delta}$ unchanged.

We know (§ 13) that the transposition $(x_1 x_2)$ changes the sign of $\sqrt{\Delta}$. In the arrangement of the factors of $\sqrt{\Delta}$ in the same Section, we might equally well have placed all the factors containing x_α or x_β in the first and second rows, $x_\alpha - x_\beta$ taking the place of $x_1 - x_2$, etc. The sign of $\sqrt{\Delta}$ may be changed by this rearrangement, but whatever this sign may be, it will be changed by the transposition $(x_\alpha x_\beta)$. Consequently $\sqrt{\Delta}$ changes its sign for every transposition.

This result is easily extended. For, if we apply successively any μ transpositions to $\sqrt{\Delta}$, its sign will be changed μ times, that is $\sqrt{\Delta}$ becomes $(-1)^\mu \sqrt{\Delta}$. If μ is even, $\sqrt{\Delta}$ is unchanged; if μ is odd, $\sqrt{\Delta}$ becomes $-\sqrt{\Delta}$. We have therefore

Theorem VIII. *All substitutions which are formed from an odd number of transpositions change the value $\sqrt{\Delta}$ into $-\sqrt{\Delta}$; all substitutions which are formed from an even number of transpositions leave $\sqrt{\Delta}$ unchanged. Similar results hold for all two-valued functions.*

§ 19. Every substitution can be reduced to a series of transpositions in a great variety of ways, as is readily seen, and as will be shown in detail in the following Chapter. But from the preceding theorem it follows that the number of transpositions into which a substitution is resolvable is always even, or always odd, according as the substitution leaves $\sqrt{J}$ unchanged or changes its sign.

Theorem IX. *If a given substitution reduces in one way to an even (odd) number of transpositions, it reduces in every way to an even (odd) number of transpositions.*

§ 20. **Theorem X.** *Every two-valued function is the root of an equation of the second degree of which the coefficients are rational symmetric functions of the elements $x_1, x_2, \ldots x_n$.*

From the equations of § 17,

$$\varphi_1 = S_1 + S_2 \sqrt{J}, \quad \varphi_2 = S_1 - S_2 \sqrt{J},$$

we have for the elementary symmetric functions of φ_1 and φ_2

$$\varphi_1 + \varphi_2 = 2 S_1,$$
$$\varphi_1 \varphi_2 = S_1^2 - J S_2^2.$$

We recognize at once that φ_1 and φ_2 are the roots of the equation

$$\varphi^2 - 2 S_1 \varphi + (S_1^2 - J S_2^2) = 0.$$

It is however to be observed here that it is not conversely true that every quadratic equation with symmetric functions of the x_λ's as coefficients has two-valued functions, in the present sense, as roots. It is further necessary that the roots should be rational in the elements x_λ, and this is not in general the case.

CHAPTER II.

MULTIPLE-VALUED FUNCTIONS AND GROUPS OF SUBSTITUTIONS.

§ 21. The preliminary explanations of the preceding Chapter enable us to indicate now the course of our further investigations, at least in their general outline. Exactly as we have treated one-valued and two-valued functions and have determined those substitutions which leave the latter class of functions unchanged, so we shall have further, either to establish the existence of functions having any prescribed number of values, or to demonstrate their impossibility; to study the algebraic form of these functions; to determine the complex of substitutions which leave a given multiple-valued function unchanged; and to ascertain the relations of the various values of these functions to one another. Further, we shall attempt to classify the multiple-valued functions; to exhibit them possibly, like the two-valued functions, as roots of equations with symmetric functions of the elements as coefficients; to discover the relations between functions which are unchanged by the same substitutions; and so on.

§ 22. At the outset it is necessary to devise a concise notation for the expression of substitutions.

Consider a rational integral function of the n independent quantities $x_1, x_2, \ldots x_n$, which we will denote by $\varphi(x_1, x_2, \ldots x_n)$. If in this expression we interchange the position of the elements x_λ in such a way that for $x_1, x_2, \ldots x_n$ we put $x_{i_1}, x_{i_2}, \ldots x_{i_n}$ respectively, where the system of numbers $i_1, i_2, \ldots i_n$ denotes any arbitrary permutation of the numbers $1, 2, \ldots n$, we obtain from the original function $\varphi(x_1, x_2, \ldots x_n)$ the new expression $\varphi(x_{i_1}, x_{i_2}, \ldots x_{i_n})$.

We consider now the manner of representing by symbols such a transition from $x_1, x_2, \ldots x_n$ to $x_{i_1}, x_{i_2}, \ldots x_{i_n}$; to this transition we have already given the name of substitution.

A. In the first place we may represent this substitution by the symbol

$$\begin{pmatrix} x_1, & x_2, & x_3, & \ldots & x_n \\ x_{i_1}, & x_{i_2}, & x_{i_3}, & \ldots & x_{i_n} \end{pmatrix},$$

which shall indicate that every element of the upper line is to be replaced by the element of the lower line immediately below it. In this mode of writing a substitution we may obviously, without loss, omit all those elements which are not affected by the substitution, that is all those for which $x_k = x_{i_k}$. In the latter case the entire number of elements is not known from the symbol, but must be otherwise given, as is also true in the case of φ itself, since, for example, it is not in any way apparent from the form of

$$\varphi = x_1 x_2 + x_3 x_4$$

whether other elements $x_5, x_6, \ldots$ may not also be under consideration, as well as those which appear in φ.

B. Secondly, we may make use of the result of the preceding Chapter, that every substitution can be resolved into a series of *transpositions*. If we denote such a transposition, *i. e.*, the interchange of *two* elements, by enclosing both in a parenthesis, every substitution may be written as a series

$$(x_a x_b)(x_c x_d) \ldots (x_p x_q).$$

This reduction can be accomplished in an endless variety of ways. For, as is shown in § 14 of the preceding Chapter, we can first bring any arbitrary element to its proper place, and then proceed with the remaining $n-1$ elements in the same way. Indeed we may introduce any arbitrary transposition into the series and cancel its effect by one or more later transpositions, which need not immediately follow it or each other.

C. Thirdly, we may also write every substitution in the form

$$(x_{a_1} x_{a_2} x_{a_3} \ldots x_{a_k})(x_{b_1} x_{b_2} x_{b_3} \ldots x_{b_m})(x_{c_1} x_{c_2} x_{c_3} \ldots x_{c_n})\ldots$$

Here each parenthesis indicates that every element contained in it except the last is to be replaced by the next succeeding, the last element being replaced by the first. The parentheses are called *cycles*, the elements contained in each of them being regarded as forming a closed system, as if they were, for example, arranged in order of succession on the circumference of a circle.

If we wish to pass from the notation A to the present one, the resulting cycles would read

$$(x_1 x_{i_1} x_{i_{l_1}} \ldots)(x_a x_{i_a} x_{i_{l_a}} \ldots) \ldots$$

Here too it is clear that those elements which are unaffected by the substitution, every one of which therefore forms a cycle by itself, may be omitted in the symbol. In the notation A these elements are the same as those immediately below them.

A fourth system of notation which is indispensable in many important special cases will be discussed later.

§ 23. It is obvious that each of the three notations, $A, B,$ and C, contains some arbitrary features. In A the order of arrangement of the elements in the first line is entirely arbitrary; in B the reduction to transpositions is possible in a great number of ways; in C the order of succession of the cycles, and again the first element of each cycle, may be taken arbitrarily.

The first of the three notations, in spite of its apparent simplicity lacks in clearness of presentation; the second is defective, in that the same element may occur any number of times, so that the important question, "by which element is a given element replaced," cannot be decided at first glance, and the equality of the two substitutions is not immediately clear from their symbols. We shall, therefore, in the following investigations, employ almost exclusively the representation of substitutions by cycles.

The following example, for the case $n = 7$, will serve as an illustration of the different notations.

It is required that the order $x_1, x_2, x_3, x_4, x_5, x_6, x_7$ shall be replaced by the order $x_3, x_7, x_5, x_4, x_1, x_6, x_2$.

The first method gives us

$$\begin{pmatrix} x_1 \ x_2 \ x_3 \ x_4 \ x_5 \ x_6 \ x_7 \\ x_3 \ x_7 \ x_5 \ x_4 \ x_1 \ x_6 \ x_2 \end{pmatrix} = \begin{pmatrix} x_1 \ x_2 \ x_3 \ x_5 \ x_7 \\ x_3 \ x_7 \ x_5 \ x_1 \ x_2 \end{pmatrix}.$$

By the second we have variously

$$(x_1 x_3)(x_1 x_5)(x_2 x_7) = (x_1 x_3)(x_1 x_5)(x_1 x_2)(x_1 x_7)(x_1 x_2)$$
$$= (x_3 x_4)(x_5 x_6)(x_1 x_7)(x_3 x_7)(x_1 x_6)(x_2 x_7)(x_4 x_5)(x_2 x_4)(x_2 x_6)$$
$$= \ldots \ldots$$

Since the given substitution resolves into 3, 5, 9, ..., transpositions, always an odd number, we have here an example of the principle of Chapter I, § 19, and it appears that this substitution changes the sign of $\sqrt{\Delta}$.

The third method gives us

$$(x_1x_3x_5)(x_2x_7)(x_4)(x_6) = (x_1x_3x_5)(x_2x_7) = (x_2x_7)(x_3x_5x_1)$$
$$= (x_7x_2)(x_5x_1x_3) = \ldots \ldots$$

§ 24. We determine now the number of all the possible substitutions by finding the entire number of possible permutations.

Two elements x_1, x_2, can form two different permutations, x_1x_2 and x_2x_1. If a third element x_3 be added to these two, it can be placed, 1) at the beginning of the permutations already present $x_3x_1x_2$, $x_3x_2x_1$, or 2) in the middle: $x_1x_3x_2$, $x_2x_3x_1$, or 3) at the end: $x_1x_2x_3$, $x_2x_1x_3$. There are therefore $2 \cdot 3 = 3!$ permutations of three elements. If a fourth element be added, it can occur in the first, second, third or fourth place of the 3! permutations already obtained, so that from every one of these proceed 4 new permutations. There are therefore in this case $2 \cdot 3 \cdot 4 = 4!$ permutations, and again, for 5 elements, 5!, in general, for n elements, $n!$ permutations.

If now, in the notation **A**, we take for the upper line the natural order of the elements, x_1, x_2, x_3, ... x_n, and for the lower line successively all the $n!$ possible permutations, we obtain all the possible distinct substitutions of the n elements.

It is to be noticed that among these there is contained that substition for which the upper and lower lines are identical. This substitution does not affect any element; it is denoted by 1, and regarded as *unity* or as the *identical substitution*.

Theorem I. *For n elements there are $n!$ possible substitutions.*

To obtain the same result from the notation **B** more elaborate investigations would be necessary for which this is not the place; in case of the notation **C** it is easy to establish the number $n!$ by the aid of induction.

We arrive in the latter case at a series of interesting relations, of which at least one may be noted here. If a substitution in the expression for which all the elements occur contains

a cycles of a elements, b cycles of β elements, ...

where
$$a\alpha + b\beta + \ldots = n, \qquad N)$$
we can obtain from this by rearrangement of the order of the cycles and by cyclical permutation of the elements of each single cycle
$$a!\,\alpha^a\,b!\,\beta^b \ldots$$
expressions for the same substitution. Consequently there are
$$\frac{n!}{a!\,\alpha^a\,b!\,\beta^b \ldots}$$
distinct substitutions which contain a cycles of α elements, b cycles of β elements, and so on. The summation of these numbers with respect to all possible modes, N), of distributing the number n gives us all the possible $n!$ substitutions. Hence
$$\sum \frac{1}{a!\,\alpha^a\,b!\,\beta^b \ldots} = 1. \ast$$

§ 25. If now we apply all the $n!$ substitutions to the function $\varphi(x_1, x_2, \ldots x_n)$, i. e., if we perform these substitutions, which may be denoted by
$$s_1 = 1,\ s_2, s_3,\ \ldots s_a,\ \ldots s_{n!},$$
among the $x_1, x_2, \ldots x_n$ in the expression φ, we obtain $n!$ expressions, including that produced by the substitution $s_1 = 1$. These expressions we may denote by
$$\varphi_{s_1} = \varphi_1,\ \varphi_{s_2}, \varphi_{s_3} \ldots \varphi_{s_a} \ldots \varphi_{s_{n!}},$$
or simply, where no confusion is likely to occur, by
$$\varphi_1,\ \varphi_2,\ \varphi_3,\ \ldots \varphi_a,\ \ldots \varphi_n.$$

These values are not necessarily all different from one another. Some of them may coincide with the original value $\varphi(x_1, x_2, \ldots x_n)$. We direct our attention at first to the complex of those substitutions which do not change the value of φ. If φ is symmetric this complex will comprise all the $n!$ substitutions; if φ is a two-valued function, the complex will contain all substitutions which are composed of an even number of transpositions, and only these.

Again, for example, consider the case of four elements x_1, x_2, x_3, x_4, and suppose
$$\varphi(x_1, x_2, x_3, x_4) = x_1 x_2 + x_3 x_4.$$

---

* Cauchy: Exercices d'analyse, III, 173.

This function is unchanged by 8 of the possible 24 substitutions, namely by ·

$$s_1 = 1, \quad s_2 = (x_1 x_2), \quad s_3 = (x_3 x_4), \quad s_4 = (x_1 x_2)(x_3 x_4),$$
$$s_5 = (x_1 x_3 x_2 x_4), \quad s_6 = (x_1 x_4 x_2 x_3), \quad s_7 = (x_1 x_3)(x_2 x_4),$$
$$s_8 = (x_1 x_4)(x_2 x_3).$$

By the remaining $4! - 8 = 16$ substitutions φ is changed, and, in fact, is converted into either

$$x_1 x_3 + x_2 x_4 \quad \text{or} \quad x_1 x_4 + x_2 x_3.$$

We note then, in passing, that we have found here a three-valued function of four elements which is unchanged by 8 of the 24 possible substitutions of the latter.

§ 26. Those substitutions which leave a function $\varphi(x_1, x_2, \ldots x_n)$ unchanged, the number of which we shall always denote by r, we shall indicate by

$$G) \qquad s_1 = 1, s_2, s_3, \ldots s_r;$$

$s_1 = 1$ is of course contained among them. Following the notation of the preceding Section we have then

$$\varphi = \varphi_{s_1} = \varphi_{s_2} = \varphi_{s_3} = \ldots = \varphi_{s_r}.$$

By supposition there is no substitution s' different from $s_1, s_2, s_3, \ldots s_r$, which leaves the value of φ unchanged; *i. e.*, we have always

$$\varphi_{s'} \neq \varphi, \quad \text{if } s' \neq s_\lambda \quad (\lambda = 1, 2, 3, \ldots r).$$

If now we apply two substitutions s_α, s_β of our series $s_1, s_2, \ldots s_r$ successively to φ, and denote the result, as above in the case of a single substitution by $\varphi_{s_\alpha s_\beta}$, then since $\varphi_{s_\alpha} = \varphi$, the result of the two operations will be

$$\varphi_{s_\alpha s_\beta} = (\varphi_{s_\alpha})_{s_\beta} = \varphi_{s_\beta} = \varphi;$$

and from this we conclude that $s_\alpha s_\beta$ also occurs in the above series G). Every substitution therefore which is produced by the successive application of two substitutions of G) occurs itself in G). What is true of two substitutions of the series is further clearly true for any number whatever.

The substitution which results from the successive application of two or more substitutions we call their *product*, and we write the substitution σ which produces the same effect on the order of the elements $x_1, x_2, \ldots x_n$ as the successive application of s_α and s_β, as the product $\sigma = s_\alpha s_\beta$.

The product of any number of the operations s occurs again in the series G) $s_1, s_2, s_3, \ldots s_r$. The succession of the operations in a product $\sigma = s_\alpha s_\beta s_\gamma \ldots$ is to be reckoned from left to right.

§ 27. The expression of such a product in the cycle notation which we have adopted is obtained as follows:

If the two factors of a product are

$$s_\alpha = (x_a\, x_{a_1}\, x_{a_2} \ldots)\, (x_h\, x_{h_1}\, x_{h_2} \ldots)\ldots,$$
$$s_\beta = (x_b\, x_{b_1}\, x_{b_2} \ldots)\, (x_k\, x_{k_1}\, x_{k_2} \ldots)\ldots,$$

then in $s_\alpha s_\beta$ that element will follow x_a by which s_β replaces x_{a_1}. Suppose, for instance, that this element is x_{h_1}. Again in $s_\alpha s_\beta$ that element will follow x_{h_1} by which s_β replaces x_{h_2}. Let this be for example, x_{k_2}, etc. We obtain

$$s_\alpha s_\beta = (x_a\, x_{h_1}\, x_{k_2} \ldots) \ldots$$

If the substitution s_α be such that it replaces every index g of the elements $x_1, x_2, \ldots x_g, \ldots x_n$, by i_g, and if s_β be such that it replaces every index g by k_g, or, in formulae, if

$$s_\alpha = (x_1 x_{i_1} x_{i_{i_1}} \ldots)\, (x_a\, x_{i_a} \ldots) \ldots,$$
$$s_\beta = (x_1 x_{k_1} x_{k_{k_1}} \ldots)\, (x_b x_{k_b} \ldots) \ldots,$$

then the product will be of the form.

$$s_\alpha s_\beta = (x_1\, x_{k_{i_1}} \ldots)\, (x_s\, x_{k_{i_s}} \ldots) \ldots$$

The following may serve as an example:

$$s_\alpha = (x_1 x_3 x_5)\, (x_2 x_7), \quad s_\beta = (x_2 x_4 x_6)\, (x_3 x_5),$$
$$s_\alpha s_\beta = (x_1 x_5)\, (x_2 x_7 x_4 x_6)\, (x_3) = (x_1 x_5)\, (x_2 x_7 x_4 x_6).$$

We have introduced here the expression "product." The question now arises how far the fundamental rules of algebraic multiplication

$$a \cdot b = b \cdot a, \quad a \cdot (b \cdot c) = (a \cdot b) \cdot c$$

remain valid in this case. An examination of this matter will show that the former, the commutative law, in general fails, while the second, the associative law, is retained. In fact the multiplication of

CORRELATION OF FUNCTIONS AND GROUPS.

$$s_\alpha = (x_1 x_{i_1} x_{i_1} \ldots) \ldots, \quad s_\beta = (x_1 x_{k_1} x_{k_1} \ldots) \ldots,$$

as performed above shows that it is only in the special case where, for every α, $i_{k_\alpha} = k_{i_\alpha}$, that the order of the two factors s_α and s_β is indifferent. This occurs, for example, as is *à priori* clear, if the expressions for s_α and s_β contain no common elements.

We may therefore interchange the individual cycles of a substitution in any way, since these contain no common elements. In the notation B of page 19, on the other hand, this is not allowable.

Passing to the associative law, however, if

$$s_\alpha = (x_s x_{i_s} \ldots) \ldots, \quad s_\beta = (x_s x_{k_s} \ldots) \ldots, \quad s_\gamma = (x_s x_{l_s} \ldots) \ldots,$$

we have the following series of products,

$$s_\beta s_\gamma = (x_s x_{l_{k_s}} \ldots) \ldots, \quad s_\alpha s_\beta = x_s x_{k_{i_s}} \ldots) \ldots$$
$$s_\alpha (s_\beta s_\gamma) = (x_s x_{l_{k_{i_s}}} \ldots) \ldots, \quad (s_\alpha s_\beta) s_\gamma = (x_s x_{l_{k_{i_s}}} \ldots) \ldots,$$

from which follows

Theorem II. *In the multiplication of substitutions a collection of the factors into sub-products without change in the order of the factors, is permissible. An interchange of the factors, on the contrary, generally alters the result. Such an interchange is however permissible if the factors contain no common elements.*

§ 28. From the preceding developments it appears that those substitutions

$G)$ $\qquad s_1 = 1, s_2, s_3, \ldots s_a, \ldots s_r$

which leave a given function $\varphi(x_1, x_2, \ldots x_n)$ unchanged, form a closed group in this respect, that the multiplicative combination of its substitutions with one another leads only to operations already contained in the group.

The name "*group*" * we shall always use to denote a system of substitutions which possesses this characteristic property of reproducing itself by multiplication of its individual members. The number of elements operated on is called the *degree* of the group. It is not however necessary that all the elements should actually occur in the cycles of the substitutions. Thus

---

*Cauchy, who gave the first systematic presentation of the Theory of Substitutions in the Exercices d'Analyse et de Physique Mathématique, employs the name "*system of conjugate substitutions.*" Serret retains this name in his Algebra. The shorter name, "*group*," was introduced by Galois.

$$s_1 = 1, \quad s_2 = (x_1 x_2)(x_3 x_4)$$

form a group, for we have $s_1 s_2 = s_2 s_1 = s_2$, $s_2 s_2 = s_1 = 1$. This group is of degree 4, if only the elements x_1, x_2, x_3, x_4 be under consideration. But we might also regard the group as affecting, for example, 6 elements $x_1, x_2, \ldots x_6$, in which case we may, if desired, write for s_2

$$s_2' = (x_1 x_2)(x_3 x_4)(x_5)(x_6).$$

The degree of the group is then 6.

The number of substitutions contained in a group is called its *order*; as already stated, this number will always be denoted by r.

The entire system of substitutions which leave the value of a function $\varphi(x_1, x_2, \ldots x_n)$ unchanged is called *the group of substitutions belonging to the function φ*, or, more briefly, *the group of φ*. The degree of the group expresses the number of elements $x_1, x_2, \ldots x_n$ under consideration; its order gives the number of substitutions which leaves the function φ unchanged.

Thus, given the four elements x_1, x_2, x_3, x_4, and the function

$$\varphi = x_1 x_2 + x_3 x_4,$$

the degree of the group belonging to φ is 4; its order, as shown in § 25, is 8.

For the five elements x_1, x_2, x_3, x_4, x_5, the same function φ has a group of degree 5 and of order 8, identical with the preceding one.

For the six elements $x_1, x_2, x_3, \ldots x_6$, the same φ has a group of degree 6. To the group above we must now add all those substitutions which arise from combining the former with the interchange of x_5 and x_6. The group now contains beside the eight substitutions of § 25 the following eight new ones:

$$s_9 = (x_5 x_6), \quad s_{10} = (x_1 x_2)(x_5 x_6), \quad s_{11} = (x_3 x_4)(x_5 x_6),$$
$$s_{12} = (x_1 x_2)(x_3 x_4)(x_5 x_6), \quad s_{13} = (x_1 x_3 x_2 x_4)(x_5 x_6), \quad s_{14} = (x_1 x_4 x_2 x_3)(x_5 x_6),$$
$$s_{15} = (x_1 x_3)(x_2 x_4)(x_5 x_6), \quad s_{16} = (x_1 x_4)(x_2 x_3)(x_5 x_6).$$

The order of the group of φ is therefore now $8 \cdot 2 = 16$.

It is easy to see that if we regard φ as dependent on $n > 4$ elements, the order of the corresponding group becomes $8 \cdot (n-4)!$, the group being obtained by multiplying the 8 substitutions of § 25 by all the substitutions of the elements $x_5, x_6, \ldots x_n$.

§ 29. The following theorem is obviously true:

Theorem III. *For every single- or multiple-valued function there is a group of substitutions which, applied to the function, leave it unchanged.*

To show the perfect correlation of the theory of multiple-valued functions and that of groups of substitutions we will demonstrate the converse theorem:

Theorem IV. *For every group of substitutions there are functions which are unchanged by all the substitutions of the group and by no others.*

We begin by constructing a function φ of the n independent elements $x_1, x_2, \ldots x_n$ which shall take the greatest possible number of values, viz: $n!$; φ is therefore to be changed in value by the application of every substitution different from unity.

Taking $n+1$ arbitrary and different constants $a_0, a_1, \ldots a_n$, we form the linear expression
$$\varphi = a_0 + a_1 x_1 + a_2 + \ldots + a_n x_n$$
If now two substitutions $s_\alpha = (x_i x_{i_s} \ldots) \ldots$ and $s_\beta = (x_i x_{k_s} \ldots) \ldots$, on being applied to φ, gave the same result, we should have
$$0 = \varphi_{s_\alpha} - \varphi_{s_\beta} = a_1(x_{i_1} - x_{k_1}) + a_2(x_{i_2} - x_{k_2}) + \ldots$$
But the a_λ's being arbitrary quantities, this equation can only be satisfied if each parenthesis vanishes separately (§ 2, C), that is we must have $x_{i_\lambda} = x_{k_\lambda}$. But, if this be the case, since the x_λ's are also independent quantities, the two substitutions s_α and s_β both replace every x_λ by the same element $x_{i_\lambda} = x_{k_\lambda}$, so that s_α and s_β are identically the same. It is only in this case, where s_α and s_β are identical, that they can produce from φ the same value. Accordingly φ has $n!$ distinct values.

§ 30. If now a group G be given, composed of the substitutions
$$s_1 = 1, s_2, s_3, \ldots s_\alpha, \ldots s_r,$$
which we will indicate symbolically by the equation
$$G = [s_1, s_2, s_3 \ldots s_\alpha, \ldots s_r],$$
we apply all the substitutions of G to the $n!$-valued function with $n+1$ parameters

$$\varphi = a_0 + a_1 x_1 + a_2 x_2 + \ldots + a_n x_n,$$

and denote the results of these operations as before by corresponding subscripts attached to the φ's

$$\varphi_{s_1} = \varphi_1, \varphi_{s_2}, \varphi_{s_3} \ldots \varphi_{s_r}.$$

Then the product of these functions φ

$$\Phi = \varphi_{s_1} \cdot \varphi_{s_2} \cdot \varphi_{s_3} \ldots \varphi_{s_r}.$$

will be one of the functions to which the group of substitutions G belongs.

To prove this it must be shown 1) that φ is unchanged by every substitution σ of G, and 2) that Φ is changed in value by every substitution τ which does not occur in G.

In regard to the first condition we have

$$\Phi_\sigma = \varphi_{s_1 \sigma} \cdot \varphi_{s_2 \sigma} \cdot \varphi_{s_3 \sigma} \ldots \varphi_{s_r \sigma};$$

and, from the definition of a group, $s_1 \sigma = \sigma$, $s_2 \sigma$, $s_3 \sigma$, ... $s_r \sigma$ are again contained in G. Moreover, these products are all different from one another; for if $s_\alpha \sigma$ and $s_\beta \sigma$ applied to φ have the same effect, this must also be the case with s_α and s_β alone, and therefore s_α and s_β are identical. Accordingly the substitutions $s_1 \sigma = \sigma$, $s_2 \sigma$, $s_3 \sigma$, ... $s_r \sigma$ are identical, apart from their order, with $s_1 = 1$, s_2, s_3, ... s_r, and hence the functions

$$\varphi_{s_1 \sigma}, \varphi_{s_2 \sigma}, \varphi_{s_3 \sigma}, \ldots \varphi_{s_r \sigma}$$

are identical with

$$\varphi_{s_1}, \varphi_{s_2}, \varphi_{s_3}, \ldots \varphi_{s_r}$$

and accordingly

$$\Phi_\sigma = \Phi$$

as was asserted.

As to the second condition, the substitutions $s_1 \tau$, $s_2 \tau$, $s_3 \tau$, ... $s_r \tau$ are all different from s_1, s_2, s_3, ... s_r, and consequently the functions $\varphi_{s_1 \tau}, \varphi_{s_2 \tau}, \varphi_{s_3 \tau}, \ldots \varphi_{s_r \tau}$ are all different from the factors $\varphi_{s_1}, \varphi_{s_2}, \varphi_{s_3}, \ldots \varphi_{s_r}$ of Φ. Moreover, this difference is such that no $\varphi_{s_i \tau}$ can be equal to the product of a φ_{s_k} by a constant c_i, in which case

$$\Phi_\tau = \varphi_{s_1 \tau} \varphi_{s_2 \tau} \ldots \varphi_{s_r \tau} = c_1 c_2 c_3 \ldots c_r \varphi_{s_1} \varphi_{s_2} \varphi_{s_3} \ldots \varphi_{s_r}$$

would become equal to Φ if $c_1 c_2 c_3 \ldots c_r = 1$. For, if

$$a_0 + a_1 x_{\tau_1} + a_2 x_{\tau_2} + \ldots = c_i(a_0 + a_1 x_{i_1} + a_2 x_{i_2} + \ldots),$$

then, since the a_λ's are arbitrary quantities, it would follow at once that
$$c_i = 1,$$
and consequently we should have the impossible equation $\varphi_\tau = \varphi_{s_i}$.

§ 31. In many cases the calculation of Φ is impracticable, since the multiplication soon becomes unmanageable even for moderately large values of r. There is however, another process of construction in which the product is replaced by a sum, and every difficulty of calculation is removed.

We begin by taking as the basis for further construction, instead of the linear function φ, the following function
$$\psi(x_1, x_2, x_3, \ldots x_n) = x_1^{a_1} x_2^{a_2} x_3^{a_3} \ldots x_n^{a_n}.$$

The a_β's are to be regarded here, as before, as arbitrary quantities, and, as the x_λ's are also arbitrary, it follows at once that ψ is an $n!$-valued function. For, if
$$\psi_\sigma = \psi_\tau,$$
then we must have identically
$$x_1^{\beta_1} x_2^{\beta_2} x_3^{\beta_3} \ldots x_n^{\beta_n} = x_1^{\gamma_1} x_2^{\gamma_2} x_3^{\gamma_3} \ldots x_n^{\gamma_n},$$
and, from § 2, C, this is only possible if every β_i is equal to the corresponding γ_i, that is, if the substitutions σ and τ are identical.

We denote the functions which proceed from ψ under the operation of the substitutions of G by
$$\psi_{s_1} = \psi_1, \psi_{s_2}, \psi_{s_3} \ldots \psi_{s_r},$$
and form now the sum
$$\Psi = \psi_{s_1} + \psi_{s_2} + \psi_{s_3} + \ldots + \psi_{s_r}.$$
The proof of the correlation of G and ψ proceeds then exactly as in the preceding Section in the case of G and Φ.

REMARK.—By making certain assumptions with respect to the a's we can assign to the ψ's some new properties. Thus we may select the a's in such a way that an equation between any two arbitrary systems of the a's,
$$a_{i_1} + a_{i_2} + \ldots a_{i_\lambda} = a_{k_1} + a_{k_2} + \ldots a_{k_\mu},$$
necessarily involves the equality of the separate terms on the right and left. This condition is satisfied if, for example, we take

$a_2 > a_1$, $a_3 > a_1 + a_2$, $a_4 > a_1 + a_2 + a_3$, ...,

in particular if

$a_1 = 1$, $a_2 = 2$, $a_3 = 4$, $a_4 = 8$, $a_5 = 16$, ...

E. g. If $a_{i_1} + a_{i_2} + \ldots + a_{i_\lambda} = 13$, we must have $i_1 = 1$, $i_2 = 3$, $i_3 = 4$, $\lambda = 3$.

EXAMPLE.—We will apply the two methods given above to the familiar group

$G = [1, (x_1 x_2), (x_3 x_4), (x_1 x_2)(x_3 x_4), (x_1 x_3)(x_2 x_4), (x_1 x_4)(x_2 x_3),$
$(x_1 x_3 x_2 x_4), (x_1 x_4 x_2 x_3)]$, $(n = 4, r = 8)$

taking as fundamental functions

$$\varphi = x_1 + i x_2 - x_3 - i x_4 \text{ and } \psi = x_1^0 x_2^1 x_3^2 x_4^3,$$

where, as usual, $i = \sqrt{-1}$.

We have then the following results:

$\Phi = (x_1 + i x_2 - x_3 - i x_4)(x_2 + i x_1 - x_3 - i x_4)(x_1 + i x_2 - x_4 - i x_3)$
$(x_2 + i x_1 - x_4 - i x_3)(x_3 + i x_4 - x_1 - i x_2)(x_4 + i x_3 - x_2 - i x_1)$
$(x_3 + i x_4 - x_2 - i x_1)(x_4 + i x_3 - x_1 - i x_2)$
$= [(x_1 + i x_2 - x_3 - i x_4)(x_2 + i x_1 - x_3 - i x_4)(x_1 + i x_2 - x_4 - i x_3)$
$(x_2 + i x_1 - x_4 - i x_3)]^2$
$= \{[(x_1 - x_3)^2 + (x_2 - x_4)^2][(x_1 - x_4)^2 + (x_2 - x_3)^2]\}^2$,

$\Psi = x_2 x_3^2 x_4^3 + x_1 x_3^2 x_4^3 + x_2 x_4^2 x_3^3 + x_1 x_4^2 x_3^3 + x_4 x_1^2 x_2^3 + x_3 x_2^2 x_1^3$
$+ x_4 x_2^2 x_1^3 + x_3 x_1^2 x_2^3$
$= (x_1 + x_2)(x_3^2 x_4^3 + x_4^2 x_3^3) + (x_3 + x_4)(x_1^2 x_2^3 + x_2^2 x_1^3)$
$= (x_1 + x_2)(x_3 + x_4)(x_3^2 x_4^2 + x_1^2 x_2^2)$.

Neither of the two methods furnishes simple results directly. But from Φ we may pass at once to the function

$$[(x_1 - x_2)^2 + (x_3 - x_4)^2][(x_1 - x_4)^2 + (x_2 - x_3)^2],$$

and from Ψ to the two functions

$$(x_1 + x_2)(x_3 + x_4) \text{ and } x_1 x_2 + x_3 x_4,$$

the latter being already known to us. It is clear also that by altering the exponents which occur in Ψ we can obtain a series of functions all of which belong to G. Among these are included all functions of the form

$$(x_1^a + x_2^a)(x_3^a + x_4^a), \quad x_1^a x_2^a + x_3^a x_4^a.$$

In general we perceive that to every group of substitutions there belong an infinite number of functions.

It may be observed however that we cannot obtain *all* functions belonging to a given group by the present methods. Thus the function

$$x_1 x_2 + x_3 x_4 - (x_1 x_3 + x_2 x_4)$$

belongs to the group

$$G = [1, (x_1 x_2)(x_3 x_4), (x_1 x_2)(x_3 x_4)],$$

but cannot be obtained by these methods. More generally, if the functions $\varphi', \varphi'', \varphi''', \ldots$ belong respectively to the groups $H', H'', H''', \ldots$, and if the substitutions common to these groups (*Cf.* § 44, Theorem VII) form the given group G, then the function

$$\Psi = a_1 \varphi' + a_2 \varphi'' + a_3 \varphi''' + \ldots,$$

where the a's are arbitrary, belongs to the group G.

§ 32. We now proceed to consider the case where the elements $x_1, x_2, \ldots x_n$ are no longer independent quantities.

Theorem V. *Even where any system of relations exists among the elements* $x_1, x_2, \ldots x_n$, *excluding only the case of the equality of two or more elements, we can still construct* $n!$-*valued functions of* $x_1, x_2, \ldots x_n$.[*]

Using the notation of the preceding Section, we start from the same linear function

$$\varphi = a_0 + a_1 x_1 + a_2 x_2 + \ldots + a_n x_n,$$

and form the product of the differences of the φ's

$$\prod (\varphi_\sigma - \varphi_\tau)$$

this product being taken over the $\dfrac{n!(n!-1)}{1 \cdot 2}$ possible combinations of the φ's in pairs. Expanding we have

$$\prod (\varphi_\sigma - \varphi_\tau) = \prod \left[a_1(x_{\sigma_1} - x_{\tau_1}) + a_2(x_{\sigma_2} - x_{\tau_2}) \ldots + a_n(x_{\sigma_n} - x_{\tau_n}) \right]$$

In no one of these factors can all the parentheses vanish, since otherwise either the substitutions σ_i and τ_i must be identical, or else the x's are not all different. The product, regarded as a function of the a's, therefore cannot vanish identically (§ 2, **D**). Consequently,

---

[*] *Cf.* G. Cantor: Math. Annalen V, 133; Acta Math. I. 372-3.

(§ 2), there are an infinite number of systems of values of the a's for which all the $n!$ values of φ are different from one another.

Corollary. *The only relations among the x's which can determine the equality of linear expressions of the form*

$$\varphi = a_0 + a_1 x_1 + a_2 x_2 + \ldots + a_n x_n$$

independently of the values of the a's, are the relations of equality of two or more x's.

§ 33. With the Theorems III and IV the foundation is laid for a classification of the integral functions of n variables. Every function belongs to a group of substitutions; to every group of substitutions correspond an infinite number of functions. This relationship is not the only connection between functions which are unchanged by the same substitutions; we shall find also a corresponding algebraic relation, namely, that every function which belongs to a group G can be rationally expressed in terms of every other function belonging to the same group.

It becomes then a fundamental problem of algebra to determine all the possible groups of substitutions of n elements. The general solution of this problem, however, presents difficulties as yet insuperable. The existence of functions which possess a prescribed number of values is discussed in one of the following Chapters. It will appear that there are narrow limits to the number of possible groups. For example, in the case of 7 elements, there is no function which possesses 3, 4, 5, or 6 values; and we shall deduce the general proposition that a function of n elements which has more than two values, will have at least n values, if $n > 4$. A series of other analogous results will also be obtained.

For the present we shall concern ourselves only with the construction and the properties of some of the simplest, and for our purpose, most important groups. *

§ 34. First of all we have the group of order $n!$, composed of all the substitutions. This group belongs to the symmetric functions, and is called the *symmetric group*.

In Chapter I we have seen that every substitution is reducible to a series of transpositions. Accordingly, if a group contains all

---

* *Cf.* Serret: Cours d'algèbre supérieure. II, §§ 416-420. Cauchy: loc. cit.

the transpositions, it contains all the possible substitutions and is identical with the symmetric group. To secure this result it is however sufficient that the group should contain all those transpositions which affect any one element, for example x_1, that is the transpositions

$$(x_1 x_2), (x_1 x_3), (x_1 x_4), \ldots (x_1 x_n).$$

For every other transposition can be expressed as a combination of these $n - 1$; in fact every $(x_\alpha x_\beta)$ is equivalent to a series of three of the system above,

$$(x_\alpha x_\beta) = (x_1 x_\alpha)(x_1 x_\beta)(x_1 x_\alpha),$$

(where it is again to be noted that the order of the factors is not indifferent). We have then

Theorem VI. *A group of n elements $x_1, x_2, \ldots x_n$ which contains the $n - 1$ transpositions*

$$(x_\alpha x_1), (x_\alpha x_2), \ldots (x_\alpha x_{\alpha-1}), (x_\alpha x_{\alpha+1}), \ldots (x_\alpha x_n)$$

is identical with the symmetric group.

Corollary. *A group which contains the transpositions*

$$(x_\alpha x_\beta), (x_\alpha x_\gamma), \ldots (x_\alpha x_\vartheta)$$

contains all the substitutions of the symmetric group of the elements $x_\alpha, x_\beta, x_\gamma, \ldots x_\vartheta$.

§ 35. We know further a group composed of all those substitutions which are equivalent to an even number of transpositions. For all these substitutions, and only these, leave every two-valued function unchanged, and they therefore form a group. We will call this group the *alternating group*. Its order r is as yet unknown, and we proceed to determine it. Let

I) $\qquad s_1 = 1, s_2, s_3, \ldots s_r$

be all the substitutions of the alternating group, and let

II) $\qquad s_1', s_2', s_3', \ldots s_t'$

be all the substitutions which are not contained in I), and which are therefore composed of an odd number of transpositions. We select now any transposition σ, for example $\sigma = (x_1 x_2)$, and form the two series

I') $\qquad s_1 \sigma, s_2 \sigma, s_3 \sigma, \ldots s_r \sigma,$

II') $\qquad s_1' \sigma, s_2' \sigma, s_3' \sigma, \ldots s_t' \sigma.$

Then every substitution of I') is composed of an odd number of transpositions, and every substitution of II') of an even number. Consequently every substitution of I') is contained in II), and every one of II') is contained in I). Moreover, $s_\alpha \sigma \neq s_\beta \sigma$, and $s_\alpha' \sigma \neq s_\beta' \sigma$, for otherwise we should have

$$s_\alpha = s_\alpha (\sigma \cdot \sigma) = (s_\alpha \sigma) \sigma = (s_\beta \sigma) \sigma = s_\beta(\sigma \cdot \sigma) = s_\beta,$$
$$s_\alpha' = s_\alpha'(\sigma \cdot \sigma) = (s_\alpha' \sigma) \sigma = (s_\beta' \sigma) \sigma = s_\beta'(\sigma \cdot \sigma) = s_\beta',$$

since $\sigma \cdot \sigma = (x_1 x_2)(x_1 x_2) = 1$.

It follows from this that $r \leq t$ and $r \geq t$, that is $r = t$. Again, since I) and II) contain all the substitutions, $r + t = n!$. Hence

$$r = \frac{n!}{2}.$$

We will note here that there is no other group l' of order $\frac{n!}{2}$. For a function φ_1 belonging to such a group would be unchanged by $\frac{n!}{2}$ substitutions, and would be changed by all others. It would therefore possess other values beside φ_1. Suppose φ_2 to be one of these values, and let σ be a substitution which converts φ_1 into $\varphi_2 = \varphi_\sigma$. If now φ_1 is unchanged by the group

III) $\qquad\qquad l' = [1, s_2', s_3', \ldots s'_{\frac{1}{2}n!}],$

then φ_1 must be converted into φ_2 by all the substitutions

IV) $\qquad\qquad \sigma, s_2'\sigma, s_3'\sigma, \ldots s'_{\frac{1}{2}n!}\sigma;$

for s_λ' leaves φ_1 unchanged and σ converts φ_1 into φ_2, consequently $s_\lambda'\sigma$ will also convert φ_1 into φ_2. Again all the substitutions $s_\alpha'\sigma$ of the series IV) are different from one another. For, if $s_\alpha'\sigma = s_\beta'\sigma$, it would follow that $s_\alpha' = s_\beta'$. The substitutions $s_\lambda'\sigma$ are also different from the s_λ''s, for the latter have a different effect on φ_1 from the former. Consequently III) and IV) exhaust all the possible substitutions, and φ_1 is therefore a two-valued function, for there is no substitution remaining which could convert φ_1 into a third value. The group l' is therefore the alternating group.

Theorem VII. *For n elements there is only one group of order $\frac{n!}{2}$. This is the alternating group. It belongs to the two-valued functions.*

We can generalize this proposition. The proof, being exactly parallel to the preceding, may be omitted.

Theorem VIII. *Either all, or exactly half of the substitutions of every group belong to the alternating group.*

Corollary. *Those substitutions of any given group of order r which belong to the alternating group, form a group within the given group, the order of which is either r or $\frac{r}{2}$.*

The simplest substitutions belonging to the alternating group contain three elements in a single cycle, $(x_\alpha x_\beta x_\gamma)$. They are equivalent to two transpositions, $(x_\alpha x_\beta x_\gamma) = (x_\alpha x_\gamma)(x_\beta x_\gamma)$.

A substitution containing only one cycle $(x_{i_1}, x_{i_2}, \ldots x_{i_m})$ we shall call a *circular substitution* of order m.

Theorem IX. *If a group of n elements contains the $n-2$ circular substitutions*

$$(x_1 x_2 x_3), (x_1 x_2 x_4), \ldots (x_1 x_2 x_n),$$

it is either the alternating or the symmetric group.

For since

$$(x_\alpha x_\beta x_\gamma) = (x_1 x_2 x_\beta)(x_1 x_2 x_\beta)(x_1 x_2 x_\gamma)(x_1 x_2 x_\alpha)(x_1 x_2 x_\alpha)(x_1 x_2 x_\beta),$$

it follows that the given group contains every circular substitution of the third order. And again, since

$$(x_1 x_2 x_3)(x_1 x_4 x_3) = (x_1 x_2)(x_3 x_4), \quad (x_1 x_3 x_2)(x_1 x_4 x_2) = (x_1 x_3)(x_2 x_4),$$

it follows that all substitutions occur in the given group which are composed of two, and consequently of four, six, or any even number of transpositions. The theorem is therefore proved.

We add the following theorem:

Theorem X. *If a group contains all circular substitutions of order $m+2$, it will contain also all those of order m, and consequently it will contain either the alternating or the symmetric group, according as m is odd or even.*

For we have

$$(x_1 x_2 \ldots x_m x_a x_b)(x_1 x_2 \ldots x_m x_a x_b)(x_m x_{m-1} \ldots x_2 x_b x_1 x_a)$$
$$= (x_1 x_2 \ldots x_{m-1} x_m).$$

Finally, we can now give the criterion for determining whether a given substitution, expressed in cycles, belongs to the alternating group, or not. The proof is hardly necessary.

Theorem XI. *If a substitution contains m elements in k cycles, it does or does not belong to the alternating group according as $m - k$ is even or odd.*

§ 36. Any single substitution at once gives rise to a group, if we multiply it by itself *i. e.*, if we form its successive *powers*. The meaning of the term "power" is already fully defined by the developments of § 27. We must have

$$s^m = s^{m-1} \cdot s = s \cdot s^{m-1} = s^{m-2} \cdot s^2 = s^2 \cdot s^{m-2} = s^a \cdot s^{m-a}$$
$$= s^a \cdot s^\beta \cdot s^{m-a-\beta} = \ldots$$

The process of calculation of the powers of a substitution is also clear from the preceding Sections. If we wish to form the second, third, fourth, ... a^{th} power of a cycle, or of any substitution, we write after each element the second, third, fourth, ... a^{th} following element of the corresponding cycle, the first element of each cycle being regarded as following the last. Thus, from the cycle $(x_1 x_2 x_3 x_4 x_5 \ldots)$ we obtain for the second power $(x_1 x_3 x_5 \ldots)$, for the third $(x_1 x_4 x_7 \ldots)$, for the fourth $(x_1 x_5 x_9 \ldots)$, etc. It is obvious that in this process a cycle may break up into several. This will occur when and only when the number of elements of the cycle and the exponent of the power have a common divisor $d > 1$. The number of resulting cycles is then equal to d.

For example,
$$(x_1 x_2 x_3 x_4 x_5 x_6)^2 = (x_1 x_3 x_5)(x_2 x_4 x_6),$$
$$(x_1 x_2 x_3 x_4 x_5 x_6)^3 = (x_1 x_4)(x_2 x_5)(x_3 x_6),$$
$$(x_1 x_2 x_3 x_4 x_5 x_6)^4 = (x_1 x_5 x_3)(x_2 x_6 x_4),$$
$$(x_1 x_2 x_3 x_4 x_5 x_6)^5 = (x_1 x_6 x_5 x_4 x_3 x_2).$$

If the number of elements of a cycle be m, then the mth, $(2m)^{th}$, $(3m)^{th}$, ... powers of the cycle, and no others, will be equal to 1.

E. g. $(x_1 x_2 x_3 x_4 x_5 x_6)^6 = (x_1 x_2 x_3 x_4 x_5 x_6)^{12} = \ldots = 1.$

If a substitution contains several cycles with $m_1, m_2, m_3, \ldots$ elements respectively, the lowest power of the substitution which is equal to 1 is that of which the exponent r is the least common multiple of $m_1, m_2, m_3, \ldots$ Thus

$$[(x_1 x_2 x_3)(x_4 x_5)(x_6 x_7)]^6 = 1.$$

This same exponent r is also the order of the group formed by the powers of the given substitution. For if we calculate

$$s^2, s^3, \ldots s^{r-1}, s^r = 1,$$

a further continuation of the series gives merely a repetition of the same terms in the same order:

$$s^{r+1} = s, \ s^{r+2} = s^2, \ s^{r+3} = s^3, \ldots s^{2r-1} = s^{r-1}, \ s^{2r} = s^r = 1, \ldots$$

Moreover the powers of s from s^1 to s^r are different from one another, for if

$$s^\lambda = s^{\lambda+\mu} = s^\lambda \cdot s^\mu \quad (\lambda + \mu \leqq r),$$

then we should have contrary to hypothesis

$$s^\mu = 1 \quad (\mu < r).$$

The extension of the definition of a power to include the case of negative exponents is now easily accomplished. We write

$$s^{-k} = s^{r-k} = s^{2r-k} = \ldots$$

so that we have

$$s^k s^{-k} = 1.$$

The substitution s^k therefore cancels the effect of the substitution s^{-k}, and vice versa. The negative powers of a substitution are formed in the same way as the positive powers, only that in forming $(-1)^{\text{st}}, (-2)^{\text{d}}, (-3)^{\text{d}}, \ldots$ powers, we pass backward in each cycle $1, 2, 3, \ldots$ elements, the last element being regarded as next preceding the first.

It may be noted that $(st)^{-1} = t^{-1} s^{-1}$. For $(st)^{-1}(st) = 1$, and by multiplying the members of this equation first into t^{-1} and then into s^{-1}, we obtain the result stated.

The simplest function belonging to the cycle $(x_1 x_2 \ldots x_m)$ is

$$\varphi = x_1 x_2^2 + x_2 x_3^2 + \ldots + x_{m-1} x_m^2 + x_m x_1^2.$$

§ 37. Given two substitutions s_α and s_β, if we wish to determine the group of lowest order which contains s_α and s_β, we have not only to form all the powers s_α^λ, s_β^μ and to multiply these together, but we must form all the combinations

$$1, \ s_\alpha^\lambda, \ s_\beta^\mu, \ s_\alpha^\lambda s_\beta^\mu, \ s_\beta^\mu s_\alpha^\lambda, \ s_\alpha^\lambda s_\beta^\mu s_\alpha^\nu, \ s_\beta^\mu s_\alpha^\lambda s_\beta^\nu, \ldots$$

Of the substitutions thus formed we retain those which are different from one another, and proceed with the construction until *all* substitutions which arise from a product of m factors are contained among the preceding ones. For then every product of $m+1$ fac-

38   THEORY OF SUBSTITUTIONS.

tors is obviously reducible to one of m factors, and is consequently also contained among those already found. The group is then complete.

In case $s_\beta s_\alpha = s_\alpha s_\beta^\mu$, *the corresponding group is exhausted by all the substitutions of the form* $s_\alpha^\kappa s_\beta^\lambda$. For we have in this case

$$s_\beta^2 s_\alpha = s_\beta \cdot s_\alpha s_\beta^\mu = s_\beta s_\alpha \cdot s_\beta^\mu = s_\alpha s_\beta^{2\mu},$$
$$s_\beta^3 s_\alpha = s_\beta \cdot s_\alpha s_\beta^{2\mu} = s_\alpha s_\beta^{3\mu},$$
$$\cdot\ \cdot\ \cdot\ \cdot\ \cdot\ \cdot\ \cdot$$
$$s_\beta^m s_\alpha = s_\alpha s_\beta^{m\mu},$$
$$s_\beta^m s_\alpha^2 = s_\beta^m s_\alpha \cdot s_\alpha = s_\alpha s_\beta^{m\mu} s_\alpha = s_\alpha^2 s_\beta^{m\mu^2},$$
$$s_\beta^m s_\alpha^3 = s_\alpha^2 s_\beta^{m\mu^2} s_\alpha = s_\alpha^3 s_\beta^{m\mu^3},$$
$$\cdot\ \cdot\ \cdot\ \cdot\ \cdot\ \cdot\ \cdot$$
$$s_\beta^m s_\alpha^k = s_\alpha^k s_\beta^{m\mu^k}.$$

Consequently any product of three factors is reducible to a product of two. Thus

$$s_\alpha^\rho s_\beta^\sigma s_\alpha^\tau = s_\alpha^{(\rho+\tau)} s_\beta^{\mu^\tau \sigma},$$
$$s_\beta^\rho s_\alpha^\sigma s_\beta^\tau = s_\alpha^\sigma s_\beta^{(\tau+\rho)\mu^\sigma},$$

and the theorem is proved.

For example, let
$$s_1 = (x_1 x_2 x_3 x_4 x_5), \quad s_2 = (x_2 x_3 x_5 x_4);$$
then
$$s_2 s_1 = (x_1 x_2 x_4 x_3) = s_1^3 s_2.$$

The group of lowest order which contains s_1 and s_2 contains therefore at the most $5 \cdot 4 = 20$ substitutions. To determine whether the number is less than this, we examine whether it is possible that

$$s_1^\alpha s_2^\beta = s_1^\gamma s_2^\delta.$$

If this were the case, it would follow that

$$s_1^{\alpha-\gamma} = s_2^{\delta-\beta}.$$

But in the series of powers of s_2 there is only one which is also a power of s_1, and this is the zero power. Consequently we must have $\alpha = \gamma$ and $\beta = \delta$. The group therefore actually contains 20 substitutions. These are the following, where for the sake of simplicity we write only the indices:

CORRELATION OF FUNCTIONS AND GROUPS. 39

$s_1^0 = 1$, $s_2 = (2354)$, $s_2^2 = (25)(34)$, $s_2^3 = (2453)$,
$s_1^1 = (12345)$, $s_1 s_2 = (1325)$, $s_1 s_2^2 = (15)(24)$, $s_1 s_2^3 = (1435)$,
$s_1^2 = (13524)$, $s_1^2 s_2 = (1534)$, $s_1^2 s_2^2 = (14)(23)$, $s_1^2 s_2^3 = (1254)$,
$s_1^3 = (14253)$, $s_1^3 s_2 = (1243)$, $s_1^3 s_2^2 = (13)(45)$, $s_1^3 s_2^3 = (1523)$,
$s_1^4 = (15432)$, $s_1^4 s_2 = (1452)$, $s_1^4 s_2^2 = (12)(35)$, $s_1^4 s_2^3 = (1342)$.

Analogous results may be obtained, for example, for the case

$$s_1 = (x_1 x_2 x_3 x_4 x_5 x_6 x_7), \quad s_2 = (x_2 x_4 x_3 x_7 x_5 x_6).$$

In case every $s_\beta^\mu s_\alpha$ ($\mu = 1, 2, 3, \ldots$) can be reduced to the form $s_\alpha^\kappa s_\beta^\lambda$, the group of lowest order which contains s_α and s_β is exhausted by the substitutions of the form $s_\alpha^\kappa s_\beta^\lambda$.

For by processes similar to those above we can bring every substitution $s_\beta^\mu s_\alpha^\nu$ to the form $s_\alpha^\kappa s_\beta^\lambda$. The proof is then reduced to the preceding.

Furthermore if s_β^q is the lowest power in the series $s_\beta, s_\beta^2, \ldots$ which occurs among the power of s_α, then the group contains q times as many substitutions as the order k of s_α. For in the first place, if the exponent λ in $s_\alpha^\kappa s_\beta^\lambda$ is greater than $q-1$, we can replace s_β^λ by a $s_\alpha^\mu s_\beta^\nu$, where $\nu \leq q-1$. There are therefore at the most $q \cdot k$ different substitutions $s_\alpha^\kappa s_\beta^\lambda$. Again if

$$s_\alpha^\kappa s_\beta^\lambda = s_\alpha^\mu s_\beta^\nu \quad (\lambda, \nu \leq q-1),$$

then we must have, if we suppose $\lambda > \nu$,

$$s_\beta^{\lambda-\nu} = s_\alpha^{\mu-\kappa} \quad (\lambda - \nu < q-1),$$

and consequently $\lambda = \nu, \mu = \kappa$. There are therefore actually $q \cdot k$ different substitutions. It is readily seen that q is a divisor of the order r of s_β.

If three substitutions $s_\alpha, s_\beta, s_\gamma$, are such that for every μ

$$s_\beta^\mu s_\alpha = s_\alpha^\delta s_\beta^\epsilon, \quad s_\gamma^\mu s_\alpha = s_\alpha^\zeta s_\beta^\eta s_\gamma^\vartheta, \quad s_\gamma^\mu s_\beta = s_\alpha^\iota s_\beta^\kappa s_\gamma^\lambda,$$

and if k be the order of s_α, and s_β^q the lowest power of s_β which is equal to a power of s_α, s_α^ν, finally if s_γ^t is the lowest power of s_γ which is equal to $s_\alpha^\kappa s_\beta^\nu$, then the group of lowest order which contains $s_\alpha, s_\beta, s_\gamma$ is of order kqt, and its substitutions are of the form

$$s_\alpha^\delta s_\beta^\epsilon s_\gamma^\zeta \quad (\delta = 0, 1, \ldots k-1;\ \epsilon = 0, 1, \ldots q-1;\ \zeta = 0, 1, \ldots t-1).$$

The proof is simple and so clearly analogous to the preceding that we may omit it.

§ 38. With this set of propositions belongs also the following:
If
$$G = [1, s_2, s_3, \ldots s_r]$$
$$H = [1, t_2, t_3, \ldots t_{r'}]$$
be two groups of substitutions between which the relation
$$s_\alpha t_\beta = t_\gamma s_\delta$$
holds for all values of α and β, and if furthermore G and H have no substitution in common except identity, then all the combinations $s_\alpha t_\beta$, or $t_\beta s_\alpha$, ($\alpha = 1, 2, \ldots r$; $\beta = 1, 2, \ldots r'$) form a group of order rr' which contains G and H as subgroups.

For since
$$s_\alpha t_\beta \cdot s_\gamma t_\delta = s_\alpha(t_\beta s_\gamma) t_\delta = s_\alpha(s_\epsilon \cdot t_\zeta) t_\delta = s_\mu t_\nu,$$
the substitutions $s_\alpha t_\beta$ form a group of, at the most, rr' substitutions. And we will show also that all of these are different from one another. For if, for example,
$$s_\alpha t_\beta = s_\gamma t_\delta$$
then if we multiply both sides of this equation by s_γ^{-1} at the left and t_β^{-1} at the right we obtain
$$s_\gamma^{-1} s_\alpha = t_\delta t_\beta^{-1}.$$
But $s_\gamma^{-1} s_\alpha$ is a substitution of G and $t_\delta t_\beta^{-1}$ a substitution of H, and consequently they can only be equal if both are equal to 1. Hence
$$s_\gamma^{-1} s_\alpha = 1, \qquad t_\delta t_\beta^{-1} = 1,$$
$$s_\alpha = s_\gamma, \qquad t_\delta = t_\beta.$$

The substitutions of the new group are then all different, and the order of the group is therefore rr'. We denote the group by
$$K = \{G, H\}.$$

We add without proof, the following generalization of the last theorem.

Under the same assumption $s_\alpha t_\beta = t_\gamma s_\delta$, if the two groups G and H have λ substitutions in common there is a group of order $\frac{rr'}{\lambda}$ which contains G and H as subgroups.*

§ 39. In later developments a group will frequently be required

---

*F. Giudice. Palermo Rend. I, pp. 222-223.

CORRELATION OF FUNCTIONS AND GROUPS. 41

the order of which is a power of a prime number p. The existence of such a group will be demonstrated by the proof of the following proposition, from which the nature of the group will also be apparent.

Theorem XII. *If p^f be the highest power of the prime number p which is a divisor of the product $n! = 1 \cdot 2 \cdot 3 \ldots n$, then there is a group of degree n and of order p.*

In the first place suppose $n < p^2$, so that $n = ap + b$ $(a, b < p)$. Then, of the numbers $1, 2, 3, \ldots n$, only $p, 2p, 3p, \ldots ap$ are divisible by p, so that $f = a$. We select now from the n elements a systems of p letters each, and form from each system a cycle, as follows:

$$s_1 = (x_{i_1} x_{i_2} x_{i_3} \ldots x_{i_p}), \quad s_2 = (x_{k_1} x_{k_2} x_{k_3} \ldots x_{k_p}), \quad s_a = (x_{t_1} x_{t_2} x_{t_3} \ldots x_{t_p})$$

Then the group which arises from these is the group required:

$$K_1 = [s_1, s_1^2, \ldots s_2, s_2^2, \ldots s_a, s_a^2, \ldots] \doteq \{s_1, s_2, \ldots s_a\}.*$$

For every s_λ with its various powers forms a subgroup of order p, and since no two of these a subgroups have any element in common, it follows from Theorem II that

$$s_\lambda^\rho s_\mu^\sigma = s_\mu^\sigma s_\lambda^\rho,$$

Accordingly every possible combination of substitutions $s_1^a, s_2^\beta, \ldots$ belonging to K can be brought to the form

$$s_1^a s_2^\beta s_3^\gamma \ldots s_a^\nu \quad (a, \beta, \gamma, \ldots \nu = 0, 1, 2, \ldots p-1).$$

The group K therefore contains at the most p^a substitutions. And it actually contains this number, for all these p^a operations are different from one another. For if

$$s_1^a s_2^\beta s_3^\gamma \ldots s_a^\nu = s_1^{a'} s_2^{\beta'} s_3^{\gamma'} \ldots s_a^{\nu'},$$

it would follow that

$$s_1^{-a'} s_1^a = s_1^{a-a'} = s_2^{\beta'} s_3^{\gamma'} \ldots s_a^{\nu'} s_a^{-\nu} s_{a-1}^{-\mu} \ldots s_3^{-\gamma} s_2^{-\beta} = s_2^{\beta'-\beta} s_3^{\gamma'-\gamma} \ldots$$

and therefore, since s_1, has no element in common with $s_2, s_3, \ldots$, we must have $a = a'$, etc.

Again, if $n = p^2$, we shall have $f = p+1$, since in the series

---
*In the designation of a group the brace, $\{\ \}$, as distinguished from the bracket, $[\ ]$, indicates that the group referred to is the smallest group which contains the included substitutions. The bracket contains *all* the substitutions of the group considered, while the brace contains only the *generating* substitutions. The latter can generally be selected in many ways. *Cf.* the notation at the close of the last Section.

3a

1, 2, 3, ... p^2, the numbers p, $2p$, $3p$, ... $(p-1)p$, p^2 are divisible by p. We form now again the substitutions

$$s_1, s_2, s_3, \ldots s_p,$$

as before, and in addition to these the substitution s_{p+1} which affects all the p^2 elements

$$s_{p+1} = (x_{i_1} x_{k_1} x_{l_1} \ldots x_{i_1} x_{i_2} x_{k_2} \ldots x_{t_2} x_{l_2} \ldots x_{l_a} \ldots x_{l_p}).$$

Then the required group is

$$K_2 = \{s_1, s_2, \ldots s_p, s_{p+1}\}.$$

For in the first place we can readily show that

$$s_1 s_{p+1} = s_{p+1} s_2, \qquad s_a s_{p+1} = s_{p+1} s_{a+1}, \qquad s_p s_{p+1} = s_{p+1} s_1,$$
$$s_1^\lambda s_{p+1} = s_{p+1} s_2^\lambda, \qquad s_a^\lambda s_{p+1} = s_{p+1} s_{a+1}^\lambda, \qquad s_p^\lambda s_{p+1} = s_{p+1} s_1^\lambda,$$
$$s_1^\lambda s_{p+1}^2 = s_{p+1}^2 s_3^\lambda, \qquad s_a^\lambda s_{p+1}^2 = s_{p+1}^2 s_{a+2}^\lambda, \qquad s_p^\lambda s_{p+1}^2 = s_{p+1}^2 s_2^\lambda,$$
$$s_1^\lambda s_{p+1}^\mu = s_{p+1}^\mu s_{\mu+1}^\lambda, \qquad s_a^\lambda s_{p+1}^\mu = s_{p+1}^\mu s_{a+\mu}^\lambda, \qquad s_p^\lambda s_{p+1}^\mu = s_{p+1}^\mu s_\mu^\lambda.$$

Accordingly every combination of the substitutions $s_1, s_2, \ldots s_{p+1}$ can be brought, as in § 37, to the form

$$s_1^\alpha s_2^\beta s_3^\gamma \ldots s_p^\iota s_{p+1}^\kappa \qquad (\alpha, \beta, \gamma, \ldots \iota, \kappa = 0, 1, 2, \ldots p-1).$$

But we must also show in the present case that we need only take the powers of s_{p+1} as far as the $(p-1)^{\text{th}}$. We find that

$$s_{p+1}^p = (x_{i_1} x_{i_2} x_{i_3} \ldots x_{i_p})(x_{k_1} x_{k_2} \ldots x_{k_p}) \ldots (x_{l_1} x_{l_2} \ldots x_{l_q})$$
$$= s_1 s_2 \ldots s_p,$$
$$s_{p+1}^{ap} = s_1^a s_2^a \ldots s_p^a.$$

Consequently, if $k > p$, we can replace the highest power of s_{p+1}^κ which occurs in s_{p+1}^k by powers of $s_1, s_2, \ldots s_p$, and these can then be written in the order above.

The question then remains whether the $p^{p+1} = p^{p'}$ substitutions thus obtained are all distinct. If two of them were equal

$$s_1^\alpha s_2^\beta \ldots s_p^\iota s_{p+1}^\kappa = s_1^{\alpha'} s_2^{\beta'} \ldots s_p^{\iota'} s_{p+1}^{\kappa'},$$

we should have

$$s_{p+1}^{\kappa-\kappa'} = s_1^{\alpha'-\alpha} s_2^{\beta'-\beta} \ldots s_p^{\iota'-\iota}.$$

But the substitution on the right does not affect the first subscripts $i, k, \ldots$, while that on the left does, unless $\kappa = \kappa'$. The proof then proceeds as before.

If $n > p^2$ but $< p^3$, that is, if $n = ap^2 + bp + c$ ($a, b, c < p$), we select from the n elements x_λ any a systems of p^2 elements each,

CORRELATION OF FUNCTIONS AND GROUPS. 43

and any other b systems of p elements each. With the former we construct a groups K_2, and with the latter b groups K_1. The combination of these $a+b$ groups gives the required group K_3. For the product of the numbers

$$(a-1)p^2+1, (a-1)p^2+2, \ldots (a-1)p^2+p, \ldots$$
$$(a-1)p^2+p^2, \quad (a<p)$$

is divisible by only the same power of p as the product of

$$1, 2, \ldots p, \ldots p^2.$$

Again, if $n = p^3$, then the exponent f of p^f is increased by 1 on account of the last term of the series

$$(p-1)p^2+1, (p-1)p^2+2, \ldots (p-1)p^2+p, \ldots$$
$$(p-1)p^2+p^2 = p^3,$$

so that in this case the multiplication of the p partial groups K_2 is not sufficient. In this case, exactly as for $n = p^2$, we add another substitution which contains all the $p \cdot p^2$ elements in a single cycle, and the pth power of which breaks up into the p substitutions as inr the case of $s_{\mu+1}$ above. Then, as in that case, we can show that the new group satisfies all the requirements. At the same time it is clear that the method here followed is perfectly general, and accordingly the theorem at the beginning of the Section is proved.

§ 40. Since all the groups $K_1, K_2, K_3, \ldots$ enter into the formation of the group K, we have the following

Corollary. *If p^f is the highest power of p which is a divisor of $n!$, then we can construct a series of groups of n elements*

$$1, K_1, K_2, \ldots K_\lambda, K_{\lambda+1}, \ldots K_f$$

which are of order respectively

$$1, p, p^2, \ldots p^\lambda, p^{\lambda+1}, \ldots p^f.$$

Every group K_λ is contained as a subgroup in the next following $K_{\lambda+1}$.

CHAPTER III.

THE DIFFERENT VALUES OF A MULTIPLE-VALUED FUNCTION AND THEIR ALGEBRAIC RELATION TO ONE ANOTHER.

§ 41. We have shown in the preceding Chapter, that to every function of n elements $x_1, x_2, x_3, \ldots x_n$ there belongs a group of substitutions, and that conversely to every group of substitutions there correspond an infinite number of functions of the elements. The examination of the relations between different functions which belong to the same group we reserve for a later Chapter. The problem which we have first to consider is the determination of the connection between the several values of a multiple-valued function and the algebraic relations of these values to one another.

If $\varphi(x_1, x_2, \ldots x_n)$ is not a symmetric function, or, in other words, if the substitutions $s_1 = 1, s_2, s_3, \ldots s_r$ of the group G belonging to φ do not exhaust all the possible substitutions (*i.e.* $r < n!$), then φ, on being operated upon by any one of the remaining substitutions σ_2, will take a new value $\varphi_2 = \varphi_{\sigma_2}$.

We proceed to construct a table, the first line of which consists of the various substitutions of the group G:

$$s_1 = 1, s_2, s_3, \ldots s_r; \quad G; \quad \varphi_1.$$

The second line is obtained from the first by right hand multiplication of all the substitutions s_λ by σ_2. This gives us

$$\sigma_2, s_2\sigma_2, s_3\sigma_2, \ldots s_r\sigma_2; \quad G \cdot \sigma_2; \quad \varphi_2.$$

We show then, as in Chapter II, § 35, 1) that all substitutions of this line convert φ_1 into φ_2; for since $\varphi_{s_a} = \varphi_1$, it follows that $\varphi_{s_a\sigma_2} = \varphi_{\sigma_2} = \varphi_2$; 2) that no other substitutions except those of this line convert φ_1 into φ_2; for if τ is a substitution which has this effect, then we shall have

$$\varphi_{\tau\sigma_2^{-1}} = \varphi_{\sigma_2\sigma_2^{-1}} = \varphi_1$$

so that $\tau\sigma_2^{-1}$ leaves the function φ_1 unchanged; consequently $\tau\sigma_2^{-1} = s_\lambda$ and $\tau = (\tau\sigma_2^{-1})\sigma_2 = s_\lambda\sigma_2$, and therefore τ is contained in the second line; 3) that all substitutions of this line are distinct; for if $s_\alpha\sigma_2 = s_\beta\sigma_2$, it follows that $s_\alpha = s_\beta$; 4) that the substitutions of this line are all different from those of the first; for the latter all leave φ_1 unchanged, while the former all convert φ_1 into φ_2.

If the $2r$ substitutions s_λ and $s_\lambda\sigma_2$ do not yet exhaust all the possible $n!$ substitutions, then any remaining substitution σ_3 will convert φ_1 into a new function $\varphi_{\sigma_3} = \varphi_3$; for all the substitutions which produce φ_1 and φ_2 are already contained in the first two lines. By the aid of σ_3 we form the third line of our table

$$\sigma_3,\ s_2\sigma_3,\ s_3\sigma_3,\ \ldots s_r\sigma_3;\quad G \cdot \sigma_3;\quad \varphi_3.$$

The substitutions of this line have again the four properties just discussed. They are all the substitutions and the only ones that convert φ_1 into φ_3, and they are all different from one another and from those of the preceding lines.

If these $3r$ substitutions do not exhaust all the possible $n!$, we proceed in the same way, until finally all the $n!$ substitutions are arranged in lines containing r each.

We shall frequently have occasion to construct tables of this kind. All these tables will possess the properties: 1) that all the substitutions in any line will have a special character; 2) that only the substitutions of this line will have this character; 3) that all the substitutions of any line are different from one another; and frequently, but not always, the fourth property will also appear: 4) that all the substitutions of any line are different from those of any other line.*

Summarizing the preceding results we have the following

Theorem I. *If the multiple-valued function $\varphi(x_1, x_2, \ldots x_n)$ has in all p values $\varphi_1, \varphi_2, \varphi_3, \ldots \varphi_p$ and if φ is converted successively into these values by certain substitutions, for example $1, \sigma_2, \sigma_3, \ldots \sigma_p$, furthermore, if G, the group of φ, is of order r and contains the substitutions $s_1 = 1,\ s_2, s_3, \ldots s_r$, we can arrange all the possible $n!$ substitutions as in the following table:*

---

*Such tables were given by Cauchy: Exercices d'analyse et de physique mathematique, III., p. 184.

$$\begin{array}{llllll}
\varphi_1; & 1, & s_2, & s_3, & \ldots s_r; & G_1 \\
\varphi_2; & \sigma_2, & s_2\sigma_2, & s_3\sigma_2, & \ldots s_r\sigma_2; & G_1 \cdot \sigma_2 \\
\varphi_3; & \sigma_3, & s_2\sigma_3, & s_3\sigma_3, & \ldots s_r\sigma_3; & G_1 \cdot \sigma_3 \\
\cdot & \cdot & \cdot & \cdot & \cdot & \cdot \\
\varphi_\rho; & \sigma_\rho, & s_2\sigma_\rho, & s_3\sigma_\rho, & \ldots s_r\sigma_\rho; & G_1 \cdot \sigma_\rho
\end{array}$$

in which every line contains all and only those substitutions which convert φ into the value φ_a prefixed to the line.

The several values $\varphi_1, \varphi_2, \ldots \varphi_\rho$ of the function φ are called *conjugate* values.

§ 42. From the circumstance that all the substitutions of this table are different from one another, and that the ρ lines of the table exhaust all the possible substitutions we deduce the following theorems:

Theorem II. *The order r of a group G of n elements is a divisor of $n!$*

Theorem III. *The number ρ of the values of an integral function of n elements is a divisor of $n!$*

Theorem IV. *The product of the number ρ of the values of an integral function by the order r of the corresponding group is equal to $n!$*

The third theorem imposes a considerable limitation on the possible number of values of a multiple-valued function. Thus, for example, there can be no seven- or nine-valued functions of five elements. But the limits thus obtained are still far too great, as the investigations of Chapter V will show.

§ 43. Precisely the same method as that of § 41 can be applied to the more general case where all the substitutions of the group G belonging to φ are contained in a group H belonging to another function ψ, so that G is a part or subgroup of H, just as in the special case above G was a subgroup of the entire or symmetric group. We see at once that all the substitutions of H can be arranged in a series of lines, each line containing r substitutions of the form $s_\lambda \sigma_\mu$ $(\lambda = 1, 2, \ldots r)$. And we pass directly from the preceding to the present case by reading everywhere for "all possible substitutions" simply "all the r_1 substitutions of H". We have then

Theorem V. *If all the r substitutions of the group G are contained among those of a group H of order r_1, then r is a divisor of r_1.*

Theorem VI. *Given two functions $\varphi\,(x_1, x_2, \ldots x_n)$ and $\psi\,(x_1, x_2, \ldots x_n)$, if ψ retains the same values for all substitutions which leave φ unchanged, the total number of values ρ of φ is a multiple of the total number of values ρ_1 of ψ.*

For we have
$$\rho = \frac{n!}{r}, \quad \rho_1 = \frac{n!}{r_1}; \quad \therefore \frac{\rho}{\rho_1} = \frac{r_1}{r}.$$

Corollary. *If a function $\varphi\,(x_1, \ldots x_n)$ belongs to a subgroup G of the group H, and if r is the order of G and r_1 that of H, then φ on being operated upon by the substitutions of H takes exactly $\dfrac{r_1}{r}$ values.*

§ 44. By a still further extension of the subject we may include the case where two groups G and H contain any substitutions in common. This case can be at once reduced to that of the preceding Section. For this purpose we employ the following proposition:

Theorem VII. *The substitutions common to two groups form a new group, the order of which is accordingly a divisor of the orders of both the given groups.*

For if σ and τ belong to both G_1 and G_2 then $\sigma \cdot \tau$ also belongs to both G_1 and G_2 and occurs among the common substitutions. The same result can also be obtained as follows. If φ_1 and φ_2 be functions belonging to G_1 and G_2 respectively, then the function

$$\psi = \alpha\,\varphi_1 + \beta\,\varphi_2,$$

where α and β are arbitrary constants, remains unchanged for those substitutions which leave both φ_1 and φ_2 unchanged, that is, which are common to G_1 and G_2. These, being all the substitutions which belong to ψ, form a group.

Corollary I. *Two groups whose orders are prime to each other can have no substitutions in common except the identical substitution.*

Corollary II. *The order of every group H which consists of all or a part of the substitutions common to the two groups G_1 and G_2 is a divisor of both r_1 and r.*

§ 45. We proceed next to determine and tabulate the groups which belong to the various values φ_i of φ, $(i = 1, 2, 3, \ldots \rho)$.

The group $G = G_1$ of φ_1 contained the substitutions

$$s_1 = 1, s_2, s_3, \ldots s_r.$$

The value φ_2 of φ was obtained from φ_1 by the substitution σ_2; consequently σ_2^{-1} converts φ_2 back into φ_1. If then we apply to φ_2 successively the substitutions $\sigma_2^{-1}, s_a, \sigma_2$, the first of these will convert φ_2 into φ_1, the second will leave φ_1 unchanged, and the third will convert φ_1 back into φ_2. It appears therefore that φ_2 is unchanged by every substitution of the form $\sigma_2^{-1} s_a \sigma_2$, $(a = 1, 2, \ldots r)$. For the second line of our table we take therefore

$$\sigma_2^{-1} s_1 \sigma_2 = 1, \quad \sigma_2^{-1} s_2 \sigma_2, \quad \sigma_2^{-1} s_3 \sigma_2, \quad \ldots \quad \sigma_2^{-1} s_r \sigma_2.$$

We can then show that this line contains all the substitutions belonging to the group of φ_2. For if τ be any substitution which leaves φ_2 unchanged, then $\tau \sigma_2^{-1}$ will convert φ_2 into φ_1, that is

$$(\varphi_2)_{\tau \sigma_2^{-1}} = (\varphi_{\sigma_2})_{\tau \sigma_2^{-1}} = \varphi_{\sigma_2 \tau \sigma_2^{-1}} = \varphi_1.$$

Consequently the substitution $\sigma_2 \tau \sigma_2^{-1}$ belongs to the group of φ_1, and we may write it equal to s_a. But from the equation $\sigma_2 \tau \sigma_2^{-1} = s_a$ follows $\tau = \sigma_2^{-1}(\sigma_2 \tau \sigma_2^{-1})\sigma_2 = \sigma_2^{-1} s_a \sigma_2$, as was asserted.

Again it is easily seen that all the substitutions of the second line are different from each other. For if

$$\sigma_2^{-1} s_a \sigma_2 = \sigma_2^{-1} s_\beta \sigma_2,$$

it follows that

$$s_a = s_\beta.$$

We may note however that the substitutions of the second line are not necessarily different from those of the first. In fact the identical substitution is of course always common to both and other substitutions may also occur in common. (*Cf.* § 50).

From the three properties of the second line obtained above, it follows that the r substitutions of this line form the group of φ_2. We will denote this group by G_2. That these substitutions from a group can also be shown formally; for

MULTIPLE-VALUED FUNCTIONS — ALGEBRAIC RELATIONS.   49

$$(\sigma_2^{-1} s_a \sigma_2)(\sigma_2^{-1} s_\beta \sigma_2) = \sigma_2^{-1} s_a (\sigma_2 \sigma_2^{-1}) s_\beta \sigma_2 = \sigma_2^{-1}(s_a s_\beta) \sigma_2,$$

so that, if s_a, s_β, ... form a group, as was assumed, the same is true of the new substitutions.

Similar results hold for all the other values, $\varphi_3, \varphi_4, \ldots \varphi_\rho$ of φ, and we have therefore

Theorem VIII. *If the values $\varphi_1, \varphi_2, \ldots \varphi_\rho$ of a ρ-valued function φ proceed from φ by the application of the substitutions $\sigma_1 = 1, \sigma_2, \sigma_3, \ldots \sigma_\rho$, then the groups $G_1, G_2, \ldots G_\rho$ of $\varphi_1, \varphi_2, \ldots \varphi_\rho$ respectively are*

$$G_1 = [s_1 = 1, s_2, s_3, \ldots s_r],$$
$$G_2 = [\sigma_2^{-1} s_1 \sigma_2 = 1, \sigma_2^{-1} s_2 \sigma_2, \sigma_2^{-1} s_3 \sigma_2, \ldots \sigma_2^{-1} s_r \sigma_2] = \sigma_2^{-1} G_1 \sigma_2,$$
$$\cdots\cdots\cdots\cdots\cdots\cdots\cdots\cdots\cdots\cdots\cdots\cdots$$
$$G_\rho = [\sigma_\rho^{-1} s_1 \sigma_\rho = 1, \sigma_\rho^{-1} s_2 \sigma_\rho, \sigma_\rho^{-1} s_3 \sigma_\rho, \ldots \sigma_\rho^{-1} s_r \sigma_\rho] = \sigma_\rho^{-1} G_1 \sigma_\rho.$$

§ 46. The functions $\varphi_1, \varphi_2, \ldots \varphi_\rho$ are of precisely the same form and only differ in the order of arrangement of the x_λ's which enter into them. Such functions we have called (§ 3) *similar* or *of the same type*. Accordingly the corresponding groups $G_1, G_2, \ldots$ must also be *similar* or *of the same type*, that is they produce the same system of rearrangement of the elements x_λ and only differ in the order in which the elements are numbered. This is clear à priori, but we can also prove it from the manner of derivation of $\sigma_i^{-1} s_a \sigma_i$ from s_a, and in fact we can show that not only the groups $G_1, G_2, \ldots$, but also the individual substitutions s_a and $\sigma_i^{-1} s_a \sigma_i$ are similar; that is, these two substitutions have the same number of cycles, each containing the same number of elements. The process of deriving the substitutions $\sigma_i^{-1} s_a \sigma_i$ from the s_a's is called *transformation*. The substitution $\sigma_i^{-1} s_a \sigma_i$ is called the *conjugate of s_a with respect to σ_i*, and similarly the group G_i is the *conjugate of G_1 with respect to σ_i*. We shall denote this latter relation frequently, as above, by the equation

$$G_i = \sigma_i^{-1} G_1 \sigma_i.$$

To prove the similarity of s_a and $\sigma^{-1} s_a \sigma_i$, let us suppose that $(x_1, x_2, \ldots x_a)$ is any one of the cycles of s_a and that σ_i replaces $x_1, x_2, \ldots x_a$ by $x_{i_1}, x_{i_2}, \ldots x_{i_a}$, so that, in the notation A of § 22, σ may be written

4

50   THEORY OF SUBSTITUTIONS.

$$\sigma_i = \begin{pmatrix} x_1 & x_2 & \ldots & x_a \\ x_{i_1} & x_{i_2} & \ldots & x_{i_a} \end{pmatrix} \ldots$$

Then the factors of the substitution $\sigma_i^{-1} s_a \sigma_i$ will replace x_{i_1} successively by x_1, x_2, x_{i_2}; similarly x_{i_2} will be replaced by x_{i_3}, x_{i_3} by x_{i_4}, and finally x_{i_a} by x_{i_1}. Accordingly $\sigma_i^{-1} s_a \sigma_i$ contains the cycle $(x_{i_1} x_{i_2} \ldots x_{i_a})$, and this is obtained from the corresponding cycle $(x_1 x_2 \ldots x_a)$ of s_a by regarding this cycle, so to speak, as a function of the elements x_λ and applying to it the substitution σ_i. In the same way every cycle of $\sigma_i^{-1} s_a \sigma_i$ proceeds from the corresponding cycle of s_a. The two substitutions have therefore the same number of cycles, each containing the same number of elements, as was to be proved.

Example: The function of four elements

$$\varphi_1 = x_1 x_2 + x_3 x_4$$

has, as we have seen, three values, and its group G_1 is of order $\frac{4!}{3} = 8$. The substitutions $\sigma_2 = (x_2 x_3)$ and $\sigma_3 = (x_2 x_4)$, which are not contained in G_1, convert φ_1 into

$$\varphi_2 = x_1 x_3 + x_2 x_4,$$
$$\varphi_3 = x_1 x_4 + x_2 x_3,$$

respectively. By transposition with respect to σ_2 and σ_3, we obtain from

$G_1 = [1, (x_1 x_2), (x_3 x_4), (x_1 x_2)(x_3 x_4), (x_1 x_3)(x_2 x_4), (x_1 x_4)(x_2 x_3), (x_1 x_3 x_2 x_4)$
$(x_1 x_4 x_2 x_3)],$

the two groups belonging respectively to φ_2 and φ_3;

$$G_2 = \sigma_2^{-1} G_1 \sigma_2 =$$
$[1, (x_1 x_3), (x_2 x_4), (x_1 x_3)(x_2 x_4), (x_1 x_2)(x_3 x_4), (x_1 x_4)(x_2 x_3), (x_1 x_2 x_3 x_4),$
$(x_1 x_4 x_3 x_2)],$

$$G_3 = \sigma_3^{-1} G_1 \sigma_3 =$$
$[1, (x_1 x_4), (x_2 x_3), (x_1 x_4)(x_2 x_3), (x_1 x_3)(x_2 x_4), (x_1 x_2)(x_3 x_4), (x_1 x_3 x_4 x_2),$
$(x_1 x_2 x_4 x_3)].$

§ 47. **Corollary I.** *If a group of substitutions is transformed with respect to any substitution whatever, the transformed substitutions form a group.*

Corollary II. *The two, generally different, substitutions $s_a s_\beta$ and $s_\beta s_a$ are similar. For $s_a s_\beta = s_\beta^{-1}(s_\beta s_a) s_\beta$.*

Corollary III. *The substitution $s_\alpha s_\beta s_\alpha^{-1}$ is conjugate to s_β with respect to s_α^{-1}.*

Corollary IV. *If the substitution s_α is of order r and if s_β be such that its qth and no lower power occurs among the powers of s_α, and if furthermore the conjugate of s_β with respect to s_α is a power of s_β, then the smallest group containing s_α and s_β is of order $q \cdot r$.* (cf. §§ 37, 38, Chapter II).

Corollary V. *All substitutions which transform a given substitution s into its powers, s^a, form a group.*

Corollary VI. *All substitutions which transform a given group into itself form a group.*

Corollary VII. *If two substitutions, or two groups, are similar, substitutions can always be found which transform the one into the other. In the case of two substitutions the transforming substitution is found at once from § 46. In the case of groups, we have only to construct the corresponding functions and determine the substitution σ_t which converts the one into the other.*

Corollary VIII. *Two powers s^α and s^β of the same substitution are similar when, and only when, α and β have the same greatest common divisor with the order of s.*

§ 47. We turn now to a series of developments relating to the existence of certain special types of groups analogous to those of § 39, Chapter II.

Given any group G of order g, let H_1 of order h_1 and K_1 of order k_1 be subgroups of G, and let φ_1 and ψ_1 be functions belonging to H_1 and K_1 respectively. These functions, on being operated on by all the substitutions of G, take respectively $\frac{g}{h_1} = h_0$ and $\frac{g}{k_1} = k_0$ values in all (§ 43, Corollary); let these be denoted by

$$\varphi_1, \varphi_2, \ldots \varphi_{h_0} \text{ and } \psi_1, \psi_2, \ldots \psi_{k_0}.$$

The group of any one of the functions ψ_a will be $K_a = \sigma_a^{-1} K_1 \sigma_a$, where σ_a is any substitution of G which converts ψ_1 into ψ_a.

We form now the entire system of values

$$a\,\varphi_\lambda + b\,\psi_\mu \quad (\lambda = 1, 2, \ldots h_0;\ \mu = 1, 2, \ldots k_0).$$

where a and b are arbitrary parameters, and divide these $\frac{g^2}{h_1 k_1}$ functions into classes, such that all the functions of each class proceed from any one among them by the operations of G.

If $a\varphi_1 + b\psi_a$ be one of the values above, and if we apply to it all the substitutions of G, the resulting values are not necessarily all distinct. In particular some of them may coincide with the given value. The number of the latter is equal to the number of substitutions common to H_1 and $K_a = \sigma_a^{-1} K_1 \sigma_a$. Let this number be d_a. Then all the g values arising from $a\varphi_1 + b\psi_a$ will coincide in sets of d_a each, as is easily seen. The number of distinct values thus obtained is therefore $\frac{g}{d_a}$.

If these do not exhaust all the possible values $a\varphi_\lambda + b\psi_\mu$, let $a\varphi_\sigma + b\varphi_\tau$ be any remaining value. Then in the same class with this belongs also

$$a\varphi_{\sigma\sigma-1} + b\psi_{\tau\sigma-1} = a\varphi_1 + b\psi_\beta.$$

From the latter value we can, as before, deduce a class containing in this case $\frac{g}{d_\beta}$ distinct values, where d_β is the number of substitutions common to H_1 and $K_\beta = \sigma_\beta^{-1} K_1 \sigma_\beta$.

Proceeding in this way, we must finally exhaust all the $\frac{g^2}{h_1 k_1}$ values of the functions $a\varphi_\lambda + b\psi_\mu$. Writing then the two numbers of values equal to each other, we have, after dividing through by g,

$$A) \qquad \frac{g}{h_1 k_1} = \frac{1}{d_1} + \frac{1}{d_2} + \ldots + \frac{1}{d_m},$$

where m denotes the number of classes of values of $a\varphi_\lambda + b\psi_\mu$ with respect to the group G. Since h_1 is a multiple of every d_a, we may write $\frac{d}{d_a} = f_a$, and consequently

$$B) \qquad \frac{g}{k_1} = f_1 + f_2 + \ldots + f_m,$$

where the f's are all integers. *

§ 49. From Theorem V of § 43 it follows as a special case that if a group contains a substitution of prime order p, the order r of

---

*Formulas $A)$ and $B)$ were obtained by G. Frobenius, Crelle CI. p. 281, as an extension of a result given by the Author, Math. Annalen XIII.

the group is a multiple of p, and if a group contains a subgroup of order p^a where p is a prime number, the order of the group is a multiple of p^a. By the aid of the results of the preceding Section we can now also prove the converse proposition:

Theorem IX. *If p^a be the highest power of the prime number p which is a divisor of the order h of a group H, then H contains subgroups of order p^a.* *

In the demonstration we take for the G of the preceding Section the symmetric group, so that $g = n!$. For H_1 we take the present group H, and for K_1 the group of order p^f of § 39, Chapter II, p^f being the highest power of p which is a divisor of $n!$. The formula B) of § 48 then becomes

$$\frac{n!}{p^f} = f_1 + f_2 + \ldots + f_m.$$

The left member of this equation is no longer divisible by p; consequently there must be at least one $f_\beta = \dfrac{h_1}{d_\beta}$ which is also not divisible by p; that is d_β is divisible by p^a, and therefore H and K_β, the latter being a conjugate of K, have exactly p^a substitutions in common.† These form the required subgroup of H.

Corollary. *At the same time it appears that the group K contains among its subgroups every type of groups of order p^γ. For we need only take any group of order p^γ for H in the above demonstration.*

§ 50. The last theorem admits of the following extension:

Theorem X. *If the order h of a group H is divisible by p^β, then H contains subgroups of order p^β.*

The proof follows at once from Theorem XI, as soon as we have proved

Theorem XI. *Every group H of order p^a contains a subgroup of order p^{a-1}.*

The corollary of the preceding Section permits us to limit the

---

*Cauchy, loc. cit., proved this theorem for the case $a = 1$. The extension to the case of any a was given by L. Sylow, Math. Annalen V, pp. 584-594.

†For every subgroup of K, and consequently every subgroup of K_β, has for its order a power of p.

proof to the case of groups of order p^a which occur as subgroups in the group of § 39, Chapter II. The group $K = K_f$ there obtained was constructed by the aid of a series of subgroups (§ 40)

$$1, K_1, K_2, \ldots K_\lambda, K_{\lambda+1}, \ldots K_{f-1},$$

of orders

$$1, p^1, p^2, \ldots p^\lambda, p^{\lambda+1}, \ldots p^{f-1},$$

every one of which is contained in the following one. If the group H occurs in this series, the theorem is already proved; if not, then let $K_{\lambda+1}$ be the lowest group which still contains H. K_λ then does not contain all the substitutions of H. We apply now the formula B) of § 48 to the groups $K_{\lambda+1}$, K_λ, and H, taking these in the place of G, K_1, and H. We find

$$p = f_1 + f_2 + \ldots + f_m.$$

This equation has two solutions, since the f's, being divisors of $h = p^a$, are powers of p: either we must have $f_1 = f_2 = \ldots = 1$ and $m = p$, or else $m = 1$ and $f_1 = p$. In the former case it would follow that $h = d_1$, i. e., that H is a subgroup of K_λ, which is contrary to hypothesis. Consequently $f_1 = p$, i. e. H and K_λ have a common subgroup of order p^{a-1}. This is the required group. *

To Theorem X we can now add the following

Corollary. *Every group of order $p^a \cdot p_1^{a_1} \cdot p_2^{a_2} \ldots$ can be constructed by the combination of subgroups, one of each of the orders $p^a, p_1^{a_1}, p_2^{a_2}, \ldots$*

A smaller number of subgroups is of course generally sufficient.

A further extension of the theory in this direction is not to be anticipated. Thus, for example, the alternating group of four elements, which is composed of the twelve substitutions

$$1, (x_1 x_2)(x_3 x_4), (x_1 x_3)(x_2 x_4), (x_1 x_4)(x_2 x_3),$$
$$(x_1 x_2 x_3), (x_1 x_2 x_4), (x_1 x_3 x_4), (x_2 x_3 x_4),$$
$$(x_1 x_3 x_2), (x_1 x_4 x_2), (x_1 x_4 x_3), (x_2 x_4 x_3),$$

has no subgroup of order 6.

§ 51. We insert here another investigation based on the construction of tables as in § 41.

Let H be a group of order h affecting the n elements $x_1, x_2, \ldots x_n$.

---
* G. Frobenius: Crelle CI. p. 283-4.

From these n elements we arbitrarily select any k, as $x_1, x_2, \ldots x_k$, and let H' be the subgroup of H which contains all the substitutions of the latter that do not affect $x_1, x_2, \ldots x_k$. Suppose h' to be the order of H', and $t_1 = 1\, t_2, \ldots t_{h'}$ to be its several substitutions. We proceed then to tabulate the substitutions of H as follows:

Given any substitution s_a of H, suppose that this converts $x_1, x_2, \ldots x_k$ into $x_{a_1}, x_{a_2} \ldots x_{a_k}$, in the order as written. Then all the substitutions

$$s_a, \; t_2 s_a, \; t_3 s_a, \; \ldots t_{h'}\, s_a$$

also convert $x_1, x_2, \ldots x_k$ into $x_{a_1}, x_{a_2}, \ldots x_{a_k}$ respectively, and these are the only substitutions of H which have this effect. We take these various sets of h substitutions for the lines of our table, which is accordingly of the form

$$T) \quad \begin{array}{llll} 1, & t_2, & t_3, & \ldots t_{h'}, \\ s_2, & t_2 s_2, & t_3 s_2, & \ldots t_{h'} s_2, \\ s_3, & t_2 s_3, & t_3 s_3, & \ldots t_{h'} s_3, \\ \cdot\cdot\cdot\cdot\cdot\cdot\cdot\cdot\cdot\cdot\cdot\cdot \\ s_\mu, & t_2 s_\mu, & t_3 s_\mu, & \ldots t_{h'} s_\mu. \end{array}$$

The substitutions of the table are obviously all different, and consequently $\mu h' = h$.

Again, suppose that v_1 is any substitution of H which contains among its cycles one of order k, say

(1) $$\omega_1 = (x_1 x_2 x_3 \ldots x_k).$$

Then all the (necessarily distinct) substitutions

(2) $$v_1, \; t_2 v_1, \; t_3 v_1, \ldots t_{h'} v_1$$

will contain the same cycle (1) and these will be the only substitutions of H which contain this cycle. We wish now to determine how many substitutions of H contain either the cycle (1) or any cycle obtainable from (1) by transformation with respect to the substitutions of H, say

(1_a) $$\omega_a = (x_{a_1} x_{a_2} x_{a_3} \ldots x_{a_k}).$$

Now since all the substitutions of the same line of T) convert the elements of the cycle (1) into one and the same system of elements, it follows that if we write

$$s_a^{-1} v_1 s_a = v_a \quad (a = 1, 2, \ldots \mu),$$

where each s_a is the same as that of the table T), then all the conjugates of the cycle (1) which occur in the substitutions of H are contained in $v_1, v_2, \ldots v_\mu$. Suppose the notation so chosen that ω_a is contained in v_a. If now we denote the substitutions of H which do not affect $x_{a_1}, x_{a_2}, \ldots x_{a_k}$, i. e. those of the group $s_a^{-1} H' s_a = H_a$ by

$$1, t_2^{(a)}, t_3^{(a)}, \ldots t_{h'}^{(a)},$$

and by right hand multiplication by v_a form the line

(3) $\qquad v_a, \; t_2^{(a)} v_a, \; t_3^{(a)} v_a, \; \ldots \; t_{h'} v_a,$

then (3) contains all the substitutions of H which involve the cycle ω_a. The μ lines of the following table

$$T_0) \quad \begin{array}{c} v_1, \; t_2 v_1, \; t_3 v_1, \; \ldots \; t_{h'} v_1, \\ v_2, \; t_2^{(2)} v_2, \; t_3^{(2)} v_2, \; \ldots \; t_{h'}^{(2)} v_2, \\ \cdots \cdots \cdots \cdots \cdots \cdots \\ v_\mu, \; t_2^{(\mu)} v_\mu, \; t_3^{(\mu)} v_\mu, \; \ldots \; t_{h'}^{(\mu)} v_\mu, \end{array}$$

therefore contain all the required substitutions.

The question then arises; how many of the lines of T_0) give the same cycle; for example, in how many lines the cycle (1) occurs. If this cycle occurs in $v_\tau = s_\tau^{-1} v_1 s_\tau$, then s_τ must permute the elements $x_1, x_2, \ldots x_k$ cyclically, and must therefore contain a power of (1) as a cycle. Consequently we must have s_τ equal to one of the substitutions $v_1, v_1^2, v_1^3, \ldots v_1^{k-1}$ or one of these multiplied by some t_r. But the substitutions $v_1, v_1^2, \ldots v_1^{k-1}$ belong to different lines of the table T_0). It appears therefore that the μ lines of T_0) coincide in sets of k each. We have then

Theorem XII. *Every individual cycle of order k which occurs in* (2), *and consequently every one which occurs in H, gives rise by transformation with respect to all the h substitutions of H, to $\dfrac{\mu}{k}$ cycles. The h' distinct conjugate cycles of order k which occur in H therefore give rise to $\dfrac{\mu h'}{k} = \dfrac{h}{k}$ cycles. The number of letters occurring in these cycles is therefore equal to h, the order of H. From this it follows that the number of letters in all the cycles of order k is a multiple of the order of H. The multiplier is the number of sets of non-conjugate cycles of order k.*

MULTIPLE-VALUED FUNCTIONS — ALGEBRAIC RELATIONS.   57

Corollary. *The number of elements which remain unchanged in the several substitutions of H is a multiple of the order of H. If every element can be replaced by every other one by the substitutions of H, this number is exactly equal to the order of H.*[*]

EXAMPLE. Consider the alternating groups of four elements,

$(x_1)(x_2)(x_3)(x_4)$, $(x_1x_2)(x_3x_4)$, $(x_1x_3)(x_2x_4)$, $(x_1x_4)(x_2x_3)$,
$(x_1)(x_2x_3x_4)$, $(x_2)(x_1x_3x_4)$, $(x_3)(x_1x_2x_4)$, $(x_4)(x_1x_2x_3)$,
$(x_1)(x_2x_4x_3)$, $(x_2)(x_1x_4x_3)$, $(x_3)(x_1x_4x_2)$, $(x_4)(x_1x_3x_2)$.

Here the number of cycles with one element is 12, which is equal to the order of the group. The number of elements which occur in cycles of the second order is also 12. But, for $k = 3$, the number of elements is $24 = 2 \cdot 12$. Correspondingly it is readily shown that the group permits of replacing any element by any other one; that the cycles of order 2 are all conjugate; but that the cycles of order 3 divide into two sets of four each:

$(x_1x_2x_3)$, $(x_1x_3x_4)$, $(x_1x_4x_2)$, $(x_2x_4x_3)$,
$(x_1x_3x_2)$, $(x_1x_4x_3)$, $(x_1x_2x_4)$, $(x_2x_3x_4)$,

the second set being non-conjugate with the first.

§ 52. We return now to the table constructed in § 45. This table did not possess the last of the four properties noted in § 41; the substitutions of one line were not necessarily all different from those of the other lines. For every group certainly contains the identical substitution 1, which therefore occurs ρ times; and again in the example of § 46 three other substitutions

$(x_1x_2)(x_3x_4)$, $(x_1x_3)(x_2x_4)$, $(x_1x_4)(x_2x_3)$

occur in each of the three groups. We have now to determine in general *when it is possible that one and the same substitution shall occur in all the groups* $G_1, G_2, \ldots G_\rho$ *belonging respectively to the several values* $\varphi_1, \varphi_2, \ldots \varphi_\rho$ *of* φ. We shall find that the example just cited is a remarkable exception, in that there is in general no substitution except 1 which leaves *all* the values of a function unchanged. [†]

---
[*] In connection with this Section *Cf.* Frobenius, Crelle CI. p. 273, followed by an article by the Author, *ibid.* CIII p. 321.

[†] L. Kronecker: Monatsberichte d. Berl. Akad. 1879, 208.

If we apply to the series of functions $\varphi_1, \varphi_2, \ldots \varphi_\rho$ any arbitrary substitution σ, we obtain

$$\varphi_\sigma, \varphi_{\sigma_2\sigma}, \varphi_{\sigma_3\sigma}, \ldots \varphi_{\sigma_\rho\sigma}.$$

These values must coincide, apart from the order of succession, with the former set, for $\varphi_1, \varphi_2, \ldots \varphi_\rho$ are all the possible values of φ, and the ρ values just obtained are all different from one another. The groups which belong to the latter

$$\sigma^{-1}G_1\sigma, \;\sigma^{-1}G_2\sigma, \;\sigma^{-1}G_3\sigma, \ldots \sigma^{-1}G_\rho\sigma$$

are therefore also, apart from the order of succession, identical with $G_1, G_2, G_3, \ldots G_\rho$, that is, the system of the G's, regarded as a whole, is unaltered by transformation with respect to any substitution whatever. If now we denote by H the group composed of those substitutions which are common to $G_1, G_2, \ldots G_\rho$, then H is also the group of those substitutions which are common to $\sigma^{-1}G_1\sigma, \sigma^{-1}G_2\sigma$, $\ldots \sigma^{-1}G_\rho\sigma$. But the latter group is also of course expressed by $\sigma^{-1}H\sigma$; consequently we have

$$\sigma^{-1}H\sigma = H;$$

that is, the group H is unaltered by transformation with respect to any substitution; it includes therefore all the substitutions which are similar to any one contained in it.

We proceed now to examine the nature of a group H of this character. We consider in particular those substitutions of H which affect the least number of elements, the identical substitution excepted. It is clear that these can contain only cycles of the same number of elements, since otherwise some of their powers would contain fewer elements, without being identically 1.

We prove with regard to these substitutions, first that no one of their cycles can contain more than three elements. For if H contains, for example, the substitution

$$s = (x_1x_2x_3x_4\ldots)\ldots,$$

and if we take $\sigma = (x_3x_4)$, then, since $\sigma^{-1}H\sigma = H$, the substitution

$$\sigma^{-1}s\sigma = (x_1x_2x_4x_3\ldots)\ldots = s_1$$

will also occur in H. Now s_1 only differs from s in the order of the two elements x_3 and x_4. Consequently their product, which must also occur in H, since H is a group,

$$s \cdot s_1 = (x_3)(x_1 x_4 \ldots) \ldots$$

will certainly not affect the element x_3, but cannot be the identical substitution, because it contains at least the cycle $(x_1 x_4 \ldots)$. This product therefore affects fewer elements than s_1 which is contrary to hypothesis.

Secondly we prove that, if $n > 4$, the substitution of H which affects the least number of elements cannot contain more than one cycle. For otherwise H would contain substitutions of the form

$$s_\alpha = (x_1 x_2)(x_3 x_4) \ldots, \quad s_\beta = (x_1 x_2 x_3)(x_4 x_5 x_6) \ldots,$$

and therefore the corresponding conjugate substitutions with respect to $\sigma = (x_4 x_5)$

$$s_\alpha' = (x_1 x_2)(x_3 x_5) \ldots, \quad s_\beta' = (x_1 x_2 x_3)(x_5 x_4 x_6) \ldots$$

Consequently the corresponding products

$$s_\alpha^{-1} s_\alpha' = (x_1)(x_2)(x_3 \ldots) \ldots, \quad s_\beta^{-1} s_\beta' = (x_1)(x_2)(x_3)(x_4 x_5 x_6) \ldots,$$

which are not 1, but affect fewer elements than s, must also occur in H, which would again be contrary to hypothesis.

If then $n > 4$, either H consists of the identical substitution 1, or H contains a substitution $(x_\mu x_\nu)$, or a substitution $(x_\lambda x_\mu x_\nu)$. In the second case H must contain all the transpositions, that is H is the symmetric group. In the third case H must contain all the circular substitutions of the third order, that is H is the alternating group. (*Cf.* §§ 34–35).

Returning from the group H to the group G, it appears that if $G_1, G_2, \ldots G_\rho$ have any substitution, except 1, common to all, then either the second or the third case occurs. H, which is contained in G, includes in either case the alternating group; G is therefore either the alternating or the symmetric group, and $\rho = 2$, or $\rho = 1$.

If, however, $n = 4$ we might have, beside $s_1 = 1$, another substitution

$$s_2 = (x_1 x_2)(x_3 x_4)$$

in the group. With this its conjugates, of which there are only two,

$$s_3 = (x_1 x_3)(x_2 x_4), \quad s_4 = (x_1 x_4)(x_2 x_3),$$

must also occur. The group H cannot contain any further substitution without becoming either the alternating or the symmetric group. We have then the exceptional group

$$H = [s_1 = 1, s_2, s_3, s_4],$$

and this actually does transform into itself with respect to every substitution. Returning to the group G it follows from § 43, Theorem II, that the order of G is a multiple of that of H, that is, a multiple of 4; again from Theorem II the order of G must be a divisor of $4! = 24$. The choice is therefore restricted to the numbers 4, 8, 12, and 24. The last two numbers lead to the general case already discussed where $\rho = 2$, or 1. The first gives $G = H$, $\rho = 6$, and for example,

$$\varphi_1 = (x_1x_2 + x_3x_4) - (x_1x_3 + x_2x_4), \quad \varphi_2 = (x_1x_2 + x_3x_4) - (x_1x_4 + x_2x_3)$$
$$\varphi_3 = (x_1x_3 + x_2x_4) - (x_1x_4 + x_2x_3), \quad \varphi_4 = (x_1x_3 + x_2x_4) - (x_1x_2 + x_3x_4)$$
$$\varphi_5 = (x_1x_4 + x_2x_3) - (x_1x_2 + x_3x_4), \quad \varphi_6 = (x_1x_4 + x_2x_3) - (x_1x_3 + x_2x_4).$$

In the second case, $r = 8$, G contains H as a subgroup. To obtain G we must add other substitutions to those of H. None of these can be cyclical of the third order, for in this case we should have $r = 12$ or 24. If we select any other substitution, we obtain the group of § 46, which is included among those treated in § 39. For this group $\rho = 3$, and, for example,

$$\psi_1 = x_1x_2 + x_3x_4, \quad \psi_2 = x_1x_3 + x_2x_4, \quad \psi_3 = x_1x_4 + x_2x_3.$$

Theorem XI. *If $n \gtreqless 4$ there is no function, except the alternating and symmetric functions, of which all the ρ values are unchanged by the same substitution (excluding the case of the identical substitution). If $n = 4$, all the values of any function belonging to the same group with*

$$\varphi = (x_1x_2 + x_3x_4) - (x_1x_3 + x_2x_4) \quad \text{or with} \quad \psi = x_1x_2 + x_3x_4$$

are unchanged by the substitutions of the group

$$H = [1, (x_1x_2)(x_3x_4), (x_1x_3)(x_2x_4), (x_1x_4)(x_2x_3)].$$

---

§ 53. We have thus far examined the connection between the values of a ρ-valued function from the point of view of the theory of substitutions. We turn now to the consideration of the algebraic relations of these values.

We saw at the beginning of the preceding Section that the system of values $\varphi_1, \varphi_2, \ldots \varphi_\rho$ belonging to a function φ is unchanged as a whole by the application of any substitution, only the order of succession of the several values being altered. All integral symmetric functions of $\varphi_1, \varphi_2, \ldots \varphi_n$ are therefore unchanged by any substitution, and are consequently symmetric not only in the φ's but also in the x_λ's. They can therefore be expressed as rational integral functions of the elementary symmetric functions c_λ of the x_λ's. If we write then

$$S(\varphi_1) = \varphi_1 + \varphi_2 + \ldots + \varphi_\rho = R_1(c_1, c_2, \ldots c_n),$$
$$S(\varphi_1 \varphi_2) = \varphi_1\varphi_2 + \varphi_1\varphi_3 + \ldots = R_2(c_1, c_2, \ldots c_n),$$
$$\cdots\cdots\cdots\cdots\cdots\cdots\cdots\cdots\cdots\cdots$$
$$S(\varphi_1\varphi_2 \ldots \varphi_\rho) = \varphi_1\varphi_2\varphi_3 \ldots \varphi_\rho = R_\rho(c_1, c_2, \ldots c_n),$$

the R's are the coefficients of an algebraic equation of which $\varphi_1, \varphi_2, \ldots \varphi_\rho$ are the roots.

Theorem XII. *The ρ values $\varphi_1, \varphi_2, \ldots \varphi_\rho$ of a ρ-valued integral rational function φ are the roots of an equation of degree ρ*

$$\varphi^\rho - R_1 \varphi^{\rho-1} + R_2 \varphi^{\rho-2} - \ldots \pm R_\rho = 0$$

the coefficients of which are rational integral functions of the elementary symmetric functions $c_1, c_2, \ldots c_n$ of the elements $x_1, x_2, \ldots x_n$.

§ 54. As an example we determine the equation of which the three roots are

$$\varphi_1 = x_1 x_2 + x_3 x_4, \quad \varphi_2 = x_1 x_3 + x_2 x_4, \quad \varphi_3 = x_1 x_4 + x_2 x_3,$$

where x_1, x_2, x_3, x_4 are themselves the roots of the equation

$$f(x) = x^4 - c_1 x^3 + c_2 x^2 - c_3 x + c_4 = 0.$$

We find at once

$$\varphi_1 + \varphi_2 + \varphi_3 = S(x_1 x_2) = c_2;$$

and again, by §10, Chapter I,

$$\varphi_1\varphi_2 + \varphi_2\varphi_3 + \varphi_3\varphi_1 = S(x_1^2 x_2 x_3) = \alpha c_4 + \beta c_1 c_3 + \gamma c_2^2.$$

The numerical coefficients α, β, γ are readily found from special examples. They are $\alpha = -4, \beta = 1, \gamma = 0$. Hence

$$\varphi_1\varphi_2 + \varphi_2\varphi_3 + \varphi_3\varphi_1 = c_1 c_3 - 4 c_4.$$

Finally
$$\varphi_1\varphi_2\varphi_3 = S(x_1^2 x_2^2 x_3^2) + x_1 x_2 x_3 x_4 \, S(x_1^2)$$
$$= c_1^2 c_4 - 4 c_2 c_4 + c_3^2,$$

Accordingly the required equation is

$$f(\varphi) = \varphi^3 - c_2 \varphi^2 + (c_1 c_3 - 4c_4)\varphi - (c_1^2 c_4 - 4c_2 c_4 + c_3^2) = 0.$$

We examine the discriminant of this equation, *i. e.*, of its three roots. To determine this function it is not necessary to employ the the general formula obtained in § 10, Chapter I. We have at once

$$\varphi_1 - \varphi_2 = (x_1 - x_4)(x_2 - x_3),$$
$$\varphi_2 - \varphi_3 = (x_1 - x_2)(x_3 - x_4),$$
$$\varphi_3 - \varphi_1 = (x_1 - x_3)(x_2 - x_4),$$

and consequently, if we denote the discriminant of the φ's by Δ_φ and that of the x_λ's by Δ,

$$\Delta_\varphi = (\varphi_1 - \varphi_2)^2 (\varphi_2 - \varphi_3)^2 (\varphi_3 - \varphi_1)^2$$
$$= (x_1 - x_2)^2 (x_1 - x_3)^2 (x_1 - x_4)^2 (x_2 - x_3)^2 (x_2 - x_4)^2 (x_3 - x_4)^2 = \Delta.$$

We observe here therefore that the discriminant of an equation of the fourth degree can also be represented in the form of the discriminant of an equation of the third degree. A more important consideration is that the special theorem here obtained can be generalized in another direction, to which we next turn our attention.

§ 55 We start out from the table of § 41. If φ is not single-valued, the first line of this table, *i. e.*, the group G_1 belonging to φ_1 does not contain all the transpositions. If a transposition $(x_\alpha x_\beta)$ occurs in the second line, it results from the construction of the table that $(x_\alpha x_\beta)$ converts φ_1 into φ_2. If therefore $x_\alpha = x_\beta$, then $\varphi_1 = \varphi_2$ also, and consequently $\varphi_1 - \varphi_2$, since it vanishes for $x_\alpha = x_\beta$, is divisible by $x_\alpha - x_\beta$.

If, then, any transposition $(x_\alpha x_\beta)$ does not occur in the group G of φ_1, one of the differences $\varphi_1 - \varphi_\lambda$ ($\lambda = 2, 3, \ldots \rho$) is divisible by a factor of the form $x_\alpha - x_\beta$.

Now there are in all $\dfrac{n(n-1)}{2}$ transpositions of n elements. If the first line of the table, *i. e.*, G_1, contains exactly q of these, then the other lines contain $\dfrac{n(n-1)}{2} - q$. The product $(\varphi_1 - \varphi_2)(\varphi_1 - \varphi_3) \ldots$ $(\varphi_1 - \varphi_\rho)$ is therefore divisible by $\dfrac{n(n-1)}{2} - q$ different factors of the form $x_\alpha - x_\beta$, and therefore by their product.

Instead of starting out with the value φ_1, we might equally well have taken φ_2. Since the group $G_2 = \sigma_2^{-1} G_1 \sigma_2$ belonging to $\varphi_2 = \varphi_{\sigma_2}$

is similar to the group G_1, it also contains q transpositions, and the product $(\varphi_2-\varphi_1)(\varphi_2-\varphi_3)\ldots(\varphi_2-\varphi_\rho)$ is also divisible by $\dfrac{n(n-1)}{2}-q$ factors of the form $x_\alpha-x_\beta$. The same reasoning holds if we start with $\varphi_3, \varphi_4 \ldots \varphi_\rho$. If we multiply the separate products together, we find that

$$J_\varphi = (-1)^{\frac{\rho(\rho-1)}{2}} \prod_{\lambda=1}^{\lambda=\rho} \{(\varphi_\lambda-\varphi_1)(\varphi_\lambda-\varphi_2)\ldots(\varphi_\lambda-\varphi_{\lambda-1})(\varphi_\lambda-\varphi_{\lambda+1})$$
$$\ldots(\varphi_\lambda-\varphi_\rho)\}$$

is divisible by the product of $\rho\left[\dfrac{n(n-1)}{3}-q\right]$ factors $x_\alpha-x_\beta$. But since J_φ is symmetric in the x_λ's, the presence of a factor $x_\alpha-x_\beta$ requires that of every other factor $x_\gamma-x_\delta$, and consequently of $J = \prod_{\alpha>\beta}(x_\alpha-x_\beta)^2$, the discriminant of $f(x)$. Suppose that Δ^t is the highest power of Δ which is contained as a factor in J_φ, then, as Δ contains $n(n-1)$ factors $x_\alpha-x_\beta$, and consequently Δ^t contains $n(n-1)t$ such factors, we must have

$$n(n-1)t \geq \rho\left[\dfrac{n(n-1)}{2}-q\right].$$

$$\therefore\ t \geq \dfrac{\rho}{2} - \dfrac{q\rho}{n(n-1)}.$$

The number t can be 0 only when $q = \dfrac{n(n-1)}{2}$, that is, when all the transpositions occur in G_1. φ is then symmetric and $\rho = 1$. Again q can be 0 only when G contains no transposition. One of the cases in which this occurs is that where G is the alternating group or one of its subgroups.

Theorem XII. *If φ is ρ-valued function of the n elements $x_1, x_2, \ldots x_n$, the group of which contains q transpositions, the discriminant J_φ of the ρ values of φ is divisible by*

$$J^\rho\left(\tfrac{1}{2} - \dfrac{q}{n(n-1)}\right).$$

If φ is not symmetric, the exponent of Δ is not zero. If the group of φ is contained in the alternating group, q is zero.

All multiple-valued functions therefore have some of their values coincident if two of the elements x_λ become equal.

We perceive now why it was impossible (§ 32) to obtain $n!$-valued functions when any of the elements x_λ were equal (*Cf.* also § 104).

§ 56. Returning now to the equation (§ 53) of which the roots are $\varphi_1, \varphi_1, \ldots \varphi_\rho$,

$$\varphi^\rho - R_1\varphi^{\rho-1} + R_2\varphi^{\rho+2} - \ldots \pm R_\rho = 0,$$

we endeavor to determine whether and under what circumstances this equation can become binomial, *i. e.*, whether there are any μ-valued functions whose ρ^{th} powers are symmetric. For $\mu = 2$, we already know that $\varphi = \sqrt{\Delta}$ satisfies this condition. In treating the general case we will assume that, if the required function φ contains any factors of the form $\sqrt{\Delta}$, these factors are all removed at the outset. If the resulting quotient is φ', so that

$$\varphi = (\sqrt{\Delta})^a \varphi',$$

then, since $\varphi^{2\rho}$ and $(\sqrt{\Delta})^{2\rho}$ are both symmetric, $\varphi'^{2\rho}$ is symmetric also. We write then

$$\varphi'^{2\rho} = S_1.$$

If φ'_1 be any root of this equation, and if ω be a primitive $(2\rho)^{\text{th}}$ root of unity, then all the roots are

$$\varphi'_1, \;\omega\varphi'_1, \;\omega^2\varphi'_1, \;\ldots \omega^{2\rho-1}\varphi'_1,$$

and consequently

$$\Delta_\psi = \varphi'_1{}^{n(n-1)} (1-\omega)^2 (1-\omega^2)^2 \ldots (\omega^{2\rho-2} - \omega^{2\rho-1})^2.$$

From Theorem XIII this discriminant must be divisible by Δ, unless φ' is itself symmetric. But the factors containing ω are constant and therefore not divisible by Δ, and by supposition φ'_1 does not contain $\sqrt{\Delta}$ as a factor. Consequently φ'_1 is symmetric, and we have, according as a is odd or even,

$$\varphi = S, \quad \varphi = S\sqrt{\Delta}.$$

Theorem XIV. *If the n elements $x_1, x_2, \ldots x_n$ are independent, the only unsymmetric functions of which a power can be symmetric are the alternating functions.*

§ 57. On account of the importance of this last proposition we add in the present and following Sections other proofs based on entirely different grounds.

If ω is a primitive p^{th} root of unity, and if φ_1 is any root of the equation

$$\varphi^p - S = 0,$$

then all the roots of this equation are

$$\varphi_1, \omega\varphi_1, \omega^2\varphi_1, \ldots \omega^{p-1}\varphi_1.$$

Since the ω's are constants, all the values of φ have the same group. From Theorem XI, this must be, for $n \gtrless 4$, either the symmetric or the alternating group, or the identical operation 1. The first two cases give $p = 1$ or 2. In the last case $p = n!$. All the values of a function are of the same type, and consequently there are substitutions which transform one into another. Suppose, in the case $p = n!$, that σ converts the value φ_1 into $\varphi_2 = \omega\varphi_1$; then σ^λ converts φ_1 into $\omega^\lambda \varphi_1$. Accordingly the series

$$1, \sigma, \sigma^2, \ldots \sigma^{n!-1}$$

consists of distinct substitutions, which therefore include all the $n!$ possible substitutions. σ must therefore be a substitution of degree n and order $n!$. Such a substitution does not exist if $n > 2$.

The case $n = 4$ furnishes no exception. In this case the group common to all the values of φ might be the special group (Theorem XIII)

$$G = [1, (x_1x_2)(x_3x_4), (x_1x_3)(x_2x_4), (x_1x_4)(x_2x_3)];$$

φ would then be a six-valued function, and there must be a substitution σ which converts φ_1 into $\omega\varphi_1$ and which is of order 6. But there is no such substitution in the case of four elements.

§ 58. Finally we give a proof which is based on the most elementary considerations and which moreover leads to an important extension of the theorem under discussion.

In the first place we may limit ourselves to the case where p is a prime number. For, if $p = p \cdot q$, where p is a prime number, it follows from

$$\varphi^{p \cdot q} = (\varphi^q)^p = S$$

that there is also a function φ^q of which a prime power, the pth, is symmetric.

If, accordingly, we denote by φ a function of which a prime power, the pth, is symmetric while φ itself is unsymmetric, then,

since the group of φ cannot contain all the transpositions (§ 34), suppose that $\sigma = (x_\alpha x_\beta)$ converts φ_1 into φ_σ, where $\varphi_\sigma \neq \varphi_1$. Since
$$\varphi_\sigma{}^{\prime\prime} = \varphi_1{}^{\prime\prime} = S,$$
it follows that
$$\varphi_\sigma = \omega \varphi_1,$$
where ω is a primitive p^{th} root of unity. If we apply the substitution σ again to the last equation, remembering that $\sigma^2 = 1$ and that consequently $\varphi_{\sigma^2} = \varphi_1$, we obtain the new equation
$$\varphi_1 = \omega \varphi_\sigma.$$
Multiplying these two equations together and dividing by $\varphi_1 \varphi_\sigma$, we have
$$\omega^2 = 1,$$
and consequently, since p is a prime number, $p = 2$ and $\varphi = S \sqrt{J}$.

Having shown that only the alternating functions have the property that their prime powers can be symmetric, we may next examine whether there are any functions prime powers of which can be two-valued.

Suppose that ζ' is multiple-valued, while its qth power is two-valued, q being prime. Then there is some circular substitution of the third order $\sigma = (x_\alpha x_\beta x_\gamma)$, which does not occur in the group of ζ', since, if this group contains all the substitutions of this form, it must be the alternating group (§ 35). Suppose, then, that $\zeta'_\sigma \neq \zeta'_1$, but that
$$\zeta'_\sigma{}^q = \zeta'_1{}^q = S_1 + S_2 \sqrt{J},$$
since ζ'^q, being a two-valued function, is unaltered by a circular substitution of the third order. We must therefore have
$$\zeta'_\sigma = \omega \zeta'_1,$$
where ω is not 1 and must therefore be a primitive q^{th} root of unity. If we apply to this last equation the substitutions σ and σ^2, and remember that $\sigma^3 = 1$, and that consequently $\zeta'_{\sigma^3} = \zeta'_1$, we obtain
$$\zeta'_{\sigma^2} = \omega \zeta'_\sigma$$
$$\zeta'_1 = \omega \zeta'_{\sigma^2}.$$
Multiplying these three equations together and removing the functional values, we have
$$\omega^3 = 1, \ q = 3.$$

If now we assume $n > 4$, then the group of ζ' cannot contain all the circular substitutions of the fifth order, (Theorem X, Chapter II). If τ is one of those not occurring in the group of ζ', then $\zeta'_\tau \neq \zeta'_1$, but

$$\zeta'_\tau{}^q = \zeta'_1{}^q = S_1 + S_2 \sqrt{\Delta},$$

and consequently, if ω be a qth root of unity different from 1,

$$\zeta'_\tau = \omega \zeta'_1.$$

It follows from this, precisely as above, that, since $\tau^5 = 1$,

$$\zeta'_{\tau^2} = \omega \zeta'_\tau, \quad \zeta'_{\tau^3} = \omega \zeta'_{\tau^2}, \quad \zeta'_{\tau^4} = \omega \zeta'_{\tau^3}, \quad \zeta'_1 = \omega \zeta'_{\tau^4},$$

and consequently $\omega^5 = 1$, $q = 5$.

But this is inconsistent with the first result. It follows therefore that n is not greater than 4.

Theorem XVII. *If $n > 4$, there is no multiple-valued function a power of which is two-valued, if the elements x are independent quantities.*

§ 59. We conclude these investigations by examining for $n < 4$ the possibility of the existence of functions having the property discussed above.

The case $n = 2$ requires no consideration. In the case $n = 3$ we undertake a systematic determination of the possible functions of the required kind. We begin with the type

$$\varphi_1 = \alpha x_1{}^\nu + \beta x_2{}^\nu + \gamma x_3{}^\nu,$$

and attempt to determine α, β, γ so as to satisfy the required conditions. For this purpose we make use of the circumstance that some $\sigma = (x_1 x_2 x_3)$ converts φ_1 into $\omega \varphi_1$ ($\omega^3 = 1$) so that

$$\varphi_\sigma = \alpha x_2{}^\nu + \beta x_3{}^\nu + \gamma x_1{}^\nu = \omega(\alpha x_1{}^\nu + \beta x_2{}^\nu + \gamma x_3{}^\nu),$$
$$\gamma = \omega \alpha, \ \beta = \omega \gamma = \omega^2 \alpha, \ \alpha = \omega \beta = \omega^2 \gamma = \omega^3 \alpha = \alpha.$$

The last three equations can be consistently satisfied for every value of α. We may take $\alpha = 1$; and therefore

$$\varphi_1 = x_1{}^\nu + \omega^2 x_2{}^\nu + \omega x_3{}^\nu$$

is a function of the required type.

This result is confirmed by actual calculation. We find

68   THEORY OF SUBSTITUTIONS.

$$\varphi^3 =$$
$$(x_1^{3r} + x_2^{3r} + x_3^{3r}) + 6x_1^r x_2^r x_3^r - \frac{3}{2}(x_1^{2r}x_2^r + x_1^{2r}x_3^r + x_2^{2r}x_3^r + \ldots)$$
$$\pm \frac{3}{2}\sqrt{-3}\,(x_1^{2r}x_2^r - x_1^{2r}x_3^r + x_2^{2r}x_3^r - x_2^{2r}x_1^r + x_3^{2r}x_1^r - x_3^{2r}x_2^r).$$

If $r = 1$ the result becomes simpler, in that the last parenthesis becomes equal to

$$\sqrt{\Delta} = (x_1 - x_2)(x_2 - x_3)(x_1 - x_3);$$

whereas, in general, this parenthesis is only a rational function of $\sqrt{\Delta}$. If we write, as usual,

$$x_1 + x_2 + x_3 = c_1, \quad x_1x_2 + x_2x_3 + x_3x_1 = c_2, \quad x_1x_2x_3 = c_3,$$

we have, for $r = 1$,

$$\varphi^3 = \tfrac{1}{2}\{2c_1^3 - 9c_1c_2 + 27c_3 \pm 3\sqrt{-3\Delta}\}.$$

Suppose now that $n = 4$.

It is obvious that a function of the type $ax_1^r + \beta x_2^r + \gamma x_3^r + \delta x_4^r$, in which every term contains only one element, cannot satisfy the condition of being multiplied by ω when operated on by $\sigma = (x_1 x_2 x_3)$.

We enquire then whether the required function can be of the type

$$\varphi_1 = ax_1^r x_2^r + \beta x_2^r x_3^r + \gamma x_3^r x_1^r + x_4^r(a_1 x_1^r + \beta_1 x_2^r + \gamma_1 x_3^r),$$

where every term contains two of the four elements. If this type should also fail to satisfy the requirements we should have to proceed to more complicated forms. We shall, however, obtain a positive result in the present case, and in fact we may take $r = 1$, so that

$$\varphi_\sigma = ax_2x_3 + \beta x_3x_1 + \gamma x_1x_2 + x_4(a_1x_2 + \beta_1x_3 + \gamma_1x_1).$$

From the condition $\varphi_\sigma = \omega\varphi_1$, we have the series of equations

$$\gamma = \omega a, \quad \beta = \omega\gamma = \omega^2 a, \quad a = \omega\beta = \omega^2\gamma = \omega^3 a = a,$$
$$\gamma_1 = \omega a_1, \quad \beta_1 = \omega\gamma_1 = \omega^2 a_1, \quad a_1 = \omega\beta_1 = \omega^2\gamma_1 = \omega^3 a_1 = a_1.$$

All of these equations are satisfied independently of the values of a and a_1, and we have

$$\varphi_1 = a(x_1x_2 + \omega^2 x_2x_3 + \omega x_3x_1) + a_1(x_1x_4 + \omega^2 x_2x_4 + \omega x_3x_4).$$

But again, the substitution $\tau = (x_1x_2x_4)$ converts φ_1 into φ_τ, where φ_τ is equal to the product of φ_1 by some cube root of unity, since

MULTIPLE-VALUED FUNCTIONS — ALGEBRAIC RELATIONS. 69

$\varphi_1^3 = \varphi_\tau^3$. Whether this cube root is 1, ω, or ω^2 cannot be determined beforehand. We find

$$\varphi_\tau = a x_2 x_4 + a_1 \omega^2 x_4 x_1 + a_1 x_1 x_2 + x_3(\omega a x_2 + \omega^2 a x_4 + \omega a_1 x_1),$$

and since the terms of φ_1 and φ_τ which contain $x_1 x_4$ are

$$a_1 x_1 x_4 \text{ and } a_1 \omega^2 x_1 x_4,$$

the cube root of unity by which φ_1 is multiplied must therefore be ω^2. From the equation $\varphi_\tau = \omega^2 \varphi_1$ it follows, by comparison of coefficients that

$$a = \omega^2(\omega^2 a_1) = \omega a_1, \quad a_1 = \omega^2 a.$$

These two equations are consistent, since $\omega^3 = 1$. Putting $a = 1$, and arranging according to the powers of ω, we have then

$$\varphi_1 = (x_1 x_2 + x_3 x_4) + \omega(x_1 x_3 + x_2 x_4) + \omega^2(x_1 x_4 + x_2 x_3).$$

The function φ_1 is therefore a combination of the three values of a function which we have already discussed. The group of φ_1 is

$$G = [1, \ (x_1 x_2)(x_3 x_4), \ (x_1 x_3)(x_2 x_4), \ (x_1 x_4)(x_2 x_3)].$$

That φ_1^3 is two-valued is also readily shown, if we write

$$x_1 x_2 + x_3 x_4 = y_1, \ x_1 x_3 + x_2 x_4 = y_2, \ x_1 x_4 + x_2 x_3 = y_3.$$

For then φ_1 coincides with the expression obtained above for the case $n = 3$; and since y_1, y_2, y_3, are the roots of the equation

$$y^3 - c_2 y^2 + (c_1 c_3 - 4 c_4) y - (c_1^2 c_4 - 4 c_2 c_4 + c_3^2) = 0,$$

where the c's are the coefficients of the equation of which the roots x_1, x_2, x_3, x_4, (§ 54), we can translate the expression obtained for $n = 3$ directly into a two-valued function of the four elements x_1, x_2, x_3, x_4, since we have (§ 54) $\Delta_y = \Delta_x$.

CHAPTER IV.

TRANSITIVITY AND PRIMITIVITY. SIMPLE AND COMPOUND GROUPS. ISOMORPHISM.

§ 60. The two familiar functions

$$x_1x_2 + x_3x_4, \quad x_1x_2 - x_3x_4$$

differ from each other in the important particular that the group belonging to the former

$$G = [1, (x_1x_2), (x_3x_4), (x_1x_3)(x_3x_4), (x_1x_3)(x_2x_4), (x_1x_4)(x_2x_3),$$
$$(x_1x_3x_2x_4), (x_1x_4x_2x_3)]$$

contains substitutions which replace x_1 by x_2, x_3, or x_4, while in the group belonging to the latter

$$G_1 = [1, (x_1x_2), (x_3x_4), (x_1x_2)(x_3x_4)]$$

there is no substitution present which replaces x_1 by x_3 or x_4. The general principle of which this is a particular instance is the basis of an important classification. We designate a group as *transitive*, if its substitutions permit us to replace any selected element x_1 by every other element. Otherwise the group is *intransitive*. Thus G, above, is transitive, while G_1 is intransitive.

It follows directly from this definition that the substitutions of a transitive group permit of replacing *every* element x_i by every element x_k. For a transitive group contains some substitution $s = (x_1x_i \ldots) \ldots$ which replaces x_1 by x_i, and also some substitution $t = (x_1x_k \ldots) \ldots$ which replaces x_1 by x_k. Consequently the product $s^{-1}t$, which also occurs in the group, replaces x_i by x_k.

The same designations, transitive and intransitive, are applied to functions as to their corresponding groups.

It appears at once that the elements of an *intransitive* group are distributed in systems of transitively connected elements. For example, in the group G_1 above x_1 and x_2, and again, x_3 and x_4, are transitively connected. Suppose that in a given intransitive group there are contained substitutions which connect $x_1, x_2, \ldots x_a$ transitively, others which connect $x_{a+1}, x_{a+2}, \ldots x_{a+b}$, and so on, but none which, for instance, replace x_1 by $x_{a+\lambda}$ ($\lambda \geq 1$), and so on. The maximum possible number of substitutions within the several systems is $a!, b!, \ldots$, and consequently the maximum number in the given group, if $a, b, \ldots$ are known, is $a!\,b!\ldots$ If only the sum $a+b+\ldots = n$ is known, the maximum number of substitutions in an intransitive group of degree n is determined by the following equations:

$$(n-1)!\,1! = \frac{n-1}{2}(n-2)!\,2! > (n-2)!\,2!, \quad (n>3)$$

$$(n-2)!\,2! = \frac{n-2}{3}(n-3)!\,3! > (n-3)!\,3!, \quad (n>5).$$

Theorem I. *The maximum orders of intransitive groups of degree n are*

$(n-1)!,\, \tfrac{1}{2}(n-1)!,\, (n-2)!\,2!,\, (n-2)!,\, (n-3)!\,3!,\, (n-3)!\,2!, \ldots$

The first two orders here given correspond to the symmetric and the alternating groups of $(n-1)$ elements, so that in these cases one element is unaffected. The third corresponds to the combination of the symmetric group of $(n-2)$ elements with that of the remaining two elements. The fourth belongs either to the combination of the alternating group of $(n-2)$ with the symmetric group of the remaining two, or to the symmetric group of $(n-2)$ elements alone, the other two elements remaining unchanged; and so on.

The construction of intransitive from transitive groups will be treated later, (§ 99).

§ 61. We proceed now to arrange the substitutions of a transitive group in a table. The first line of the table is to contain all those substitutions

$$s_1 = 1,\, s_2,\, s_3,\, \ldots s_m$$

which leave the element x_1 unchanged, each substitution occurring only once. From the definition of transitivity, there is in the given group a substitution σ_2 which replaces x_1 by x_2. For the second line of the table we take

$$\sigma_2,\, s_2\sigma_2,\, s_3\sigma_2,\, \ldots s_m\sigma_2.$$

We show then, 1) that all the substitutions of this line replace x_1 by x_2; for every s_λ leaves x_1 unchanged and σ_2 converts x_1 into x_2; 2) that all the substitutions which produce this effect are contained in this line; for if τ replaces x_1 by x_2, $\tau\sigma_2^{-1}$ leaves x_1 unchanged and is therefore contained among the s_λ's; but from $\tau\sigma_2^{-1} = s_\lambda$ it follows that $\tau = s_\lambda \sigma_2$; 3) that all the substitutions of the line are distinct; for if $s_a \sigma_2 = s_\beta \sigma_2$, we obtain by right hand multiplication by σ_2^{-1}, $s_a = s_\beta$; 4) that the substitutions of the second line are all different from those of the first; for the latter leave x_1 unchanged, while the former do not.

We select now any substitution σ_3 which converts x_1 into x_3 and form for the third line of the table

$$\sigma_3, \; s_2\sigma_3, \; s_3\sigma_3 \ldots s_m\sigma_3.$$

The substitutions of this line may then be shown to possess properties similar to those of the second. We continue in this way until all the substitutions of the group are arranged in n lines of m substitutions each. We have then

Theorem II. *If the number of substitutions of a transitive group, which leave any element x_1 unchanged, is m, the order r of the group is mn, i. e., always a multiple of n.*

The following extension of this theorem is readily obtained:

Corollary. *If $x_a, x_b, x_c, \ldots$ are any arbitrary elements of the group G, and if m is the order of the subgroup of G which does not affect these elements, then the order of G is mn', where n' is the number of distinct systems of elements $x_\alpha, x_\beta, x_\gamma, \ldots$ by which the substitutions of G can replace $x_a, x_b, x_c, \ldots$*

§ 62. A group is said to be k-*fold transitive* when its substitutions permit of replacing k given elements by any k arbitrary ones. It can be readily shown that any k arbitrary elements can then be replaced by any k others. The definition includes the case where any number of the k elements remain unchanged. A k-fold transitive group must contain one or more substitutions involving any arbitrarily chosen cycle of the kth or any lower order. Thus in a four-fold transitive group there must be substitutions which leave x_1 and x_2 unchanged but interchange x_3 and x_4, and which are therefore of the form $(x_1)(x_2)(x_3x_4)\ldots$ The same group must also contain a substitution which leaves x_1, x_2, x_3, x_4 all unchanged; this may of course be the identical substitution.

For an example of a three-fold transitive group we may take the alternating group of 5 elements. If, for instance, we require a substitution which leaves x_2 unchanged and replaces x_1 and x_5 by x_5 and x_3 respectively, the circular substitution $S = (x_1 x_5 x_3)$ satisfies these conditions, and, being equivalent to the two transpositions $(x_1 x_5)(x_1 x_3)$, belongs to the alternating group. This same alternating group cannot however, be four-fold transitive, for it must then contain a substitution which converts x_1, x_2, x_3, x_4 into x_1, x_2, x_3, x_5 respectively; this could only be the transposition $(x_4 x_5)$, and this cannot occur in the alternating group.

In general, we can show *that the alternating group of n elements is always $(n-2)$-fold transitive*. The requirement that any $(n-2)$ elements shall be replaced by $(n-2)$ others may take any one of three forms. In the first place it may be required that $(n-2)$ given elements shall be replaced by the same elements in a different order, so that two elements are not involved. Secondly, the requirement may involve $(n-1)$ elements, or, thirdly, all the n elements.

In the first case suppose that σ is a substitution which satisfies the conditions, and let τ be the transposition of the two remaining elements. Then $\sigma\tau$ also satisfies the conditions, and one of the two substitutions σ, $\sigma\tau$ belongs to the alternating group.

If $(n-1)$ elements are involved, suppose that the remaining element is x_n, so that neither the element which replaces x_n nor that which x_n replaces is assigned. The elements which are to replace $x_1, x_2, \ldots x_{n-1}$ are all known with the exception of one. Suppose that it is not known which element replaces x_{n-1}. Then from the elements $x_1, x_2, \ldots x_{n-1}$ we can construct one substitution which satisfies the requirements, say $\sigma = (\ldots x_a\, x_{n-1}\, x_b \ldots)\ldots$, and from the n elements a second one, only distinguished from the first in the fact that x_{n-1} is followed by x_n, thus $\tau = (\ldots x_a\, x_{n-1} x_n\, x_b \ldots)$ $= \sigma \cdot (x_b x_n)$. Then either σ or τ belongs to the alternating group.

Finally, if all the n elements are involved, there are two elements for which the substituted elements are not assigned. Suppose these to be x_{n-1} and x_n. If now the elements are arranged in cycles in the usual manner, there will be two cycles which are not closed, the one ending with x_{n-1}, the other with x_n. We can then construct two substitutions σ and τ which satisfy the requirements, the one being obtained by simply closing the two incomplete cycles, the other by uniting the latter in a single parenthesis. From Chapter II, Theorem XI, it then follows that either σ or τ belongs to the alternating group.

The alternating group of n elements is therefore at least $(n-2)$-fold transitive. It cannot be $(n-1)$-fold transitive, since it contains no substitution which leaves $x_1, x_2, \ldots x_{n-2}$ unchanged, and converts x_{n-1} into x_n.

§ 63. If G is a k-fold transitive group, the subgroup G' of G which does not affect x_1 will be $(k-1)$-fold transitive; the subgroup G'' of G' which does not affect x_2 will be $(k-2)$-fold transitive, and so on. Finally the subgroup $G^{(k-1)}$ which does not affect $x_1, x_2, \ldots x_{k-1}$ will be simply transitive. Applying Theorem II successively to $G^{(k-1)}, \ldots G'', G', G$, we obtain

Theorem III. *The order r of a k-fold transitive group is equal to $n(n-1)(n-2)\ldots(n-k+1)m$, where m is the order of any subgroup which leaves k elements unchanged.*

§ 64. A simply transitive group is called *non-primitive* when its elements can be divided into systems, each including the same number, such that every substitution of the group replaces all the elements of any system either by the elements of the same system or by those of another system. The substitutions of the group can therefore be effected by first interchanging the several systems as units, and then interchanging the elements within each separate system.

A simply transitive group which does not possess this property is called *primitive*.

For example, the groups

$$G_1 = [1, (x_1x_2), (x_3x_4), (x_1x_2)(x_3x_4), (x_1x_3)(x_2x_4), (x_1x_4)(x_2x_3),$$
$$(x_1x_3x_2x_4), (x_1x_4x_2x_3)],$$
$$G_2 = \{1, (x_1x_2x_3), (x_4x_5x_6), (x_7x_8x_9), (x_1x_5x_9x_2x_6x_7x_3x_4x_8)\}$$

are both non-primitive. G_1 has two systems of elements, x_1, x_2 and x_3, x_4; G_2 has three systems, $x_1, x_2, x_3, \quad x_4, x_5, x_6$, and x_7, x_8, x_9.

The powers of a circular substitution of prime order form a primitive group, e. g. $G_3 = [1, (x_1x_2x_3), (x_1x_3x_2)]$.

The powers of a circular substitution of composite order form a non-primitive group. If the degree of the substitution is $n = p_1^{a_1} \cdot p_2^{a_2} \cdot p_3^{a_3} \ldots$, where $p_1, p_2, p_3, \ldots$ are the different prime factors of n, the corresponding systems of elements can be selected in $[(a_1+1)(a_2+1)(a_3+1) \ldots -2]$ different ways, as is readily seen. For example, in the case of the group

$$G_4 = [1, (x_1x_2x_3x_4x_5x_6), (x_1x_3x_5)(x_2x_4x_6), (x_1x_4)(x_2x_5)(x_3x_6),$$
$$(x_1x_5x_3)(x_2x_6x_4), (x_1x_6x_5x_4x_3x_2)],$$

we may take either two systems of three elements each, x_1, x_2, x_5 and x_2, x_4, x_6, or three systems of two elements each, $x_1, x_4, \quad x_2, x_5,$ and x_3, x_6.

A theorem applies here, the proof which may be omitted on account of its obvious character:

Theorem IV. *If, for a non-primitive group, the division of the elements into systems is possible in two different ways such that one division is not merely a subdivision of the other, then a third mode of division can also be obtained by combining into a new system the elements common to a system of the first division and one of the second.*

It must be observed that a single element is not to be regarded as a "system" in the present sense. Thus the group G_4 above admits of only two kinds of systems.

§ 65. The elements of a non-primitive group G can be arranged in a table, as follows. The first line contains all and only those substitutions

$$s_1 = 1, s_2, s_3, \ldots s_m$$

which leave the several systems unchanged as units, and which accordingly only interchange the elements within the systems. (The line will of course vary with the particular distribution of the elements in systems.)

From the definition of transitivity, (for the names "primitive" and "non-primitive" apply only to simply transitive groups), there must be in the given group a substitution σ_2 which converts any element x_a of one system into an element x_b of another system, and which consequently interchanges the several systems in a certain way. For the second line of the table we take

$$\sigma_2,\; s_2\sigma_2,\; s_3\sigma_2,\; \ldots s_m\sigma_2.$$

We show then, 1) that all the substitutions of this line produce the same rearrangement of the order of the systems as σ_2; for every s_λ leaves this order unchanged; 2) that all substitutions which produce the same rearrangement of the systems as σ_2 are contained in this line; for if τ is one of these, then $\tau\sigma_2^{-1} = s_\lambda$, so that $\tau = s_\lambda\sigma_2$; 3) that all the substitutions of the second line are different from one another; and 4) that they are all different from those of the first line.

If there is then still another substitution σ_3 which produces a new arrangement of the systems, this gives rise to a third line which possesses similar properties, and so on.

Theorem V. *If a non-primitive group G contains a subgroup G_1 of order m which does not interchange the several systems of elements, the order r of G is equal to mq, where q is a divisor of $\mu!$, μ being the number of systems.*

§ 66. If we denote the several systems, regarded, so to speak, as being themselves elements, by $A_1, A_2, \ldots A_\mu$, then all the substitutions of any one line of the table above, and only these, produce the same rearrangement of the $A_1, A_2, \ldots A_\mu$. To every line of the table corresponds therefore a substitution of the A's, the first line, for example, corresponding to the identical substitution, etc. These new substitutions we denote by $\mathfrak{s}_1 = 1, \mathfrak{s}_2, \ldots \mathfrak{s}_q$. It is readily seen that they form a new group $\mathfrak{G}$. For the successive application of $\mathfrak{s}_\alpha$ and $\mathfrak{s}_\beta$ to the elements A produces the same rearrangement of these elements as if the corresponding σ_α and σ_β were successively applied to the elements x. Accordingly, since $\sigma_\alpha\sigma_\beta = \sigma_\gamma$, we have also $\mathfrak{s}_\alpha\mathfrak{s}_\beta = \mathfrak{s}_\gamma$, where $\mathfrak{s}_\gamma$ corresponds to the line of the table above which contains σ_γ. The system of $\mathfrak{s}$'s therefore possesses the characteristic property of a group.

We perceive here a peculiar relation between the two groups G and $\mathfrak{G}$. To every substitution s of the former corresponds one substitution $\bar{s}$ of the latter, and again to every $\bar{s}$ of $\mathfrak{G}$ corresponds either one substitution, or a certain constant number of substitutions s of G. And this correspondence is moreover of such a nature that to the product of any two s's corresponds the product of the two corresponding $\bar{s}$'s.

If to every $\bar{s}$ there corresponds only one s, then there is only one substitution, identity, in $\mathfrak{G}$ which leaves the order of the systems A unchanged. The two following groups may serve as an example of this type. Suppose that

$$G = [1, (x_1x_2)(x_3x_4)(x_5x_6), (x_1x_3)(x_2x_6)(x_4x_5), (x_1x_4x_6)(x_2x_5x_3),$$
$$(x_1x_5)(x_2x_4)(x_3x_6), (x_1x_6x_4)(x_2x_3x_5)].$$

Here the systems A_1, A_2, and A_3 are composed respectively of x_1 and x_2, x_3 and x_6, and x_4 and x_5. The corresponding substitutions of the A's form the group

$$\mathfrak{G} = [1, (A_2A_3), (A_1A_2), (A_1A_3A_2), (A_1A_3), (A_1A_2A_3)].$$

§ 67. We examine more closely the subgroup

$$G_1 = [s_1, s_2, s_3, \ldots s_m]$$

of the group G of § 65. Since G_1 cannot replace any element of one system by an element of another system, it follows that G_1 is intransitive. Any arbitrary substitution t of G transforms G_1 into $^{-1}G_1 t = G'_1$. The latter is also a subgroup of G; it is similar to G_1; and it evidently does not interchange the systems of G. It follows that $G'_1 = G_1$.

Suppose that any system of a non-primitive group consists of the elements $x'_1, x'_2, x'_3, \ldots$ The subgroup G_1 therefore permutes the elements x' among themselves. We proceed to examine whether these elements are transitively connected with one another by the group G_1, or whether this is the case only when substitutions of G are added which interchange the systems of elements. Suppose that $x'_1, x'_2, \ldots x'_a$, and again $x'_{a+1}, \ldots x'_\beta$, etc., are transitively connected by G_1. Then G contains a substitution of the form $t = (x'_1 x'_{a+1} \ldots) \ldots$, and since $t^{-1}G_1 t = G_1$, it follows that t replaces all the elements $x'_1, x'_2, \ldots x'_a$ by $x'_{a+1} \ldots x'_\beta$. Further consideration then shows that $x'_1, x'_2, \ldots x'_a$ form a system of non-primitivity.

Accordingly if the systems of non-primitivity are chosen at the outset as small as possible, then the group G_1 connects all the elements of every system transitively.

Assuming the systems to be thus chosen, we direct our attention to those cycles of the several substitutions of G_1 which interchange the elements $x'_1, x'_2, \ldots$ of any one system of non-primitivity. These form a transitive group H'. Similarly the components of the several substitutions of G_1 which interchange the elements $x_1^{(a)}, x_2^{(a)} \ldots$ of any second system form a group $H^{(a)}$. The groups $H', H'', \ldots$ are similar, for if $t = (x'_1 x_1^{(a)} \ldots) \ldots$ is a substitution of G, then the transformation $t^{-1} G_1 t = G_1$ will convert H' into $H^{(a)}$. The order of H' is a multiple of $\dfrac{n}{\mu}$ and a divisor of $\dfrac{n}{\mu}!$, where μ is the number of systems of non-primitivity.

§ 68. The following easily demonstrated theorems in regard to to primitive and non-primitive groups may be added here:

Theorem VI. *If from the elements $x_1, x_2, \ldots x_n$ of a transitive group G any system $x'_1, x'_2, \ldots$ can be selected such that every substitution of G which replaces any x'_a by an x'_β permutes the x''s only among themselves, then G is a non-primitive group.*

Theorem VII. *If from the elements $x_1, x_2, \ldots x_n$ of a transitive group G two systems $x'_1, x'_2, \ldots$ and $x''_1, x''_2, \ldots$ can be selected such that any substitution which replaces any element x'_a by an x''_β replaces all the x''s by x'''s, then G is a non-primitive group.*

Theorem VIII. *Every primitive group G contains substitutions which replace an element x'_a of any given system $x'_1, x'_2, \ldots$ by an element of the same system, and which at the same time replace any second element of the system by some element not belonging to the system.* *

§ 69. The preceding discussion has led us to two general properties of groups which, together with transitivity and primitivity, are of fundamental importance.

In § 67 the subgroup G_1 of G possessed the property of being reproduced by transformation with respect to every substitution t of

---

* Rudio: Ueber primitive Gruppen. Crelle CI. p. 1.

G, so that for every t we have $t^{-1}G_1 t = G_1$. We may conveniently indicate this property of G_1 by the equation

$$G^{-1}G_1 G = G_1 \text{ or } G_1 G = G G_1.$$

It is to be observed however that this notation must be cautiously employed. For example, if G_1 is *any* subgroup of G, we have always $G_1^{-1} G G_1 = G$, and from this equation would apparently follow $G G_1 = G_1 G$, and consequently $G_1 = G^{-1} G_1 G$. But this last equation holds only for a special type of subgroup G_1. The reason for this apparent inconsistency lies in the fact, that in the equation $G^{-1} G_1 G = G_1$ the two G's represent the *same* substitution and the two G_1's in general *different* substitutions, while in the equation $G_1^{-1} G G_1 = G$ the reverse is the case.

We introduce here the following definitions.

1) *Two substitutions s_1 and s_2 are commutative* * *if*

$$s_1 s_2 = s_2 s_1.$$

2) *A substitution s_1 and a group H are commutative if*

$$s_1 H = H s_1.$$

3) *Two groups H and G are commutative if*

$$HG = GH.$$

The last equation is to be understood as indicating that the product of any substitution of H into any substitution of G is equal to the product of some substitution of G into some substitution of H, so that, if the substitutions of G are denoted by s and those of H by t, then

$$s_\alpha t_\beta = t_\gamma s_\delta$$

for every α and β.

Under 2) s_1 may be a substitution of H; for s_1 and H are then always commutative. Under 3) a case of special importance is that, an instance of which we have just considered, for which H is a subgroup of G. In this case s_α and s_δ of the equation $s_\alpha t_\beta = t_\gamma s_\delta$ are always to be taken equal.

A subgroup H of any group G, for which $G^{-1}HG = H$, is called a self-conjugate subgroup of G.

---

* German: "vertauschbar"; French: "échangeable", retained as "interchangeable" by Bolza: Amer. Jour. Math. XIII, p. 11.

The following may serve as examples:

1) The substitutions $s_1 = (x_1 x_2 x_3)(x_4 x_5 x_6)$, $s_2 = (x_1 x_4)(x_2 x_5)(x_3 x_6)$ are commutative; for their product is $(x_1 x_5 x_3 x_4 x_2 x_6)$, independently of the order of the factors.

Every power s^α of any substitution s is commutative with every other power s^β of the same substitution.

Two substitutions which have no common element are commutative.

2) The group $H = [1, (x_1 x_2)(x_3 x_4), (x_1 x_3)(x_2 x_4), (x_1 x_4)(x_2 x_3)]$ is commutative with every substitution of the four elements x_1, x_2, x_3, x_4.

The alternating group of n elements is commutative with every substitution of the same elements.

3) The group H of 2), being commutative with the symmetric group of the four elements x_1, x_2, x_3, x_4, is a self-conjugate subgroup of the latter.

The alternating group of n elements is a self-conjugate subgroup of the corresponding symmetric group.

Every group G of order r, which is not contained in the alternating group A, contains as a self-conjugate subgroup the group H of order $\frac{1}{2}r$ composed of those substitutions of G which are contained in A (Theorem VIII, Chapter II).

The identical substitution is, by itself, a self-conjugate subgroup of every group.

§ 70. We may employ the principle of commutativity to further the solution of the problem of the construction of groups begun in Chapter II (§§ 33–40).

All substitutions of n elements which are commutative with any given substitution of the same elements, form a group.

For if $t_1, t_2 \ldots$ are commutative with s, it follows from
$$t_1^{-1} s t_1 = s, \quad t_2^{-1} s t_2 = s,$$
that
$$(t_1 t_2)^{-1} s (t_1 t_2) = s,$$
so that the product $t_1 t_2$ also occurs among the substitutions t.

All substitutions of n elements which are commutative with a *given group G of the same elements form a group which contains G as a self-conjugate subgroup.*

For from
$$t_1^{-1}Gt_1 = G, \quad t_2^{-1}Gt_2 = G,$$
follows
$$(t_1t_2)^{-1}G(t_1t_2) = G;$$
and among the t's are included all the substitutions of G.

If two commutative groups G and H have no substitution, except the identical substitution, in common, then the order of the smallest group
$$K = \{G, H\}$$
is equal to the product of the orders of G and H.

§ 71. *If a group G of order $2r$ contains a subgroup H of of order r, then H is a self-conjugate subgroup of G.*

For if the substitutions of H are denoted by $1, s_2, s_3, \ldots s_r$, and if t is any substitution of G which is not contained in H, then $t, ts_2, ts_3, \ldots ts_r$ are the remaining r substitutions of G. But in the same way, $t, s_2t, s_3t, \ldots s_rt$ are also these remaining substitutions. Consequently every substitution $s_\alpha t$ is equal to some ts_β, that is, we have in every case $t^{-1}s_\beta t = s_\alpha$, and therefore $G^{-1}HG = H$.

If a group G contains a self-conjugate subgroup H and any other subgroup K, then the greatest subgroup L common to H and K is a self-conjugate subgroup of K. If the orders of G, H, K, L are respectively g, h, k, l, then $\dfrac{g}{h}$ is a multiple of $\dfrac{k}{l}$.

For if s is any substitution of K, then $s^{-1}Ls$ is contained in K, since all the separate factors s^{-1}, L, s are contained in K. But $s^{-1}Ls$ is also contained in H, for L is a subgroup of H and $s^{-1}Hs = H$. Consequently $s^{-1}Ls$ is contained in L, and, as these two groups have the same number of substitutions, we must have $s^{-1}Ls = L$, and L is a self-conjugate subgroup of K.

The relation between the orders of the four groups follows at once from the formula of Frobenius (§ 48). We have only to take for the K of this formula the present group H, and to put all the $d_1, d_2, \ldots d_m$ equal to l. We have then $\dfrac{g}{hk} = \dfrac{m}{l}$.

A self-conjugate subgroup of a transitive group either affects every element of the latter, or else it consists of the identical substitution alone.

For if $H = G^{-1}HG$ is a self-conjugate subgroup of the transitive group G, and if H does not affect the element x_1, then, since G contains a substitution s_λ which replaces x_1 by x_λ, it would follow that $s_\lambda^{-1} H s_\lambda = H$ would also not affect x_λ, that is, that H would not affect any element.

If a self-conjugate subgroup of a transitive group G is intransitive, then G is non-primitive and H only interchanges the elements within the several systems of non-primitivity.

For suppose that x_1 and x_λ belong to two different systems of intransitivity with respect to H. Then G contains a substitution s_λ which replaces x_1 by x_λ, and since $s_\lambda^{-1} H s_\lambda = H$, it follows that $s_\lambda^{-1} H s_\lambda$ must replace x_λ only by elements transitively connected with x_λ with respect to H. But s_λ^{-1} replaces x_λ by x_1 and H replaces x_1 by every element of the same system of intransitivity with x_1. Consequently the remaining factor s_λ must replace every element of the system containing x_1 by an element of the system containing x_λ. The systems of intransitivity of H are therefore the systems of non-primitivity of G.

§ 72. Another important property is that of the correspondence of two groups, of which an instance has already been met with in § 66. The two groups G and $\mathfrak{G}$ of this Section were so related that to every substitution s of G corresponded one substitution $\mathfrak{s}$ of $\mathfrak{G}$, and to every $\mathfrak{s}$ corresponded a certain number of s's. The correspondence was moreover such that to the product of any two s's corresponded the product of two corresponding $\mathfrak{s}$'s.

We may consider at once the more general type of correspondence,* where to every substitution of either group correspond a certain number of substitutions of the other, and to every product $s_\alpha s_\beta$ corresponds every product $\mathfrak{s}_\alpha \mathfrak{s}_\beta$ of corresponding $\mathfrak{s}$'s and *vice versa*. We may then readily show that to every substitution of the one group correspond the same number of substitutions of the other. For if to 1 of the group G correspond $1, \mathfrak{s}_2, \mathfrak{s}_3, \ldots \mathfrak{s}_q$ of $\mathfrak{G}$,

---

* A. Capelli: Battaglini Gior. 1878, p. 32 *seq.*

then, if $\bar{s}$ corresponds to s, all the substitutions $\bar{s}, \bar{s}\bar{s}_2, \bar{s}\bar{s}_3, \ldots \bar{s}\bar{s}_q$ correspond to s, by definition. Conversely, if any substitution $\bar{s}'$ corresponds to s, then $\bar{s}^{-1}\bar{s}'$ corresponds to $s^{-1}s = 1$, and therefore $\bar{s}^{-1}\bar{s}'$ is contained in the series $1, \bar{s}_2, \bar{s}_3, \ldots \bar{s}_q$. Consequently the series $\bar{s}, \bar{s}\bar{s}_2, \bar{s}\bar{s}_3, \ldots \bar{s}\bar{s}_q$ contains all the substitutions of $\mathfrak{G}$ which correspond to s, and the number q is constant for every s. Similarly to every $\bar{s}$ correspond the same number p of the substitutions s.

It is evident at once that

the substitutions of $G\,(\mathfrak{G})$ which correspond to the identical substitution of $\mathfrak{G}\,(G)$ form a group $H\,(\mathfrak{H})$ which is a self-conjugate subgroup of $G\,(\mathfrak{G})$.

The correspondence of two groups as just defined is called *isomorphism*. If to every substitution of G correspond q substitutions of $\mathfrak{G}$, and to every substitution of $\mathfrak{G}$ p substitutions of G, then G and $\mathfrak{G}$ are said to be $(p-q)$-*fold isomorphic*, or if p and q are not specified, *manifold isomorphic*. If $p = q = 1$, the groups are said to be *simply isomorphic*. *

EXAMPLES.

I. The groups

$G = [1, (x_1x_2)(x_3x_6)(x_4x_5), (x_1x_3)(x_2x_5)(x_4x_6), (x_1x_4)(x_2x_6)(x_3x_5),$
$(x_1x_5x_6)(x_2x_3x_4), (x_1x_6x_5)(x_2x_4x_3)],$
$\varGamma = [1, (\xi_1\xi_2), (\xi_1\xi_3), (\xi_2\xi_3), (\xi_1\xi_2\xi_3), (\xi_1\xi_3\xi_2)]$

are simply isomorphic, the substitutions corresponding in the order as written. For if any two substitutions of G, and the corresponding substitutions of $\varGamma$, are multiplied together, the resulting products again occupy the same positions in their respective groups.

II. The groups

$G = [1, (x_1x_2)], \quad \varGamma = [1, (\xi_1\xi_2)(\xi_3\xi_4), (\xi_1\xi_3)(\xi_2\xi_4), (\xi_1\xi_4)(\xi_2\xi_3)]$

are $(1-2)$-fold isomorphic. Corresponding to 1 of G we may take, beside 1, any other arbitrary substitution of $\varGamma$. It follows that $\varGamma$ is simply isomorphic with itself in different ways.

---

* *Cf.* Camille Jordan: Traité etc., § 67-74, where the names "holoedric" and 'meriedric" isomorphism are employed. These have been retained by Bolza: Amer. Jour. Vol. XIII. The "simply, manifold, $(p-q)$-fold isomorphic" above represent the "einstufig, mehrstufig, $(p-q)$-stufig isomorph" of the German edition.

III. The groups
$$G = [1, (x_1x_2)(x_3x_4), (x_1x_3)(x_2x_4), (x_1x_4)(x_2x_3)],$$
$$\Gamma = [1, (\xi_1\xi_2\xi_3), (\xi_1\xi_3\xi_2), (\xi_1\xi_2), (\xi_1\xi_3), (\xi_2\xi_3)]$$
are $(2-3)$-fold isomorphic. To the substitution 1 of G correspond $1, (\xi_1\xi_2\xi_3), (\xi_1\xi_3\xi_2)$ of Γ, and conversely to 1 of Γ correspond $1, (x_1x_2)(x_3x_4)$ of G.

§ 73. *If G and Γ are $(m-n)$-fold isomorphic, then their orders are in the ratio of $m:n$.*

If L is a self-conjugate subgroup of G, and if Λ is the corresponding subgroup of Γ, then Λ is a self-conjugate subgroup of Γ.

For from $G^{-1}LG = L$ follows at once $\Gamma^{-1}\Lambda\Gamma = \Lambda$. In the case of $(p-1)$-fold isomorphism, it may however happen that the group Λ consists of the identical substitution alone.

§ 74. Having now discussed the more elementary properties of groups in reference to transitivity, primitivity, commutativity, and isomorphism, we turn next to certain more elaborate investigations devoted to the same subjects.

The m substitutions of a transitive group G which do not affect the element x_1 form a subgroup G_1 of G. Similarly the substitutions of G which do not affect x_2 form a second subgroup G_2, and so on to the subgroup G_n which does not affect x_n. All these subgroups are similar; for if σ_a is any substitution of G which replaces x_1 by x_a, we have $\sigma_a^{-1} G_1 \sigma_a = G_2$. The groups G_a are therefore all of order m.

If now we denote by $[q]$ the number of those substitutions of G_1 which affect exactly q elements but leave the remaining $(n-q-1)$ unchanged, then $[q]$ is also the corresponding number for each of the other groups $G_2, G_3, \ldots G_n$. It follows then from the meaning of the symbol $[q]$ that
$$m = [n-1] + [n-2] + \ldots + [q] + \ldots + [2] + [0],$$
where the symbol $[1]$ does not of course occur, and $[0] = 1$. $G_1, G_2, \ldots G_n$ therefore possess together $n[n-1]$ substitutions which affect exactly $(n-1)$ elements. These are all different, for any substitution which leaves only x_a unchanged occurs in G_a, but cannot also occur in G_β. But this is not the case with substitutions

which affect exactly $(n-2)$ elements; for if any one of these leaves both x_a and x_β unchanged, it will occur in both G_a and G_β. Accordingly every one of these $n[n-2]$ substitutions is counted twice, and G therefore contains $\frac{1}{2}n[n-2]$ substitutions which affect exactly $(n-2)$ elements. Similarly every one of the $n[q]$ substitutions of q elements which occur in $G_1, G_2, \ldots G_n$ is counted $(n-q)$ times, and there are therefore only $\dfrac{n}{n-q}[q]$ different substitutions in G which affect exactly q elements. We have then for the total number of substitutions in G, which affect less than n elements

$$\frac{n}{1}[n-1]+\frac{n}{2}[n-2]+\ldots+\frac{n}{n-q}[q]+\ldots+\frac{n}{n}[0].$$

If this number is subtracted from that of all the substitutions in G, the remainder gives the number of substitutions in G which affect exactly n elements. But from Theorem II

$$r=mn=n[n-1]+n[n-2]+\ldots+n[q]+\ldots+n[0],$$

and consequently the required difference N is

$$n\Big(\frac{1}{2}[n-2]+\frac{2}{3}[n-3]+\ldots+\frac{n-q-1}{n-q}[q]+\ldots+\frac{n-1}{n}[0]\Big).$$

No term in the parenthesis is negative. The last one is equal to $\dfrac{n-1}{n}$ since $[0]=1$. Consequently $N \geq (n-1)$.

Theorem IX. *Every transitive group contains at least $(n-1)$ substitutions which affect all the n elements. If there are more than $(n-1)$ of these, then the group also contains substitutions which affect less than $(n-1)$ elements.* *

Corollary. *A k-fold transitive group contains substitutions which affect exactly n elements, and others which affect exactly $(n-1), (n-2), \ldots (n-k+1)$ elements.*

Those substitutions which affect exactly k elements we shall call *substitutions of the kth class*. We have just demonstrated the existence of substitutions of the nth, or highest class.

If we consider a non-primitive group G, there is (§ 66) a second group $\mathfrak{G}$ isomorphic with G, the substitutions of which interchange

---

*C. Jordan: Liouville Jour. (2), XVII, p 351.

the elements $A_1, A_2, \ldots A_n$ exactly as the corresponding substitutions of G interchange the several systems of non-primitivity. Since G is transitive, $\mathfrak{G}$ is also transitive. From Theorem IX follows therefore

Theorem X. *Every non-primitive group G contains substitutions which interchange all the systems of non-primitivity.*

§ 75. We construct within the transitive group G the subgroup H of lowest order, which contains all the substitutions of the highest class in G, and prove that this group H is also transitive.

H is evidently a self-conjugate subgroup of G. If H were intransitive, G must then be non-primitive (Theorem VI). If this is the case, let $\mathfrak{G}$ be the group of § 66 which affects the systems $A_1, A_2, \ldots A_\mu$ regarded as elements. $\mathfrak{G}$ is transitive. To substitutions of the highest class in $\mathfrak{G}$ correspond substitutions of the highest class in G. (The converse is not necessarily true). Suppose that $\mathfrak{H}$ is the subgroup of the lowest order which contains all the substitutions of the highest class in $\mathfrak{G}$. To $\mathfrak{H}$ then corresponds either H or a subgroup of H. If $\mathfrak{H}$ is transitive in the A's, H is transitive in the x's. The question therefore reduces to the consideration of the groups $\mathfrak{G}$ and $\mathfrak{H}$. $\mathfrak{H}$ can be intransitive only if $\mathfrak{G}$ is non-primitive and G accordingly contains more comprehensive systems of non-primitivity. If this were the case, we should again start out in the same way from $\mathfrak{G}$ and $\mathfrak{H}$, and continue until we arrive at a primitive group. The proof is then complete.

Theorem XI. *In every transitive group the substitutions of the highest class form by themselves a transitive system.*

§ 76. Suppose a second transitive group G' to have all its substitutions of the highest class in common with G of the preceding Section. If then we construct the subgroup H' for G', corresponding to the subgroup H of G, we have $H' = H$.

Moreover the number N_1 of the substitutions of the highest class in H is

$$n\left(\frac{1}{2}[n-2]_1 + \frac{2}{3}[n-3]_1 + \cdots + \frac{n-q-1}{n-q}[q]_1 + \cdots + \frac{n-1}{n}[0]_1\right),$$

where $[q]_1$ has the same relation to H as $[q]$ to G. But the number N_1 is, as we have just seen equal to the N of § 74. Consequently

$$n\{\tfrac{1}{2}([n-2]-[n-2]_1)+\tfrac{2}{3}([n-3]-[n-3]_1)+\cdots$$
$$\cdots+\frac{n-q-1}{n-q}([q]-[q]_1)+\cdots\}=0.$$

But, since H is entirely contained in G, it follows that $[q] \geq [q]_1$, and therefore the left hand member of the equation above can only vanish if each parenthesis is 0. Consequently G and G' can only differ in respect to substitutions of the $(n-1)^{\text{th}}$ class.

Theorem XII. *If two transitive groups have all their substitutions of the highest class in common, they can only differ in those substitutions which leave only one element unchanged.*

§ 77. Let G be any transitive group and $G_1, G_2, \ldots G_n$ those subgroups of G which do not affect $x_1, x_2, \ldots x_n$ respectively. These groups are, as we have seen, all similar. If now G_1, and consequently $G_2, \ldots G_n$, are k-fold transitive, then G is at least $(k+1)$-fold transitive. For if it be required that the $(k+1)$ elements $x_1, x_2, \ldots x_{k+1}$ shall be replaced by $x_{i_1}, x_{i_2}, \ldots x_{i_{k+1}}$ respectively, we can find in G some substitution s which replaces $x_1, x_2, x_3, \ldots x_{k+1}$ by $x_{i_1}, x_{h_2}, x_{h_3}, \ldots x_{h_{k+1}}$, where $x_{h_2}, x_{h_3}, \ldots$ may be any elements whatever. Again G_{i_1} contains some substitution t which replaces $x_{h_2}, x_{h_3}, \ldots$ by $x_{i_2}, x_{i_3}, \ldots$ Consequently the substitution st of G satisfies the requirement.

From this follows the more general

Theorem XIII. *If a group G is at least k-fold transitive, and if the subgroup of G which leaves k given elements unchanged is still h-fold transitive, then G is at least $(k+h)$-fold transitive.*[*]

§ 78. Suppose that those substitutions of a k-fold transitive group G, which, excluding the identical substitution, affect the smallest number of elements, are of the qth class, *i. e.*, that they affect exactly q elements. The question arises whether there is any connection between the numbers k and q.

In the first place suppose $k \geq q$, and let one of the substitutions of the qth class contained in G be $s = (x_1 x_2 \ldots)\ldots(\ldots x_{q-1} x_q)$. Then, on account of its k-fold transitivity, G also contains a substitu-

---
[*] G. Frobenius: Ueber die Congruenz nach einem aus zwei endlichen Gruppen gebildeten Doppelmodul. Crelle CI. p. 290.

88   THEORY OF SUBSTITUTIONS.

tion σ, which replaces $x_1, x_2, \ldots x_{q-1}, x_q$ by $x_1, x_2, \ldots x_{q-1}, x_\varkappa$ ($\varkappa > q$), and which is therefore of the form

$$\sigma = (x_1)(x_2) \ldots (x_{q-1})(x_q x_\varkappa \ldots).$$

We have then

$$(\sigma^{-1} s \sigma) s^{-1} = [(x_1 x_2 \ldots) \ldots (\ldots x_{q-1} x_\varkappa)] s^{-1} = (x_{q-1} x_\varkappa x_q),$$

and since this substitution affects only 3 elements, it follows that $q \leq 3$.

Secondly, suppose $k < q$. The substitutions of the qth class may then be of either of the forms

$$s_1 = (x_1 x_2 \ldots) \ldots (\ldots x_{k-1} x_k \ldots) \ldots (\ldots x_q),$$
$$s_2 = (x_1 x_2 \ldots) \ldots (\ldots x_{k-1})(x_k \ldots) \ldots (\ldots x_q').$$

In the first case we take

$$\sigma_1 = (x_1)(x_2) \ldots (x_{k-1})(x_k x_\varkappa \ldots) \quad (k+1) \leq \varkappa \leq q,$$

and in the second

$$\sigma_2 = (x_1)(x_2) \ldots (x_{k-1})(x_k x_\lambda \ldots) \quad \lambda > q.$$

It is evident that both are possible, if in the latter case it is remembered that $n > q$. We obtain then

$$\sigma_1^{-1} s_1 \sigma_1 = (x_1 x_2 \ldots) \ldots (\ldots x_{k-1} x_\varkappa \ldots),$$
$$\sigma_2^{-1} s_2 \sigma_2 = (x_1 x_2 \ldots) \ldots (\ldots x_{k-1})(x_\lambda \ldots),$$

and if we form now $(\sigma_1^{-1} s_1 \sigma_1) s_1^{-1}$, the first $(k-2)$ elements are removed, and there remain, at the most, $q + (q-k) - (k-2) = 2q - 2k + 2$. Similarly, if we form $(\sigma_2^{-1} s_2 \sigma_2) s_2^{-1}$, the first $(k+1)$ elements are removed, and there remain, at the most, $q + (q-k+1) - (k-1) = 2q - 2k + 2$. By hypothesis, this number cannot be less than q. Consequently

$$q \geq 2k - 2.$$

Theorem XIV. *If a k-fold transitive group contains any substitution, except the identical substitution, which affects less than $(2k-2)$ elements, it contains also substitutions which affect at the most only three elements.*

This theorem gives a positive result only if $k > 2$. In this case, by anticipating the conclusions of the next Section, we can add the following

Corollary. *If a k-fold transitive group $k>2$ contains substitutions, different from identity, which affect not more than $(2k-2)$ elements, it is either the alternating or the symmetric group.*

We may now combine this result with the corollary of Theorem IX. If G is k-fold transitive, it contains substitutions of the class $(n-k+1)$. Accordingly $q \leq (n-k+1)$. If G is neither the alternating nor the symmetric group, $q > (2k-2)$. Consequently $(n-k+1) \geq (2k-2)$ and $k \leq \dfrac{n+3}{3}$.

Theorem XV. *If a group of degree n is neither the alternating nor the symmetric group, it is, at the most, $\left(\dfrac{n}{3}+1\right)$-fold transitive.*

That the upper limit of transitivity here assigned may actually occur is demonstrated by the five-fold transitive group of twelve elements discovered by Matthieu,

$$G = \{(x x_1 x_2 x_3)(x_4 x_5 x_6 x_7), (x x_5 x_2 x_7)(x_1 x_4 x_3 x_6),$$
$$(y_1 x)(x_1 x_6)(x_3 x_7)(x_4 x_5), (y_2 y_1)(x_1 x_3)(x_4 x_7)(x_5 x_6),$$
$$(y_3 y_2)(x_1 x_5)(x_3 x_7)(x_4 x_6), (y_1 y_3)(x_1 x_3)(x_4 x_5)(x_6 x_7)\}.$$

§ 79. **Theorem XVI.** *If a k-fold transitive group $(k>1)$ contains a circular substitution of three elements, it contains the alternating group.*

Suppose that $s = (x_1 x_2 x_3)$ occurs in the given group G. Then, since G is at least two-fold transitive, it must contain a substitution $\sigma = (x_3)(x_1 x_4 x_\lambda \ldots) \ldots$ and consequently also

$$\tau = \sigma^{-1} s \sigma = (x_3 x_4 x_\mu), \quad \tau^{-1} s \tau = (x_1 x_2 x_4).$$

In the same way it appears that G contains

$$(x_1 x_2 x_5), (x_1 x_2 x_6), \ldots$$

Consequently (§ 35) G contains the alternating group.

Theorem XVII. *If a k-fold transitive group $(k>1)$ contains a transposition, the group is symmetric.*

The proof is exactly analogous to the preceding.

For simply transitive groups the last two theorems hold only under certain limitations, as appear from the following instances

$$G_1 = [1, (x_1x_2), (x_3x_4), (x_1x_2)(x_3x_4), (x_1x_3)(x_2x_4), (x_1x_4)(x_2x_3),$$
$$(x_1x_3x_2x_4), (x_1x_4x_2x_3)],$$
$$G_2 = \{1, (x_1x_2x_3), (x_4x_5x_6), (x_7x_8x_9), (x_1x_5x_9x_2x_6x_7x_3x_4x_8)\}.$$

Both of these are transitive. But the former contains a substitution of two elements, without being symmetric, and the latter a substitution of three elements without being the alternating group.

§ 80. An explanation of this exception in the case of simply transitive groups is obtained from the following considerations.

If we arbitrarily select two or more substitutions of n elements, it is to be regarded as extremely probable that the group of lowest order which contains these is the symmetric group, or at least the alternating group. In the case of two substitutions the probability in favor of the symmetric group may be taken as about $\frac{3}{4}$, and in favor of the alternating, but not symmetric, group as about $\frac{1}{4}$. In order that any given substitutions may generate a group which is only a part of the $n!$ possible substitutions, very special relations are necessary, and it is highly improbable that arbitrarily chosen substitutions $s_i = \begin{pmatrix} x_1 \, x_2 \, \ldots \, x_n \\ x_{i_1} x_{i_2} \ldots x_{i_n} \end{pmatrix}$ should satisfy these conditions. The exception most likely to occur would be that all the given substitutions were severally equivalent to an even number of transpositions and would consequently generate the alternating group.

In general, therefore, we must regard every transitive group which is neither symmetric nor alternating, and every intransitive group which is not made up of symmetric or alternating parts, as decidedly exceptional. And we shall expect to find in such cases special relations among the substitutions of the group, of such a nature as to limit the number of their distinct combinations.

Such relations occur in the case of the two groups cited above. Both of them belong to the groups which we have designated as non-primitive. In G_1 the elements x_1, x_2 form one system, and x_3, x_4 another; it is therefore impossible that G_1 should include, for example, the transposition (x_1x_3). In G_2 there are three systems of non-primitivity x_1, x_2, x_3, x_4, x_5, x_6, and x_7, x_8, x_9, G_2 therefore cannot contain the substitution $(x_1x_4x_7)$.

It is, then, evidently of importance to examine the influence of primitivity on the character of a transitive group, and we turn our attention now in this direction.

§ 81. With the last two theorems belongs naturally

Theorem XVIII. *If a primitive group contains either of the two substitutions*

$$\sigma = (x_1 x_2 x_3), \quad \tau = (x_1 x_2),$$

it contains in the former case the alternating, in the latter the symmetric group.

The proofs in the two cases are of the same character. We give only that for the latter case.

From Theorem VIII, the given group must contain a substitution which leaves x_1 unchanged and replaces x_2 by a new element x_3, or which leaves x_2 unchanged and replaces x_1 by a new element x_3, or which replaces x_1 by x_2 or x_2 by x_1 and the latter element in either case by a new element x_3. If then we transform τ with respect to this substitution, we obtain a transposition τ' connecting either x_1 or x_2 with x_3, for example $\tau' = (x_1 x_3)$. The presence of τ and τ' in the group shows that the latter must contain the symmetric group of the three elements x_1, x_2, x_3. From Theorem VIII there must also be in the given group a second substitution which replaces one of these three elements by either itself or a second one among them, and which also replaces one of them by a new element x_4. Suppose this substitution to be, for instance,

$$s = (\ldots x_2 x_1 \ldots x_3 x_4 \ldots) \ldots$$

We obtain then

$$\tau'' = s^{-1}(x_2 x_3) s = (x_1 x_4),$$

and it follows that the given group contains the symmetric group of the four elements x_1, x_2, x_3, x_4; and so on.

§ 82. We can generalize the last theorem as follows:

Theorem XIX. *If a primitive group G with the elements $x_1, x_2 \ldots x_n$ contains a primitive subgroup H of degree $k < n$, then G contains a series of primitive subgroups similar to H,*

$$H_1, H_2, H_3, \ldots H_{n-k+}$$

such that every H_λ affects the elements $x_1, x_2, \ldots x_{k-1}, x_{k+\lambda-1}$, where $x_1, x_2, \ldots x_{k-1}$ may be selected arbitrarily.

We take $H_1 = H$ and transform H with respect to all the substitutions of G into $H_1, H'_1, H''_1, \ldots$ Now let H'_1 be that one of the transformed groups which connects the k elements $x_1, x_2, \ldots x_k$ of H_1 with other elements, but with the smallest number of these. We maintain that this smallest number is one. For if several new elements $\xi_1, \xi_2, \ldots$ occurred in H'_1, then from Theorem VIII there must be in the primitive group H'_1 a substitution which replaces one ξ by another ξ and at the same time replaces a second ξ by one of the elements $x_1, x_2, \ldots x_k$. Suppose that

$$t = (\xi_\alpha \xi_\beta \ldots \xi_\gamma x_\delta \ldots) \ldots$$

is such a substitution, the case where $\beta = \gamma$ being included. Then $H''_1 = t H_1 t^{-1}$ will still contain ξ_γ but will not contain ξ_α. H''_1, therefore contains fewer new elements ξ than H'_1. Consequently if H'_1 is properly chosen, it will contain only one new element, say x_{k+1}. It will therefore not contain some one of the elements of H_1, say x_α. We select then from H_1 a substitution $u = (\ldots x_\alpha x_k \ldots) \ldots$ and form the group $u^{-1} H'_1 u = H_2$. This group contains

$$x_1, x_2 \ldots x_{k-1}, x_{k+1},$$

but not x_k. In the same way we can form a group H_3 which affects only $x_1, x_2 \ldots x_{k-1}, x_{k+2}$, and so on.

It remains to be shown that $x_1, x_2, \ldots x_{k-1}$ can be taken arbitrarily, that is, that the assumption $H = H_1$ is always allowable. Suppose that H_1 contains $x_1, x_2, \ldots x_{k-a}$. Then in the series $H_1, H_2, \ldots$ there is a group H_0 which also contains H_{k-a+1}. Proceeding from H_0 and the elements $x_1, x_2, \ldots x_{k-a+1}$, we construct a series of groups, as before, arriving finally at the group H.

§ 83. **Theorem XX.** *If a primitive group G of degree n contains a primitive subgroup H of degree k then G is at least $(n-k+1)$-fold transitive.*

From the preceding theorem H_1 affects the elements $x_1, x_2, \ldots x_k$; $\{H_1, H_2\}$ the elements $x_1, x_2, \ldots x_k, x_{k+1}$; $\{H_1, H_2, H_3\}$ the elements $x_1, x_2, \ldots x_k, x_{k+1}, x_{k+2}$; and so on. All these groups are transitive; consequently, from Theorem XIII, $\{H_1, H_2\}$ is two-fold transitive, $\{H_1, H_2, H_3\}$ three-fold transitive, and finally

$$\Gamma = \{H_1, \ldots H_{n-k+1}\}$$

is at least $(n-k+1)$-fold transitive. Therefore G, which includes Γ, is also at least $(n-k+1)$-fold transitive. *

Corollary I. *If a primitive group of degree n contains a circular substitution of the prime order p, the group is at least $(n-p+1)$-fold transitive.*

For the powers of the circular substitution form a group H of degree p.

Corollary II. *If a transitive group of degree n contains a circular substitution of prime order $p < \dfrac{2n}{3}$, then, if the group does not contain the alternating group, it is non-primitive.*

From Theorem XV, every group which is more than $\left(\dfrac{n}{3}+1\right)$-fold transitive is either alternating or symmetric. And since the presence of a circular substitution of a prime order p in a primitive group would require the latter to be at least $(n-p+1)$-fold transitive, it would follow, if $p < \dfrac{2n}{3}$, that the group would be more than $\left(\dfrac{n}{3}+1\right)$-fold transitive and must therefore be either alternating or symmetric. As these alternatives are excluded, the group must be non-primitive.

§ 84. In the proof of Theorem XIX the primitivity of the group H was only employed to demonstrate the presence of substitutions which contained two successions of elements of a certain kind. The presence of such substitutions would also evidently be assured if H were two-fold or many-fold transitive. Theorems XIX and XX would therefore still be valid in this case. The latter then takes the form:

Theorem XXI. *If a primitive group G of degree n con-*

---

*Another proof of this theorem is given by Rudio: Ueber primitive Gruppen, Crelle CII, p. 1.

tains a k-fold transitive subgroup ($k \geq 2$) of degree q, then G is at least $(n-q+2)$ transitive.

§ 85. If the requirement that the subgroup H of the preceding Section shall be primitive or multiply transitive is not fulfilled, the the theory becomes at once far more complicated.* We give here only a few of the simpler results.

Theorem XXII. *If a primitive group G of degree n contains a subgroup H of degree $\lambda < n$, then G also contains a subgroup whose degree is exactly $n-1$; or in other words: A transitive group G of n elements, which has no subgroup of exactly $n-1$ elements, but has a subgroup of lower degree, is non-primitive.*

Suppose that the subgroup H of degree $\lambda < n$ affects the elements $x_1, x_2, \ldots x_\lambda$. In the first place if $\lambda < \dfrac{n}{2}$ then the group G, on account of its primitivity, contains a substitution s_1 which replaces one element of $x_1, x_2, \ldots x_\lambda$ by another element of the same system and at the same time replaces a second element of $x_1, x_2, \ldots x_\lambda$ by some new element. Then $H' = s_1^{-1} H s_1$ contains beside some of the old elements, also certain new ones, so that $H_1 = \{H, H'\}$ affects more than λ elements, but less than n, since H and H' together affect at the most $(2\lambda - 1) < n$. If the degree λ_1 of H_1 is still $< \dfrac{n}{2}$, we repeat the same process, until λ_1 is equal to or greater than $\dfrac{n}{2}$. Suppose that the elements of the last H_1 are $x_1, x_2, \ldots x_\lambda$. Then the primitive group G must again contain a substitution s_2 which replaces two elements not belonging to H_1 by two elements, one of which does, while the other does not, belong to H_1. Then the group $H'_1 = s_2 H_1 s_2^{-1}$ will connect new elements with those of H_1; but, from the way in which s_2 was taken, one new element is still not contained in H'_1. That some of the old elements actually occur in H'_1 follows from the fact that $\lambda_1 \geq \tfrac{1}{2} n$. Accordingly $H_2 = \{H_1, H'_1\}$ contains more elements than H_1 but less than G. Proceeding in this way, we must finally arrive at a group K which contains exactly $(n-1)$ elements.

---

*C. Jordan: Liouville Jour. (2) XVI. B. Marggraf: Ueber primitive Gruppen mit transitiven Untergruppen geringeren Grades; Giessen Dissertation, 1890.

If H is transitive, then H', and consequently $H_1 = \{H', H\}$, and so on to K, are also transitive. From Theorem XIII, G must therefore in this case be at least two-fold transitive. We have then the following

Corollary. *If a primitive group G contains a transitive subgroup of lower degree, then G is at least two-fold transitive.*

§ 86. We turn now to a series of properties based on the theory of self-conjugate subgroups.

Let $H = [1, s_2, s_3, \ldots s_m]$ be a self-conjugate subgroup of a group G of order $n = km$. The substitutions of G can be arranged (§ 41) in a table, the first line of which contains the substitutions of H.

$$
\begin{array}{cccc}
s_1 = 1, & s_2, & s_3, & \ldots s_m, \\
\sigma_2, & s_2\sigma_2, & s_3\sigma_2, & \ldots s_m\sigma_2, \\
\sigma_3, & s_2\sigma_3, & s_3\sigma_3, & \ldots s_m\sigma_3, \\
\cdot & \cdot & \cdot & \cdot \\
\sigma_k, & s_2\sigma_k, & s_3\sigma_k, & \ldots s_m\sigma_k.
\end{array}
$$

From the definition of a self-conjugate subgroup we have then

$$(s_\lambda \sigma_\alpha)(s_\mu \sigma_\beta) = s_\lambda(\sigma_\alpha s_\mu \sigma_\alpha^{-1})\sigma_\alpha \sigma_\beta = s_\lambda s_\nu \sigma_\alpha \sigma_\beta = s_\kappa \sigma_\alpha \sigma_\beta,$$

that is, the line of the table in which the product $(s_\lambda \sigma_\alpha)(s_\mu \sigma_\beta)$ occurs depends only on σ_α and σ_β, or in other words, if every substitution of the αth line is multiplied into every substitution of the βth line, the resulting products all belong to one and the same line.

If we denote the several lines, regarded as units, by $z_1, z_2, \ldots z_k$, then the line containing the product of the substitutions of z_α into those of z_β may be denoted by $z_\alpha z_\beta$. This symbol has then a definite, unambiguous meaning. Moreover, $z_\alpha z_\beta$ cannot be equal to $z_\alpha z_\gamma$ or to $z_\gamma z_\beta$. For then we should have from the last paragraph $\sigma_\alpha \sigma_\beta = \sigma_\alpha \sigma_\gamma$ or $\sigma_\alpha \sigma_\beta = \sigma_\gamma \sigma_\beta$, that is, $\sigma_\beta = \sigma_\gamma$ or $\sigma_\alpha = \sigma_\gamma$. Consequently

$$t_\alpha = \begin{pmatrix} z_1, & z_2, & \ldots z_i, & \ldots \\ z_1 z_\alpha, & z_2 z_\alpha, & \ldots z_i z_\alpha & \ldots \end{pmatrix}$$

denotes a substitution among the z's, and this substitution corresponds to all the substitutions $s_\lambda \sigma_\alpha$ $(\lambda = 1, 2, \ldots m)$ of the αth line of the table. The t's therefore form a group T, which is $(1-m)$-

fold isomorphic with the given group G. The degree and order of T are equal, and both are equal to k (*Cf.* § 97). To the identical substitution of T corresponds the self-conjugate subgroup H of G.

We shall designate T as the *quotient* of G and H, and write accordingly $T = G:H$.

§ 87. A group G which contains a self-conjugate subgroup H, different from identity, is called a *compound* group; otherwise G is a *simple* group. If G contains no other self-conjugate subgroup K which includes H, then H is a *maximal self-conjugate subgroup*.

If G is a compound group, and if the series of groups

$$G, G_1, G_2, \ldots G_\mu, 1$$

is so taken that every G_λ is a maximal self-conjugate subgroup of the preceding one, then this series is called *the series belonging to the compound group G*, or *the series of composition of G*, or, still more briefly, *the series of G*.

If the numbers

$$r, \ r_1 = r:e_1, \ r_2 = r_1:e_2, \ \ldots \ r_\mu = r_{\mu-1}:e_\mu, \ r_{\mu+1} = r_\mu:e_{\mu+1} = 1$$

are the orders of the successive groups of the series of composition of G, then $e_1, e_2, \ldots e_{\mu+1}$ are called *the factors of composition* of G; and we have $r = e_1 e_2 e_3 \ldots e_{\mu+1}$.

If, in accordance with the notation of § 86, we write,

$$G:G_1 = \varGamma'_1, \quad G_1:G_2 = \varGamma'_2, \quad \ldots \quad G_{\mu-1}:G_\mu = \varGamma'_\mu, \quad G_\mu:1 = \varGamma',$$

the order and the degree of every $\varGamma'_a$ is equal to e_a ($a = 1, 2, \ldots \mu + 1$). All the groups $\varGamma'_a$ are simple. For $\varGamma'_a$ is $(1-r_a)$-fold isomorphic with G_{a-1}, and to the identical substitution in $\varGamma'_a$ corresponds G_a in G_{a-1}. Consequently, if $\varGamma'_a$ contains a self-conjugate subgroup different from identity, then the corresponding self-conjugate subgroup of G_{a-1} (§ 73) contains and is greater than G_a. The latter would therefore not be a *maximal* self-conjugate subgroup of G_{a-1}.

The groups $\varGamma'$, which define the transition from every G_a to the following one in the series of composition, are called the *factor groups* of G.*

---

* O. Hölder; Math. Ann. XXXIV, p. 30 ff.

§ 88. Given a compound group G, it is quite possible that the corresponding series of composition is not fully determinate. It is conceivable that, if a series of composition

$$G, G_1, G_2, \ldots G_\mu, 1$$

has been found to exist, there may also be a second series

$$G, G'_1, G'_2, \ldots G'_\nu, 1$$

in which every G' is contained as a maximal self-conjugate subgroup in the preceding one. We shall find however that, in whatever way the series of composition may be chosen, the number of groups G is constant, and moreover the factors of composition are always the same, apart from their order of succession.

Suppose the substitutions of G_1 and G'_1 to be denoted by s_a and s'_a respectively. Let $r_1 = r : e_1$ be the order of G_1, and $r'_1 = r : e'_1$ that of G'_1. The substitutions common to G_1 and G'_1 form a group Γ (§ 44), the order x of which is a factor of both r_1 and r'_1. We write

$$r_1 = xy, \quad r'_1 = xy'.$$

The substitutions of Γ we denote by σ_a. All the substitutions of G_1 may then be arranged in a table, the first line of which consists of the substitutions σ_a of Γ. We obtain

$$\sigma_1 = 1, \quad \sigma_2, \quad \sigma_3, \ldots \sigma_x; \quad \Gamma,$$
$$\mathfrak{s}_2\sigma_1, \mathfrak{s}_2\sigma_2, \mathfrak{s}_2\sigma_3, \ldots \mathfrak{s}_2\sigma_x; \quad \mathfrak{s}_2\Gamma,$$
$$\cdot \quad \cdot \quad \cdot \quad \cdot \quad \cdot \quad \cdot$$
$$\mathfrak{s}_y\sigma_1, \mathfrak{s}_y\sigma_2, \mathfrak{s}_y\sigma_3, \ldots \mathfrak{s}_y\sigma_x; \quad \mathfrak{s}_y\Gamma,$$

where the $\mathfrak{s}$ belonging to any line is any substitution of G_1 not contained in the preceding lines. The group G'_1 can be treated in the same way. We will suppose that in this case, in place of $\mathfrak{s}_1, \mathfrak{s}_2, \ldots$, we have $\mathfrak{s}'_1, \mathfrak{s}'_2, \ldots$. Every substitution of G' or G'_1 can, then, be written in the form

$$s_a = \mathfrak{s}_\beta \sigma_\gamma, \quad s'_a = \mathfrak{s}'_\beta \sigma_\gamma.$$

Again, the product

$$s_a{}^{-1} s'_\beta{}^{-1} s_a s'_\beta = s'_a{}^{-1}(s'_\beta{}^{-1} s_a s'_\beta) = (s_a{}^{-1} s'_\beta{}^{-1} s_a) s'_\beta$$

belongs to G_1. For, since $G^{-1} G_1 G = G_1$, it follows that $s'_\beta{}^{-1} s_a s'_\beta$, which occurs in the second form of the product, is equal to s_γ, and the product itself is equal to $s_a{}^{-1} s_\gamma$. But, from the third form, this

7

same product belongs to G', since $G^{-1}G'G = G'$, and therefore $s_a^{-1}s'_\beta{}^{-1}s_a = s'_\gamma$, so that the product is equal to $s'_\gamma s'_\beta$. Consequently the product belongs to the group I' which is common to G_1 and G'_1. Hence

A) $\quad s_a{}^{-1}s'_\beta{}^{-1}s_a s'_\beta = \sigma_\gamma; \quad s_a s'_\beta = s'_\beta s_a \sigma_\gamma; \quad s'_\beta s_a = s_a s'_\beta \sigma_\delta.$

In particular, since the σ's belong to both the s's and the s''s, we obtain

B) $\quad \sigma_a \mathfrak{s}'_\beta = \mathfrak{s}'_\beta \sigma_\gamma; \quad \sigma_a \mathfrak{s}_\beta = \mathfrak{s}_\beta \sigma_\delta; \quad \mathfrak{s}'_a \mathfrak{s}_\beta = \mathfrak{s}_\beta \mathfrak{s}'_a \sigma_\epsilon.$

From this it follows that the substitutions of the form $\mathfrak{s}_a \mathfrak{s}'_\beta \sigma_\gamma$ form a group $\mathfrak{G}$. For, by repeated applications of the equation B), we obtain

$$(\mathfrak{s}_a \mathfrak{s}'_\beta \sigma_\gamma)(\mathfrak{s}_a \mathfrak{s}'_b \sigma_c) = \mathfrak{s}_a \mathfrak{s}'_\beta \mathfrak{s}_a \sigma_\delta \mathfrak{s}'_b \sigma_c = \mathfrak{s}_a \mathfrak{s}_n \cdot \mathfrak{s}'_\beta \sigma_\epsilon \mathfrak{s}'_b \sigma_c = \mathfrak{s}_a \mathfrak{s}_n \mathfrak{s}'_\beta \mathfrak{s}'_b \sigma_d$$
$$= \mathfrak{s}_\zeta \sigma_\epsilon \mathfrak{s}'_\eta \sigma_f \sigma_a = \mathfrak{s}_\zeta \mathfrak{s}'_\eta \sigma_\vartheta.$$

The group $\mathfrak{G}$ is commutative with G; for we have

$$G^{-1}(\mathfrak{s}_\zeta \mathfrak{s}'_\eta \sigma_\vartheta)G = G^{-1}\mathfrak{s}_\zeta G \cdot G^{-1}\mathfrak{s}'_\eta \sigma_\vartheta G = s_a s'_\beta = \mathfrak{s}_\gamma \sigma_\delta \cdot \mathfrak{s}'_\epsilon \sigma_\iota = \mathfrak{s}_\gamma \mathfrak{s}'_\epsilon \sigma_\kappa.$$

The group $\mathfrak{G}$ is more extensive than G_1 or G'; it is contained in G; consequently, from the assumption as to G_1 and G', $\mathfrak{G}$ must be identical with G.

The order of $\mathfrak{G}$ is equal to $xy \cdot y'$. For, if $\mathfrak{s}_a \mathfrak{s}'_\beta \sigma_\gamma = \mathfrak{s}_a \mathfrak{s}'_b \mathfrak{s}_c$, it is easily seen that $a = a, b = \beta, c = \gamma$. Consequently the order of G is also xyy', and since we have

$$r = r_1 e_1 = xye_1, \quad r = r'_1 e'_1 = xy'e'_1$$

it follows that

$$y' = e_1, \quad y = e'_1.$$

This last result gives us for the order of I', $x = \dfrac{r}{e_1 e'_1} = \dfrac{r_1}{e'_1} = \dfrac{r'_1}{e_1}$.

We can show, further, that I' is a maximal self-conjugate subgroup of G_1 and of G'_1, and consequently occurs in one of the series of composition of either of these groups. For in the first place I', as a part of G'_1, is commutative with G_1, and, as a part of G_1, is commutative with G'_1, so that we have

$$G_1^{-1}I'G_1 = G'_1, \quad G'_1{}^{-1}I'G'_1 = G_1.$$

But since the left member of the first equation belongs entirely to G_1, the same is true for the right member, and a similar result holds for the second equation. Consequently

GENERAL CLASSIFICATION OF GROUPS. 99

$$G_1^{-1} \mathit{I'} G_1 = \mathit{I'}, \quad G'_1{}^{-1} \mathit{I'} G'_1 = \mathit{I'}.$$

Again there is no self-conjugate subgroup of G_1 intermediate between G_1 and $\mathit{I'}$ which contains the latter. For if there were such a group H with substitutions t_α, then it would follow from $A)$ that

$$t_\alpha^{-1} s'_\beta{}^{-1} t_\alpha s'_\beta = \sigma_\gamma, \quad s'_\beta{}^{-1} t_\alpha s'_\beta = t_\alpha \cdot t_\alpha^{-1} s'_\beta{}^{-1} t_\alpha s'_\beta = t_\alpha \sigma_\gamma = t_\delta,$$

that is, H is also commutative with G'_1. And since G_1 and G'_1 together generate G, it appears that H must be commutative with G. If now we add to the t_α's the $\mathfrak{s}'_2, \mathfrak{s}'_3, \ldots$, then the substitutions $\mathfrak{s}'_\alpha t_\beta$ form a group. For since $\mathit{I'}$ is contained in H and in G_1, we have from $A)$

$$(\mathfrak{s}'_\alpha t_\beta)(\mathfrak{s}'_\gamma t_\delta) = \mathfrak{s}'_\alpha \cdot \mathfrak{s}'_\gamma t_\beta \sigma_\epsilon \cdot t_\delta = \mathfrak{s}'_\alpha t_\nu.$$

This group is commutative with G, since this is true of its component groups H and G'_1. It contains G'_1, which consists of the substitutions $\mathfrak{s}'_\alpha \sigma_\beta$. It is contained in G, which consists of the substitutions $\mathfrak{s}'_\alpha \mathfrak{s}_\beta \sigma_\gamma$. But this is contrary to the assumption that G'_1 is a maximal self-conjugate subgroup of G. We have therefore the following preliminary result:

If in two series of composition of the group G, the groups next succeeding G are respectively G_1 and G'_1, then in both series we may take for the group next succeeding G_1 or G'_1 one and the same maximal self-conjugate subgroup $\mathit{I'}$, which is composed of all the substitutions common to G_1 and G'_1. If e_1 and e'_1 are the factors of composition belonging to G_1 and G'_1 respectively, then $\mathit{I'}$ has for its factors of composition, in the first series e'_1, in the second e_1.

§ 89. We can now easily obtain the final result.

Let one series of composition for G be

1) $\qquad G, G_1, G_2, G_3, \ldots,$

$$r, \; r_1 = r : e_1, \; r_2 = r_1 : e_2, \; r_3 = r_2 : e_3, \ldots,$$

and let a second series be

2) $\qquad G, G'_1, G'_2, G'_3, \ldots,$

$$r, \; r'_1 = r : e'_1, \; r'_2 = r'_1 : e'_2, \; r'_3 = r'_2 : e'_3, \ldots$$

Then from the result just obtained, we can construct two more series belonging to G:

100   THEORY OF SUBSTITUTIONS.

3) $G, G_1, I', J, H, \ldots$ 4) $G, G'_1, I', J, H, \ldots$,
$r, r_1 = r : e_1, r_2 = r_1 : e'_1, \ldots;$ $r, r'_1 = r : e'_1, r'_2 = r'_1 : e_1, \ldots,$

and apply the same proof for the constancy of the factors of composition to the series 1) and 3), and again 2) and 4), as was employed above in the case of the series 1) and 2). The series 3) and 4) have obviously the same factors of composition.

The problem is now reduced, for while the series 1) and 2) agree only in their first terms, the series 1) and 3), and again 2) and 4), agree to two terms each. The proof can then be carried another step by constructing from 1) and 2) as before two new series, both of which now begin with G, G_1:

1') $G, G_1, G_2, \mathfrak{G}, \mathfrak{H}, \mathfrak{J}, \ldots;$
       $r, r_1, r_2 = r_1 : e_2, r''_3 = r_2 : e'_2, \ldots$
3') $G, G_1, I', \mathfrak{G}, \mathfrak{H}, \mathfrak{J}, \ldots,$
       $r, r_1, r'_2 = r_1 : e'_2, r''_3 = r'_2 : e_2, \ldots$

These series have again the same factors of composition, and 1') and 1) and again 3') and 3) agree to three terms, and so on.

We have then finally

Theorem XXIII. *If a compound group G admits of two different series of composition, the factors of composition in the two cases are identical, apart from their order, and the number of groups in the two series is therefore the same.*

§ 90. From § 88 we deduce another result. Since $G^{-1}\Gamma G$ belongs to G_1, because $G^{-1}G_1 G = G_1$, and also to G'_1 because $G^{-1}G'_1 G = G'_1$, it appears that $G^{-1}I'G$, as a common subgroup of G_1 and G'_1, must be identical with I', so that I' is a self-conjugate subgroup of G. From § 86 it follows that it is possible to construct a group Ω of order $e_1 e'_1$ which is $(1 - \dfrac{r}{e_1 e'_1})$-fold isomorphic with G, in such a way that the same substitution of Ω corresponds to all the substitutions of G which only differ in a factor σ. We will take now, to correspond to the substitutions $1, \mathfrak{s}_2, \mathfrak{s}_3, \ldots \mathfrak{s}_{e'_1}$ of G_1, the substitutions $1, \omega_2, \omega_3, \ldots \omega_{e'_1}$ of Ω, and, to correspond to the $1, \mathfrak{s}'_2, \mathfrak{s}'_3, \ldots \mathfrak{s}'_{e_1}$ of G'_1, the substitutions $1, \omega'_2, \omega'_3, \ldots \omega'_{e_1}$ of Ω. In no case is $\mathfrak{s}'_\alpha = \mathfrak{s}_\beta \mathfrak{s}_\gamma$, for the σ's form the common subgroup of G_1

and G'_1. Consequently the ω's are different from the ω''s. Both classes of substitutions give rise to groups:

$$\Omega_1 = [1, \omega_2, \omega_3, \ldots \omega_{e_1}], \quad \Omega_2 = [1, \omega'_2, \omega'_3, \ldots \omega'_{e_1}],$$

and, since $\bar{s}_a\bar{s}'_\beta = \bar{s}'_\beta \bar{s}_a \sigma_\gamma$, it follows that $\Omega_1 \Omega'_1 = \Omega'_1 \Omega_1$. Moreover every s in G is equal to $\bar{s}_a \bar{s}'_\beta \sigma_\gamma$, that is $\Omega = \Omega_1 \Omega'_1$.

We obtain Ω therefore, by multiplying every substitution of Ω_1 by every one of Ω'_1.

§ 91. We consider now two successive groups of a series of composition, or, what is the same thing, a group G and one of its maximal self-conjugate subgroups H. Suppose that s'_1 is a substitution of G which does not occur in H, and let $s'_1{}^m$ be the lowest power of s'_1 which does occur in H (m is either the order of s'_1 or a factor of the order). If m is a composite number and equal to pq, we put $s'_1{}^q = s_1$, and obtain thus a substitution s_1 which does not occur in H, and of which a prime power $s_1{}^p$ is the first to occur in H. We then transform s_1 with respect to all the substitutions of G, and obtain in this way a series of substitutions $s_1, s_2, \ldots s_\lambda$. No one of these can occur in H. For if this were the case with $s_a = \sigma^{-1} s_1 \sigma$, then $\sigma s_a \sigma^{-1} = s_1$, being the transformed of a substitution s_a of H with respect to a substitution σ^{-1} of G, would also occur in H.

We consider then the group

$$\Gamma' = \{H, s_1, s_2, \ldots s_\lambda\}.$$

This group contains H and is contained in G. If t is any arbitrary substitution of G, we have

$$t^{-1} \Gamma' t = t^{-1} \{ H s_{i_1}{}^a s_{i_2}{}^\beta \ldots \} t = t^{-1} H t \cdot t^{-1} s^a t \cdot t^{-1} s_{i_2}{}^\beta t \ldots$$
$$= H s_{i_1}{}^{\iota} s_{i_2}{}^{\kappa} \ldots = \Gamma'.$$

Γ' is therefore commutative with G. These three properties of Γ' are inconsistent with the assumption that H is a maximal self-conjugate subgroup of G, unless Γ' and H are identical.

If we remember further that all substitutions, as $s_1, s_2, \ldots s_\lambda$, which are obtained from one another by transformation, are similar, we have

Theorem XXIV. *Every group of the series of composition of any group G, is obtainable from the next following (or, every group is obtainable from any one of its maximal self-conjugate*

102   THEORY OF SUBSTITUTIONS.

subgroups) by the addition of a series of substitutions, 1) *which are similar to one another, and* 2) *a prime power of which belongs to the smaller group. The last actual group of a series of composition consists entirely of similar substitutions of prime order.*

§ 92. The following theorem is of great importance for the theory of equations:

Theorem XXV. *The series of composition of the symmetric group of n elements, consists, if $n > 4$, of the alternating group and the identical substitution. The corresponding factors of composition are therefore 2 and $\frac{1}{2}n!$ The alternating group of more than four elements is simple.*

We have already seen that the alternating group is a maximal self-conjugate subgroup of the symmetric group. It only remains to be shown that, for $n > 4$, the alternating group is simple. The proof is perfectly analogous to that of § 52, and the theorem there obtained, when expressed in the nomenclature of the present Chapter, becomes: a group which is commutative with the symmetric group is, for $n > 4$, either the alternating group or the identical substitution. It will be necessary therefore to give only a brief sketch of the proof.

Suppose that H_1 is a maximal self-conjugate subgroup of the alternating group H, and consider the substitutions of H_1 which affect the smallest number of elements. All the cycles of any one of these substitutions must contain the same number of elements (§ 52). The substitutions cannot contain more than three elements in any cycle. For if H contains the substitution

$$s = (x_1 x_2 x_3 x_4 \ldots) \ldots,$$

and if we transform s with respect to $\sigma = (x_2 x_3 x_4)$, which of course occurs in H, then $s^{-1}\sigma^{-1}s\sigma$ contains fewer elements than s.

Again the substitutions of H_1 with the least number of the elements cannot contain more than one cycle. For if either

$$s_\alpha = (x_1 x_2)(x_3 x_4) \ldots, \quad s_\beta = (x_1 x_2 x_3)(x_4 x_5 x_6) \ldots,$$

occurs in H, and if we transform with respect to $\sigma = (x_1 x_2 x_5)$, the products

$$s_\alpha \sigma^{-1} s_\alpha \sigma, \quad s_\beta \sigma^{-1} s_\beta \sigma,$$

will contain fewer elements than s_α, s_β respectively.

The substitutions which affect the smallest number of elements are therefore of one or the other of the forms

$$s_a = (x_a x_\beta), \quad t = (x_a x_\beta x_\gamma).$$

The first case is impossible, since the alternating group cannot contain a transformation. The second case leads to the alternating group itself.

If $n = 4$, we obtain the following series of composition: 1) the symmetric group; 2) the alternating group; 3) $G_2 = [1, (x_1 x_2)(x_3 x_4), (x_1 x_3)(x_2 x_4), (x_1 x_4)(x_2 x_3)]$; 3) $G_3 = [1, (x_1 x_2)(x_3 x_4)]$; 5) $G_4 = 1$. The exceptional group G_2 is already familiar to us.

§ 93. We may add here the following theorems:

Theorem XXVI. *Every group G, which is not contained in the alternating group is compound. One of its factors of composition is 2. The corresponding factor group is $[(1, z_1, z_2,)]$.*

The proof is based on § 35, Theorem VIII. The substitutions of G which belong to the alternating group form the first self-conjugate subgroup of G.

Theorem XXVII. *If a group G is of order p^a, p being a prime number, the factors of composition of G are all equal to p.*

The group K of order p^f obtained in § 30 is obviously, from the method of its construction, compound. It contains a self-conjugate subgroup L of order p^{f-1} and this again contains a self-conjugate subgroup M of order p^{f-2}, and so on. The series of composition of K consists therefore of the groups

$$K, L, M, \ldots Q, R, \ldots S, 1,$$

of orders

$$p^f, p^{f-1}, p^{f-2}, \ldots p^k, p^{k-1}, \ldots p, 1.$$

The last corollary of § 49 shows that we need prove the present theorem only for the subgroups of K. If G occurs among these and is one of the series above, the proof is already complete. If G does not occur in this series, suppose that R is the first group of the series which does not contain G, while G is a subgroup of Q. We apply then to G the second proposition of § 71. Suppose that H is the common subgroup of R and G. Then H is a self-conjugate subgroup of G, and its order is a multiple of p^{a-1} and is consequently either p^{a-1} or p^a. The latter case is impossible since then G would be contained in R. Consequently H is of order p^{a-1}, and the theorem is proved.

Theorem XXVIII. *If a group G of order r contains a self-conjugate subgroup H of order $\dfrac{r}{e}$ then no substitution of G, which does not occur in H can be of an order prime to e.* *

We construct the factor group $\Gamma = G:H$ of the order e. No one of the substitutions of Γ, except the identical substitution, is of an order prime to e. To any substitution s of G which does not occur in H corresponds a σ which is different from 1. On account of the isomorphism of G and Γ, there corresponds to every power s^k of s the same power σ^k of σ. If $\varkappa$ is the lowest power of s for which $s^\varkappa = 1$, then at the same time $\sigma^\varkappa = 1$. $\varkappa$ is to therefore a multiple of the order of σ and consequently is not prime to e.

If, in particular, e is a prime number, then the order of every substitution of G which is not contained in H is divisible by e.

§ 94. Among the various series of composition of a group G, the *principal series of composition*, or briefly, *the principal series*, is of special importance in the algebraic solution of equations. This principal series is obtained from any series of composition by retaining only those groups of the series which are themselves self-conjugate subgroups of G. Suppose the resulting series to be

$$G, H, J, \ldots K, L, M, 1.$$

The series of G may itself be the principal series. This will be the case, for example, as we shall immediately show, if all the factors of composition of the series are prime numbers.

Assuming that the principal series is not identical with the given series, suppose that the latter contains, for instance between H and J, other groups, as

$$H_1, H_2, \ldots H_{\nu-1}.$$

H_1 is therefore commutative with H, but not with G. Consequently

$$H^{-1}H_1 H = H_1, \quad G^{-1}H_1 G \neq H_1.$$

---

*A. Kneser: Ueber die algebraische Unauflösbarkeit höherer Gleichungen. Crelle CVI., pp. 59–60.

Accordingly, if we transform H_1 with respect to all the substitutions of G, we shall obtain a series of groups $H_1, H'_1, H''_1, \ldots$ All of these are contained as self-conjugate subgroups in H, for if σ is any substitution of G, then $\sigma^{-1}H_1\sigma = H'_1$ is contained in $\sigma^{-1}H\sigma = H$. Moreover

$$H^{-1}H'_1 H = H^{-1}\left(\sigma^{-1}H_1\sigma\right) H = \left(\sigma^{-1}H^{-1}\sigma\right)\left(\sigma^{-1}H_1\sigma\right)\left(\sigma^{-1}H\sigma\right)$$
$$= \sigma^{-1}(H^{-1}H_1H)\sigma = \sigma^{-1}H_1\sigma = H'_1;$$

for if τ is any substitution of H, then from $\sigma^{-1}\tau\sigma = \nu$ follows $\sigma^{-1}\tau^{-1}\sigma = \nu^{-1}$ (*cf.* § 36).

Again J is contained in every one of the groups $H_1, H'_1, H''_1, \ldots$ For J is contained in H_1, and consequently $\sigma^{-1}J\sigma = J$ is contained in $\sigma^{-1}H_1\sigma = H'_1$, and so on. Finally H'_1, like H_1, is a maximal self-conjugate subgroup of H. For if there were any self-conjugate subgroup between H and H'_1, then the same would be true of H and H_1. In fact if H'_1 is obtained from H_1 by transformation with respect to σ, then the intermediate group between H and H_1 would proceed from that between H and H'_1, by transformation with respect to σ^{-1}. In the series of G, the group H may therefore be any one of the several groups of the same type $H_1, H'_1, \ldots$ All of these belong to the same factor ϵ of composition, ϵ being the quotient of the orders of H and H_1. In accordance with the preliminary result of § 88, we can then continue the series of G by taking for the group next succeeding H_1 the substitutions common to H_1 and H'_1, or to H_1 and H''_1, or to H_1 and H'''_1, and so on. From the same result the new groups all belong to the same factor of composition ϵ. Every one of them contains J. We need of course consider only the different groups among them. If there is only one, this must coincide with J. For the entire system of groups

$$H_1, H'_1, H''_1, H'''_1, \ldots,$$

and consequently the group common to all of them, is unaltered by transformation with respect to G. The order of J is therefore obtained by dividing that of H by ϵ^2.

But if there are several different groups, we can then proceed in the same way. The substitutions common to H_1, H'_1, H''_1, for example, form a group which in the series of G can succeed the group composed of the substitutions common to H_1 and H'_1. The corresponding factor of composition is again ε.

After ν repetitions of this process we arrive at the group J. The order of J is therefore obtained by dividing that of H by ε^ν. The last system before J consists of ν groups $H_{\nu-1}, H'_{\nu-1} \ldots$ which are all similar and all belong to the factor ε, and which give

$$H = \{H_{\nu-1}, H'_{\nu-1}, H''_{\nu-1}, \ldots \}.$$

Theorem XXIX. *If a series of composition of G does not coincide with a principal series, but if, between two groups H and J of the latter, $\nu - 1$ groups $H_1, H_2, \ldots H_{\nu-1}$ of the former are inserted, then to $H_1, H_2, \ldots J$ belong the same factors of composition ε, and the order r' of G is therefore equal to the order r'' of J multiplied by ε^ν. H can be obtained from J by combining with J a series of ν groups $H_{\nu-1}, H'_{\nu-1}, \ldots$, which are all similar, and of the order $r''\varepsilon$.*

Corollary I. *If the factors of composition of a group are not all equal, the group has a principal series.*

Corollary II. *Every non-primitive group is compound if it contains any substitution except identity which leaves the several systems of non-primitivity unchanged as units. If the group contains greater (including) and lesser (included) systems of non-primitivity, it has a principal series.*

The instance of the group

$$G = [1, (x_1x_2)(x_3x_4)(x_5x_6), (x_1x_3)(x_2x_5)(x_4x_6), (x_1x_6)(x_2x_4)(x_3x_5),$$
$$(x_1x_4x_5)(x_2x_3x_6), (x_1x_5x_4)(x_2x_6x_3)]$$

shows that non-primitivity may occur in a simple group. In this case the only substitution which leaves the systems $x_1, x_2, \quad x_3, x_5$, and x_4, x_6 unchanged is the identical substitution.

Corollary III. *The groups $H_{\nu-1}, H'_{\nu-1}, H''_{\nu-1}, \ldots$ are commutative, i. e., the equations hold*

$$H_{\nu-1}^{(\alpha)} H_{\nu-1}^{(\beta)} = H_{\nu-1}^{(\beta)} H_{\nu-1}^{(\alpha)}.$$

For in the series preceding J we may assume the sequence $H_{\nu-1}^{(\alpha)}, \{H_{\nu-1}^{(\alpha)}, H_{\nu-1}^{(\beta)}\}, \ldots$ to occur. Accordingly we must have

GENERAL CLASSIFICATION OF GROUPS. 107

or
$$(H_{\nu-1}^{(\alpha)} H_{\nu-1}^{(\beta)})^{-1} H_{\nu-1}^{(\alpha)} (H_{\nu-1}^{(\alpha)} H_{\nu-1}^{(\beta)}) = H_{\nu-1}^{(\alpha)},$$

$$(H_{\nu-1}^{(\beta)})^{-1} (H_{\nu-1}^{(\alpha)})^{-1} H_{\nu-1}^{(\alpha)} H_{\nu-1}^{(\alpha)} H_{\nu-1}^{(\beta)}$$
$$= (H_{\nu-1}^{(\beta)})^{-1} H_{\nu-1}^{(\alpha)} H_{\nu-1}^{(\beta)} = H_{\nu-1}^{(\alpha)}.$$

Corollary IV. *The last actual group M of the principal series of G is composed of one or more groups similar to one another, which have no substitutions except identity in common, and which are commutative with one another.*

§ 95. We have now to consider the important special case where ε is a prime number p.

Instead of $H'_{\nu-1}, H''_{\nu-1}, \ldots$ we employ now the more convenient notation

$$H', H'', H''', \ldots H^{(\nu)}.$$

Then H' is obtained from J by adding to the latter a substitution t_1, the pth power of which is the first to occur in J. We may write (§ 91)

$$H' = t_1^a J, \quad H'' = t_2^a J, \quad H''' = t_3^a J, \ldots \quad (a = 0, 1, \ldots p-1).$$

Since J is a self-conjugate subgroup of every one of the groups $H', H'', \ldots$, we have

$$t_1^{-a} J t_1^a = J, \quad t_2^{-a} J t_2^a = J, \quad t_3^{-a} J t_3^a = J, \ldots$$

and, if we denote the substitutions of J by $i_1, i_2, i_3, \ldots$,

$$t_1^{-a} i_1^{-1} t_1^a = i_2, \quad t_2^{-a} i_1^{-1} t_2^a = i_3, \quad t_3^{-a} i_1^{-1} t_3^a = i_4, \ldots,$$
$$t_1^a i_1 = i_1 t_1^a (i_2 i_1), \quad t_2^a i_1 = i_1 t_2^a (i_3 i_1), \quad t_3^a i_1 = i_1 t_3^a (i_4 i_1), \ldots,$$

that is, the substitutions of H', of H'', of H''', and so on, are commutative among themselves, apart from a factor belonging to J.

Since we can return from J to H by combining the substitutions of H' and H'', for example, into a single group (§ 88), we have from § 94, Corollary III

$$t_2^{-1} t_1 t_2 = t_1^a i_1, \quad t_1^{-1} t_2^{-1} t_1 = t_2^\beta i_2,$$

and consequently, by combination of these two results,

$$t_1^{-1} t_2^{-1} t_1 t_2 = t_2^\beta i_2 t_2 \quad = t_2^{\beta+1} i_3,$$
$$= t_1^{-1} t_1^a i_1 = t_1^{a-1} i_1,$$
$$t_1^{a-1} i_1 = t_2^{\beta+1} i_3.$$

The left member of the last equation is a substitution of H', the right member a substitution of H''. Since these two groups have only the substitutions of J in common, the powers of t_1 and t_2 must disappear. Consequently $\alpha = 1$, $\beta = -1$, and

$$t_2^{-1} t_1 t_2 = t_1 i_1,$$
$$t_1 t_2 = t_2 t_1 i_1,$$
$$(t_1 i_1)(t_2 i_2) = (t_2 i_2)(t_1 i_1) i_3,$$
$$(t_1{}^\alpha i_1)(t_2{}^\beta i_2) = (t_2{}^\beta i_2)(t_1{}^\alpha i_1) i_4.$$

The substitutions of the group formed from J, t_1, and t_2 are therefore commutative among themselves, apart from a factor belonging to J. The same is true of the group formed from J, t_1, and t_3, or from J, t_2, and t_3, and consequently of the group $\{J, t_1, t_2, t_3\}$, and so on, to the group H itself. (It is to be noted that Corollary III of § 94 involves much less than this. There it was a question of the commutativity of groups, here of the single substitutions.)

Every two substitutions of H are, then, commutative apart from a factor belonging to J. We will prove now the converse proposition: If two substitutions of H are commutative apart from a factor belonging to J, then ε is a prime number. In fact this will be the case, if the substitutions of H' have this property. For, this being assumed, if ε were a composite number, suppose its prime factors to be $q, q', q'', \ldots$ We select from H'_1, in accordance with Theorem XXIV, § 91, a substitution t which is not contained in J. The lowest power of t which occurs in J will then be, for example, t^q. Transforming, we have

$$H'^{-1}(t^\alpha J)H' = H'^{-1} t^\alpha H' H'^{-1} J H'$$
$$= H'^{-1} t^\alpha H' J,$$

and, since by assumption, $t^\alpha H' = H' t^\alpha J$,

$$H'^{-1}(t^\alpha J)H' = t^\alpha J.$$

The group $\{t, J\}$ is therefore a self-conjugate subgroup of H', which contains J and is larger than J. Moreover, it is contained in H', and is smaller than H'. For, if t is commutative with J, then from §§ 37–8 the order of $\{t, J\}$ is $r''q < r''e$. This is contrary to the assumption that J is the group immediately following H' in the series of G.

GENERAL CLASSIFICATION OF GROUPS. 109

Theorem XXX. *If, in the principal series of composition of G, the order r' of H is obtained from the order r'' of J by multiplication by p^ν, where the prime number p is the factor of composition for the intervening groups in the series of G, then the substitutions of H are commutative among themselves apart from factors belonging to J. Conversely, if this is the case, the factors of composition of the groups between H and J are all equal to the same prime number p.*

§ 96. We turn finally to certain properties of groups in relation to isomorphism.

If L is a maximal self-conjugate subgroup of G, and Λ the corresponding group of Γ, then Λ is also a maximal self-conjugate subgroup of Γ.

For if Γ contained a self-conjugate subgroup θ, which contained Λ, then the corresponding group T of G would contain L.

The series of composition of G corresponds to that of Γ. If G and Γ are simply isomorphic, all the factors of the one group are equal to the corresponding factors of the other. But if G is multiply isomorphic to Γ, then there occur in the series for G, besides the factors of Γ, also a factor belonging to the group S which corresponds to the identical substitution of Γ.

The proof is readily found.

If G is multiply isomorphic to Γ, then G is compound, and S is a group of the series of composition of G.

§ 97. Suppose that G is any transitive group of order r, affecting the n elements $x_1, x_2, \ldots x_n$. We construct any arbitrary $n!$-valued function ξ of $x_1, x_2, \ldots x_n$, denote its different values by $\xi_1, \xi_2, \ldots \xi_{n!}$, and apply to any one of these, as ξ_1, all the substitutions of G. Let the values obtained from ξ_1 in this way be

$$\xi_1, \xi_2, \xi_3, \ldots \xi_r.$$

The r substitutions of G will not change this system of functions as a whole, but will merely interchange its individual members, producing r rearrangements of these, which we may also regard as substitutions. These substitutions of the ξ's, as we have seen, form a new group Γ. The group Γ is transitive, for G contains substi-

110   THEORY OF SUBSTITUTIONS.

tutions which convert ξ_1 into any one of the values $\xi_1, \xi_2, \ldots \xi_r$, and therefore the substitutions of l' replace ξ_1 by any element $\xi_1, \xi_2, \ldots \xi_r$. Again every substitution of G alters the order of $\xi_1, \xi_2, \ldots \xi_r$, for ξ is a $n!$-valued function. Consequently every substitution of l' also rearranges the $\xi_1, \xi_2, \ldots \xi_r$. The order of l' is therefore equal to its degree, and both are equal to r.

G and l' are simply isomorphic. For to every substitution of G corresponds one substitution of l', and conversely to every substitution of l' at least one substitution of G. And in the latter case it can be only one substitution of G, since G and l' are of the same order.

Theorem XXXI. *To any transitive group of order r corresponds a simply isomorphic transitive group, the degree and order of which are both equal to r. Such groups are called regular.*

§ 98. **Theorem XXXII.** *Every substitution of a regular group, except the identical substitution, affects all the elements. A regular group contains only one substitution which replaces a given element by a prescribed element. Every one of its substitutions consists of cycles of the same order. If two regular groups of the same degree are (necessarily simply) isomorphic, they are similar i. e., they differ only in respect to the designation of the elements. Every regular group is non-primitive.* *

The greater part of the the theorem is already proved in the preceding Section, and the remainder presents no difficulty. We need consider in particular only the last two statements.

Suppose that $\varGamma$, with elements $\xi_1, \xi_2, \ldots \xi_n$ and substitutions $\sigma_1, \sigma_2, \ldots \sigma_n$ is isomorphic to G with elements $x_1, x_2, \ldots x_n$ and substitutions $s_1, s_2, \ldots s_n$, the isomorphism being such that to every σ_λ corresponds s_λ. Then we arrange the elements x_α and ξ_β in pairs as follows. Any two of them, x_1 and ξ_1, form the first pair. If then s_λ converts x_1 into x_λ, and if the corresponding σ_λ converts ξ_1 into ξ_λ', then x_λ and ξ_λ form a second pair. No inconsistency can arise in this way, for there is only one substitution which converts x_1 into x_α. We have now to prove that, if s_λ contains the succession $x_a x_b$, then σ_λ contains the succession $\xi_a \xi_b$.

---
*A regular group of prime degree is cyclical.

We have
$$s_a = \ldots x_1 x_a \ldots, \quad s_b = \ldots x_1 x_b \ldots, \quad s_a^{-1} s_b = \ldots x_a x_b \ldots,$$
$$\sigma_a = \ldots \xi_1 \xi_a \ldots, \quad \sigma_b = \ldots \xi_1 \xi_b \ldots, \quad \sigma_a^{-1} \sigma_b = \ldots \xi_a \xi_b \ldots,$$
and since there is only one substitution which replaces x_a by x_b, it follows that
$$s_a^{-1} s_b = s_\lambda, \quad \sigma_a^{-1} \sigma_b = \sigma_\lambda.$$
If therefore s_λ contains a cycle composed of a given number of elements x_a, then σ_λ contains an equal cycle composed of the corresponding elements ξ_a. Therefore s_λ and σ_λ, and consequently G and l' are of the same type.

The last part of the theorem is proved as follows. If a regular group G contains a substitution $s = (x_1 x_2 \ldots x_m)(x_{m+1} x_{m+2} \ldots) \ldots$ then it cannot also contain $t = (x_1 x_2 \ldots)(x_4 x_{m+2} \ldots) \ldots$ For we should then have $st^{-1} = (x_1)(x_{m+1} x_4 \ldots) \ldots$, and G would not be a regular group. Consequently $x_1, x_2, \ldots x_m$, i. e., the elements of any arbitrary cycle, form a system of non-primitivity. (The remaining systems however are not necessarily formed from the remaining cycles of the same substitution).

§ 99. If the groups G and l' are isomorphic, and if G is intransitive, then, if in every substitution of G we suppress all elements which are not transitively connected with any one among them, as x_1, the remaining portions of the several substitutions form a new transitive group G_1 also isomorphic with l'. It may however happen that the order of isomorphism of l' to G_1 is increased. Again, if x_2 is any new element, not transitively connected with x_1, we can then form a second transitive group G_2, isomorphic to l' and containing x_2, and so on.

The intransitive group G can therefore be decomposed into a system of transitive groups isomorphic with l', and conversely every intransitive group can be compounded from transitive groups $G_1, G_2, \ldots$ In the case of simple isomorphism it is only necessary to multiply the several constituents directly together.

§ 100. Suppose that G is a transitive group of degree m and order $r = mm_1$, which is k-fold isomorphic with a second group l'. Let the elements of G be $x_1, x_2, \ldots x_m$, and let G_1 be the subgroup of G which does not affect x_1. The order of G_1 is therefore

m_1. If $s_2, s_3, \ldots$ are substitutions of G which convert x_1 into $x_2, x_3, \ldots$ then $G_1 \cdot s_2, G_1 \cdot s_3, \ldots$ comprise in each case all and only the substitutions which produce the same effects.

Suppose that to G_1 in G corresponds l'_1 in l', the order of l'_1 being $m_1 k$. The order of l' is rk. Consequently, if the function φ_1 belongs to l'_1, then φ_1 takes exactly $\dfrac{rk}{m_1 k} = m$ values under the operation of all the substitutions of l'. Suppose that the substitution σ_2 of l' which corresponds to s_2 of G converts φ_1 into φ_2. Then $l' \cdot \sigma_2$ contains all the substitutions which convert φ_1 into φ_2. Similar considerations hold for σ_3 and φ_3, σ_4 and φ_4, and so on to σ_m and φ_m. If we apply all the substitutions of l' to the system of values

$$\varphi_1, \varphi_2, \ldots \varphi_m,$$

we obtain rearrangements which can be regarded as substitutions of the new elements φ. The order of the new group H is equal to the quotient of kr by the number of substitutions of l' which leave all the φ's unchanged. These correspond to the substitutions of G which leave all the x's unchanged, i. e., to the identical substitution. To this correspond k substitutions of l', and consequently the order of H is r.

G and H are, then, the same degree m_1, of the same order r, and they are isomorphic and, in fact, similar. For if s is a substitution of G which replaces x_β by x_α, then s belongs to the system $s_\alpha^{-1} G_1 s_\beta$. The corresponding substitution of H is obtained by applying a substitution $\sigma_\alpha^{-1} l'_1 \sigma_\beta$ to the system $\varphi_1, \varphi_2, \ldots \varphi_m$. Every one of these substitutions replaces φ_α by φ_β. Accordingly the substitutions of H only differ from those of G in the fact that the latter contain x's where the former contain the corresponding φ's.

We can therefore construct all groups G (or H) isomorphic to l' by applying all the substitutions of l' to any function φ belonging to any arbitrary subgroup l'_1 of l', and noting the resulting group of substitutions of the elements $\varphi_1, \varphi_2, \ldots \varphi_m$.

If l'_1 is a self-conjugate subgroup of l', the resulting isomorphic group H will be regular, as is easily seen.

§ 101. In conclusion we deduce the following

Theorem XXXIII. *Given any number of mutually multiply isomorphic groups, in which the elements of any one are* *all different from those of any other one, if we multiply every substitution of the one group by every corresponding substitution of every other group and form all the possible products, the result is an intransitive group, and conversely every intransitive group can be constructed in this way.*

The first part of the theorem is sufficiently obvious. For the second part we consider the special case of an intransitive group the elements of which break up into two transitive systems. The general proof is obtained in a perfectly similar way.

Suppose that the substitutions s_λ of the intransitive group G divide into two components

$$s_\lambda = \sigma_\lambda \tau_\lambda,$$

where σ_λ affects only the elements $x_1, x_2, \ldots x_m$, and τ_λ only $\xi_1, \xi_2, \ldots \xi_\mu$. It is possible that σ_λ occurs also in other combinations

$$\sigma_\lambda \tau_\lambda, \quad \sigma_\lambda \tau'_\lambda, \quad \sigma_\lambda \tau''_\lambda, \ldots$$

Similarly τ_λ may occur in other combinations

$$\sigma_\lambda \tau_\lambda, \quad \sigma'_\lambda \tau_\lambda, \quad \sigma''_\lambda \tau_\lambda \ldots$$

We coordinate now with σ_λ all the $\tau_\lambda, \tau'_\lambda, \tau''_\lambda, \ldots$, and with τ_λ all the $\sigma_\lambda, \sigma'_\lambda, \sigma''_\lambda, \ldots$, and proceed in the same way with all the substitutions s_λ of G. The σ_λ's form a group Σ and the τ_λ's a group T. Suppose that $\sigma_\lambda, \sigma_\mu$ are coordinated with τ_λ, τ_μ. Then there are substitutions s_λ, s_μ, s_ν, such that

$$s_\lambda = \sigma_\lambda \tau_\lambda, \quad s_\mu = \sigma_\mu \tau_\mu,$$
$$s_\lambda s_\mu = s_\nu = \sigma_\nu \tau_\nu,$$

and consequently $\sigma_\lambda \sigma_\mu = \sigma_\nu$ is coordinated with $\tau_\lambda \tau_\mu = \tau_\nu$.

8

CHAPTER V.

ALGEBRAIC RELATIONS BETWEEN FUNCTIONS BELONGING TO THE SAME GROUP. FAMILIES OF MULTIPLE-VALUED FUNCTIONS.

§ 102. It has been shown that to every multiple-valued function there belongs a group composed of all those substitutions and only those which leave the value of the given function unchanged. Conversely, we have seen that to every group there correspond an infinite number of functions. The question now to be considered is whether the property of belonging to the same group is a fundamentally important relation among functions; in particular, whether this property implies corresponding algebraic relations.

An instance in point is that of the discriminant Δ_ϕ of the values of a function φ, considered in Chapter III, § 55. It was there shown simply from the consideration of the group belonging to φ, that Δ_ϕ, and therefore the corresponding discriminant of any function belonging to the same group, is divisible by a certain power of the discriminant of the elements $x_1, x_2, \ldots x_n$.

§ 103. We shall prove now another mutual relation of great importance.

Theorem I. *Two functions belonging to the same group can be rationally expressed one in terms of the other.*

Suppose φ_1 and ψ_1 to be two functions belonging to the same group of order r and degree n

$$G_1 = [s_1 = 1, s_2, s_3, \ldots s_r].$$

If σ_2 is any substitution not belonging to G_1, and if φ_2 and ψ_2 are the values which proceed from φ_1 and ψ_1 by the application of σ_2, then all the substitutions

$$\sigma_2, \ s_2\sigma_2, \ s_3\sigma_2, \ \ldots s_r\sigma_2$$

also convert φ_1 and φ_2 into ψ_1 and ψ_2 respectively, and these are the only substitutions which produce this effect. The values φ_2 and ψ_2

therefore again belong to one and the same group $G_2 = \sigma_2^{-1} G_1 \sigma_2$. Proceeding in the same way, we obtain all the ρ pairs of values of φ and ψ', together with the ρ corresponding groups.

For every integral value of λ, the function
$$\varphi_1^\lambda \psi'_1 + \varphi_2^\lambda \psi'_2 + \ldots + \varphi_\rho^\lambda \psi'_\rho = A_\lambda$$
is therefore, like $\varphi_1 + \varphi_2 + \ldots + \varphi_\rho$ or $\psi'_1 + \psi'_2 + \ldots + \psi'_\rho$, an integral symmetric function of the elements $x_1, x_2, \ldots x_n$. For this function is merely the sum of all the values which $\varphi_1^\lambda \psi'_1$ can assume and is accordingly unchanged by any substitution, only the order of the several terms being affected. Accordingly, if φ_1 and ψ'_1 are integral rational functions of the elements x_λ, then A_λ is an integral rational function of $c_1, c_2, \ldots c_n$.

Taking successively $\lambda = 0, 1, 2, \ldots \rho - 1$, we write the corresponding equations:

$$S) \quad \begin{aligned} \psi'_1 + & \psi'_2 + & \psi'_3 + \ldots + & \psi'_\rho = A_0, \\ \varphi_1 \psi'_1 + & \varphi_2 \psi'_2 + & \varphi_3 \psi'_3 + \ldots + & \varphi_\rho \psi'_\rho = A_1, \\ \varphi_1^2 \psi'_1 + & \varphi_2^2 \psi'_2 + & \varphi_3^2 \psi'_3 + \ldots + & \varphi_\rho^2 \psi'_\rho = A_2, \\ & \ldots \ldots \ldots \ldots \ldots \ldots \ldots \ldots \\ \varphi_1^{\rho-1} \psi'_1 + & \varphi_2^{\rho-1} \psi'_2 + & \varphi_3^{\rho-1} \psi'_3 + \ldots + & \varphi_\rho^{\rho-1} \psi'_\rho = A_{\rho-1}. \end{aligned}$$

If these equations are solved for $\psi'_1, \psi'_2, \ldots \psi'_\rho$, every ψ'_λ is obtained as a rational function of $\varphi_1, \varphi_2, \ldots \varphi_\rho$.

§ 104. We multiply the first $\rho - 2$ equations of the system S) successively by the undetermined quantities $y_0, y_1, y_2 \ldots y_{\rho-2}$, and the last equation by $y_{\rho-1} = 1$, and add the resulting products, writing for brevity
$$y_{\rho-1} \varphi^{\rho-1} + y_{\rho-2} \varphi^{\rho-2} + y_{\rho-3} \varphi^{\rho-3} + \ldots + y_1 \varphi + y_0 = \chi(\varphi).$$
We obtain then

(1) $\psi'_1 \chi(\varphi_1) + \psi'_2 \chi(\varphi_2) + \ldots \psi'_\rho \chi(\varphi_\rho) = A_0 y_0 + A_1 y_1 + A_2 y_2 + \ldots$
$\ldots + A_{\rho-2} y_{\rho-2} + A_{\rho-1} y_{\rho-1}.$

From this equation we can eliminate $\psi'_2, \psi'_3, \ldots \psi'_\rho$ and obtain ψ'_1. For this purpose we need only select the y's so that we have simultaneously
$$\chi(\varphi_2) = 0, \quad \chi(\varphi_3) = 0, \quad \ldots \quad \chi(\varphi_\rho) = 0; \quad \chi(\varphi_1) \neq 0.$$

In Chapter III, § 53, we have shown that $\varphi_1, \varphi_2, \ldots \varphi_\rho$ satisfy an equation of degree ρ
$$X(\varphi) = 0,$$
the coefficients of which are rational in $c_1, c_2, \ldots c_n$. Again, the quotient
$$\frac{X(\varphi)}{\varphi - \varphi_1} = (\varphi - \varphi_2)(\varphi - \varphi_3) \ldots (\varphi - \varphi_\rho)$$
$$= \varphi^{\rho-1} - \beta_2 \varphi^{\rho-2} + \beta_3 \varphi^{\rho-3} - \ldots \pm \beta_\rho$$
vanishes if $\varphi = \varphi_1, \varphi_2, \ldots \varphi_\rho$. But, if $\varphi = \varphi_1$, we have
$$(\varphi_1 - \varphi_2)(\varphi_1 - \varphi_3) \ldots (\varphi_1 - \varphi_\rho) = X'(\varphi_1).$$
The derivative $X'(\varphi_1)$ is not zero, for if $x_1, x_2, \ldots x_n$ are independent, the values $\varphi_1, \varphi_2, \ldots \varphi_\rho$ are all different.

We can therefore satisfy the requirements above by taking
$$\chi(\varphi) = \frac{X(\varphi)}{\varphi - \varphi_1},$$
that is,
$$y_{\rho-2} = -\beta_2, \quad y_{\rho-3} = \beta_3, \quad y_{\rho-4} = -\beta_4, \ldots y_0 = \pm \beta_\rho.$$
Or, if we write
$$X(\varphi) = \varphi^\rho - a_1 \varphi^{\rho-1} + a_2 \varphi^{\rho-2} - \ldots \pm a_\rho,$$
we have
$$\frac{X(\varphi)}{\varphi - \varphi_1} = \varphi^{\rho-1} + [\varphi_1 - a_1]\varphi^{\rho-2} + [\varphi_1^2 - a_1\varphi_1 + a_2]\varphi^{\rho-3} + \ldots,$$
and consequently
$$y_{\rho-2} = \varphi_1 - a_1, \quad y_{\rho-3} = \varphi_1^2 - a_1\varphi_1 + a_2, \quad y_{\rho-4} = \varphi_1^3 - a_1\varphi_1^2 + a_2\varphi_1 - a_3, \ldots$$
By substitution in (1) we obtain then
$$(2) \qquad \psi_1 X'(\varphi_1) = R(\varphi_1), \quad \psi_1 = \frac{R(\varphi_1)}{X'(\varphi_1)}.$$

The value of ψ_1 thus obtained can be reduced to a simpler form as follows. The product
$$X'(\varphi_1) X'(\varphi_2) \ldots X'(\varphi_\rho)$$
is a symmetric function of the φ's, and in fact, as appears from the expression for $X'(\varphi_1)$ above, only differs from the discriminant Δ_φ in algebraic sign. Moreover, the product
$$(3) \qquad X'(\varphi_2) X'(\varphi_3) \ldots X'(\varphi_\rho)$$
is a symmetric function of the roots of the equation

$$\frac{X(\varphi)}{\varphi - \varphi_1} = 0,$$

and can therefore be rationally expressed in terms of the coefficients of this equation, that is, in terms of $a_1, a_2, \ldots a_\rho$, and φ_1, and consequently in terms of $c_1, c_2, \ldots c_n$ and φ_1. If now we multiply numerator and denominator of the expression for ψ_1 in (2) by the product (3), we obtain

(4) $$\psi_1 = \frac{R_1(\varphi_1)}{\Delta_\phi}.$$

The denominator of this last fraction is rational and integral in $c_1, c_2, \ldots c_n$; the numerator is rational and integral in $c_1, c_2, \ldots c_n$ and φ_1.

If the numerator $R_1(\varphi_1)$ is of a degree higher than $\rho - 1$ with respect to φ_1, a still further reduction is possible. For suppose that

$$R_1(\varphi) = X(\varphi)\, Q(\varphi) + R_2(\varphi),$$

where $Q(\varphi)$ and $R_2(\varphi)$ are the quotient and remainder obtained by dividing $R_1(\varphi)$ by $X(\varphi)$. The degree of $R_2(\varphi)$ then does not exceed $\rho - 1$. Now if $\varphi = \varphi_1, \varphi_2, \ldots \varphi_\rho$, $X(\varphi) = 0$. Consequently

$$R_1(\varphi_\lambda) = R_2(\varphi_\lambda) \quad (\lambda = 1, 2, 3, \ldots \rho),$$

and therefore

$$\psi_1 = \frac{R_2(\varphi_1)}{\Delta_\phi}$$

Similar considerations hold for the values $\psi_2, \psi_3, \ldots \psi_\rho$. We have therefore

Theorem II. *If two ρ-valued functions φ_λ and ψ_λ belong to the same group G_λ, then ψ_λ can be expressed as a rational function of which the denominator is the discriminant Δ_ϕ and is therefore rational and integral in $c_1, c_2, \ldots c_n$, while the numerator is an integral rational function of φ_λ, of a degree not exceeding $\rho - 1$, with coefficients which are integral and rational in $c_1, c_2, \ldots c_n$.*[*]

§ 105. The converse of Theorem I is proved at once:

Theorem III. *If two functions can be rationally expressed one in terms of the other, they belong to the same group.*

---

[*] *Cf.* Kronecker: Crelle 91, p. 307.

In fact, given the two equations

$$\varphi = R_1(\psi), \quad \psi = R_2(\varphi),$$

it appears from the former that φ is unchanged by all substitutions which leave ψ unchanged, so that the group of φ contains that of ψ, while from the latter equation it appears in the same way that the group of ψ contains that of φ. The two groups are therefore identical.

REMARK. Apparently the proof of this theorem does not involve the requirement that φ and ψ shall be rational functions. It must however be distinctly understood that this requirement must always be fulfilled. For example, in the irrational functions

$$\sqrt{x_1^2 - x_1 x_2 + x_2^2}, \quad \sqrt{x_1^2 - 2 x_1 x_2 + x_2^2}, \quad \sqrt{x_1^2 + 2 x_1 x_2 + x_2^2},$$

the expressions under the square root sign are all unchanged by the transposition $\sigma = (x_1 x_2)$. But it remains entirely uncertain whether the algebraic signs of the irrationalities are affected by this substitution. Considerations from the theory of substitutions alone cannot determine this question, and accordingly the sphere of application of this theory is restricted to the case of rational functions. If, in the last two irrationalities above, the roots are actually extracted and written in rational form

$$\pm (x_1 - x_2), \quad \pm (x_1 + x_2),$$

it appears at once that the transposition σ changes the sign of the former expression but leaves that of the latter unchanged, while in the case of the first irrationality this matter is entirely undecided.

§ 106. Theorems I and III furnish the basis for an algebraic classification of functions resting on the theory of groups. All rational integral functions which can be rationally expressed one in terms of another, that is, which belong to the same group, are regarded as forming a *family of algebraic functions*. The number ρ of the values of the individual functions of a family is called the *order* of the family. The several families to which the different values of any one of the functions belong are called *conjugate families*.*

---

* L. Kronecker: Monatsber. d. Berl. Akad., 1879, p. 212.

The product of the order of a family by the order of the corresponding group is equal to $n!$, where n is the degree of the group.

Every function of a family of order p is a root of an equation of degree p, the coefficients of which are rational in $c_1, c_2, \ldots c_n$. The remaining $p-1$ roots of this equation are the conjugate functions.

The groups which belong to conjugate families have, if $p > 2$, $n > 4$, no common substitution except the identical substitution.

For $p = 2$ the two conjugate families are identical.

For $p = 6$, $n = 4$ there is a family which is identical with its five conjugate families.

§ 107. In the demonstration of § 104 the condition that φ and ψ should belong to the same family was not wholly necessary. It is only essential that ψ shall remain unchanged for all those substitutions which leave the value of φ unaltered. The demonstration would therefore still be valid if some of the values of ψ should coincide; but the values of φ must all be different, as appears, for example, from the presence of the discriminant $\mathcal{J}_\phi$ in the denominator of ψ. Under the more general condition that the group of ψ includes that of φ we have then the following

Theorem IV. *If a function ψ is unchanged by all the substitutions of the group of a second function φ, while the converse is not necessarily true, then ψ can be expressed as a rational function of φ, as in Theorem II.*

Under these circumstances *the family of the function ψ is said to be included in the family of the function φ*. ψ can be rationally expressed in terms of φ, but φ cannot in general be thus expressed in terms of ψ. An *including* group corresponds to an *included* family and *vice versa*. The larger the group the smaller the family, the same inverse relation holding here as between the orders r and p.

From the preceding considerations we further deduce the following theorems:

120   THEORY OF SUBSTITUTIONS.

Theorem V. *It is always possible to find a function in terms of which any number of given functions can be rationally expressed. This function can be constructed as a linear combination of the given functions. Its family includes all the families of the given functions.*

Thus any given functions $\varphi, \psi, \chi, \ldots$ can be rationally expressed in terms of
$$\omega = \alpha \varphi + \beta \psi + \gamma \chi + \ldots,$$
where $\alpha, \beta, \gamma, \ldots$ are arbitrary parameters. For the group of ω is composed of those substitutions which leave $\varphi, \psi, \chi, \ldots$ all unchanged, and which are therefore common to the groups of $\varphi, \psi, \chi, \ldots$. The group of ω is therefore contained in that of every function $\varphi, \psi, \chi, \ldots$, and the theorem follows at once.

A special case occurs when the group of ω reduces to the identical operation, ω being accordingly a $n!$-valued function. In this case every function of the n elements $x_1, x_2, \ldots x_n$ can be rationally expressed in terms of ω, and every family is contained in that of ω. The family of ω is then called the *Galois family*.

Theorem VI. *Every rational function of n independent elements $x_1, x_2, \ldots x_n$ can be rationally expressed in terms of every $n!$-valued function of the same elements; in particular, in terms of any linear function*
$$\varphi = a_1 x_1 + a_2 x_2 \ldots + a_n x_n,$$
where $a_1, a_2, \ldots a_n$ are arbitrary parameters.

§ 108. We attempt now to find a means of expressing a multiple-valued function φ in terms of a less valued function ψ, the group of the former being included in that of the latter. A rational solution of this problem is, from the preceding developments, impossible. The problem is an analogue and a generalization of that treated in Chapter III, § 53, where a ρ-valued function was expressed in terms of a symmetric function by the aid of an equation with symmetric coefficients of which the former was a root.

From the analogy of the two cases we can state at once the present result:

FUNCTIONS BELONGING TO THE SAME GROUP. 121

Theorem VII. *If the group of a mp-valued function φ is contained in that of a p-valued function ψ, and if*

$$\varphi_1, \varphi_2, \ldots \varphi_m$$

are the m values which φ takes under the application of all the substitutions which leave ψ unchanged, then these m values of φ are the roots of an equation of degree m, the coefficients of which are rational functions of ψ.

In fact, the substitutions of the group

$$G_1 = [s_1 = 1, s_2, s_3, \ldots s_r] \quad r = \frac{n!}{p}$$

of ψ_1 are applied to any symmetric function of $\varphi_1, \varphi_2, \ldots \varphi_m$, the value of this function is unchanged, only the order of the several terms being altered. In particular we have for the elementary functions

$$\varphi_1 + \varphi_2 + \ldots + \varphi_m = A_1(\psi_1),$$
$$\varphi_1\varphi_2 + \varphi_1\varphi_3 + \ldots + \varphi_{m-1}\varphi_m = A_2(\psi_1),$$
$$\cdots \cdots \cdots \cdots \cdots$$
$$\varphi_1\varphi_2 \ldots \varphi_m = A_m(\psi_1),$$

where the A's are rational, but in general not integral functions of ψ_1. We obtain therefore the equation

$$(A_1) \quad \varphi^m - A_1(\psi_1)\varphi^{m-1} + A_2(\psi_1)\varphi^{m-2} + \ldots \pm A_m(\psi_1) = 0,$$

of which the roots are $\varphi_1, \varphi_2, \ldots \varphi_m$, and in general the equation

$$(A_\lambda) \quad \varphi^m - A_1(\psi_\lambda)\varphi^{m-1} + A_2(\psi_\lambda)\varphi^{m-2} + \ldots \pm A_m(\psi_\lambda),$$

of which the roots are the m values of φ which correspond to the value ψ_λ of ψ.

The denominators of the A_λ's and, in fact, their least common denominator is always a divisor of the discriminant Δ_ϕ, as appears from the proof of Theorem II. If ψ is a symmetric function, there is no longer a discriminant, and the denominator is removed, as we have seen in Chapter III, § 53.

§ 109. One special case deserves particular notice. If the included group H of the function φ is commutative with the including group G of ψ then, if a single root of the equation (A_1) is known, the other roots are all rationally determined. For if

$$\varphi_1, \varphi_2, \varphi_3, \ldots \varphi_m$$

122   THEORY OF SUBSTITUTIONS.

are the roots of the equation (A_1), and if
$$H_1, H_2, H_3, \ldots H_m$$
are the groups belonging to these values, finally, if
$$\sigma_1 = 1, \sigma_2, \sigma_3 \ldots \sigma_m$$
are any arbitrary substitutions of G which convert φ_1 into $\varphi_1, \varphi_2, \ldots \varphi_m$ respectively, then we have (Chapter III, § 45)
$$H_1 = \sigma_1^{-1} H_1 \sigma_1, \quad H_2 = \sigma_2^{-1} H_1 \sigma_2, \ldots H_m = \sigma^{-1} H_1 \sigma_m.$$
But by supposition H is a *self-conjugate subgroup* of G, and therefore
$$G^{-1} H_1 G = H_1,$$
that is,
$$\sigma_2^{-1} H_1 \sigma_2 = H_1, \quad \sigma_3^{-1} H_1 \sigma_3 = H_1, \ldots \sigma_m^{-1} H_1 \sigma_m = H_1,$$
and consequently
$$H_1 = H_2 = H_3 = \ldots = H_m.$$
The m different values $\varphi_1, \varphi_2, \ldots \varphi_m$ therefore belong to one and the same group H, and can consequently all be rationally expressed in terms of any one among them, in accordance with Theorem I.

The family of ψ_1 is included in that of φ_1. When, as in the present case, the group H_1 of φ_1 is not merely contained in the group G of ψ_1 but is a self-conjugate subgroup of G, the family of ψ_1 is called a *self-conjugate subfamily* of the family of φ_1.

Theorem VIII. *In order that all the roots of the equation (A_1) should be rationally expressible in terms of any one among them, as φ_1, it is necessary and sufficient that the family of ψ_1 should be a self-conjugate subfamily of that of φ_1, i. e., that the group of φ_1 should be a self-conjugate subgroup of that of ψ_1. The groups of $\varphi_1, \varphi_2, \ldots \varphi_m$ are then coincident.*

We consider in particular the case where m is a *prime* number. Suppose G_1 to be the group of ψ_1 and H_1 that of φ_1. Since every substitution of G_1 produces a corresponding substitution of the values $\varphi_1, \varphi_2, \ldots \varphi_m$, the group G_1 is isomorphic with a group of the φ's. The latter group is transitive and of degree m. From Theorem II, Chapter IV, its order is divisible by m, and from Theorem X, Chapter III, it therefore contains a substitution of order m.

For m elements, where m is prime, there is only one type of such substitutions

$$t = (\varphi_1 \varphi_2 \ldots \varphi_m).$$

The corresponding substitution τ of G_1 therefore permutes $\varphi_1, \varphi_2, \ldots \varphi_m$ cyclically. Moreover, since τ^m corresponds to t^m, it follows that τ^m leaves all the functions $\varphi_1, \varphi_2, \ldots \varphi_m$ unchanged. Accordingly τ^m, and no lower power of τ, is contained among the substitutions of H_1.

Furthermore we readily show that

$$G_1 = \{H_1, \tau\}.$$

For the substitutions $H_1, H_1\tau, \ldots H_1\tau^{m-1}$ are all different and, since H_1 is of order $\dfrac{n!}{m\rho}$, there are $m \dfrac{n!}{m\rho} = \dfrac{n!}{\rho}$ of them. They are all necessarily contained in G_1, which, being itself of order $\dfrac{n!}{\rho}$, cannot contain any other substitutions. From this it appears again that τ is commutative with H_1.

Theorem IX. *If the equation* (A_1) *is of prime degree* m, *and if the group* H_1 *of* φ_1 *is a self-conjugate subgroup of the group* G_1 *of* ζ_1, *then* G_1 *contains a substitution* τ *which permutes* $\varphi_1, \varphi_2, \ldots \varphi_\rho$ *cyclically. This substitution is commutative with* H_1; *its* m^{th}, *and no lower, power is contained in* H_1; *together with* H_1 *it generates the group* G_1.

§ 110. We examine now under what circumstances (A_1) can become a *binomial* equation, again assuming the degree m to be a *prime* number. If (A_1) is binomial, its roots, $\varphi_1, \omega\varphi_1, \omega^2\varphi_1, \ldots \omega^{m-1}\varphi_1$ evidently all belong to the same group. It is therefore necessary that G_1 should be a self-conjugate subgroup of G_1.

We proceed now to show conversely that, if the group H_1 is a self-conjugate subgroup of G_1, then a function χ_1 belonging to H_1 can be found, the mth power of which belongs to G_1.

Denoting any primitive m^{th} root of unity by ω, we write

$$\chi_1 = \varphi_1 + \omega\varphi_2 + \omega^2\varphi_3 + \ldots + \omega^{m-1}\varphi_m.$$

If we apply to this expression the successive powers of t or τ, we obtain

$$\chi_2 = \varphi_2 + \omega\varphi_3 + \omega^2\varphi_4 + \ldots + \omega^{m-1}\varphi_1 = \omega^{-1}\chi_1,$$
$$\chi_3 = \varphi_3 + \omega\varphi_4 + \omega^2\varphi_5 + \ldots + \omega^{m-1}\varphi_2 = \omega^{-2}\chi_1,$$

consequently
$$\chi_1^m = \chi_2^m = \ldots = \chi_m^m.$$

We have now to prove 1) that χ_1 belongs to the group H_1, and 2) that χ_1^m belongs to the group G_1.

In the first place, since $\varphi_1, \varphi_2, \ldots \varphi_m$ are unchanged by all the substitutions of H_1, the same is true of χ_1. Moreover if there were any other substitutions which left χ_1 unchanged we should have, for example,

$$\varphi_1 + \omega\varphi_2 + \ldots + \omega^{m-1}\varphi_m = \varphi_{i_1} + \omega\varphi_{i_2} + \ldots + \omega^{m-1}\varphi_{i_m},$$

and therefore

$$\omega^{m-1}(\varphi_m - \varphi_{i_m}) + \omega^{m-2}(\varphi_{m-1} - \varphi_{i_{m-1}}) + \ldots + (\varphi_1 - \varphi_{i_1}) = 0.$$

The latter equation would then have one of its roots, and consequently all its roots, in common with the irreducible equation

$$\omega^{m-1} + \omega^{m-2} + \ldots + 1 = 0,$$

and we must therefore have

$$\varphi_1 - \varphi_{i_1} = \varphi_2 - \varphi_{i_2} = \ldots = \varphi_m - \varphi_{i_m}.$$

But we may assume the function φ_1 to have been constructed by the method of § 31 as a sum of $\dfrac{n!}{m\rho}$ terms of the form $x_1^a x_2^\beta \ldots$ with undetermined exponents. The systems of exponents in $\varphi_1, \varphi_2, \ldots \varphi_m$ will then all be different, and therefore, since the x's are independent variables, the equation

$$\varphi_1 + \varphi_{i_2} = \varphi_2 + \varphi_{i_1}$$

can hold only if $\varphi_1 = \varphi_{i_1}$ and $\varphi_2 = \varphi_{i_2}$ identically. The function χ_1 therefore belongs to H_1.

It follows at once that χ_1^m belongs to G_1. For this function is unchanged by H_1 and τ, and consequently by

$$G_1 = \{H_1, \tau\}.$$

No other substitutions can leave χ_1^m unchanged. For otherwise χ_1^m would take less than ρ values, and its mth root χ_1 less than $m\rho$ values, which would be contrary to the result just obtained.

Theorem X. *In order that the family belonging to a group H may contain functions the mth power of which belongs to the family of a group G, it is necessary and sufficient that H should be a self-conjugate subgroup of G, or, in other words, that the family of G should be a self-conjugate subfamily of that of H.*

From Theorems IX and X the following special case of the latter is readily deduced:

Theorem XI. *In order that the prime power φ^p of a $p\mu$-valued function φ may have μ values, it is necessary and sufficient that there should be a substitution τ, commutative with the group H of φ, of which the pth power is the first to occur in H.*

Finally an extension of the last theorem furnishes the following important result:

Theorem XII. *If the series of groups*

$$G, G_1, G_2, G_3, \ldots G_\nu$$

is so connected that every G_{a-1} can be obtained from the following G_a by the addition to the latter of a substitution τ_a commutative with G_a, of which a prime power, the p_a^{th}, is the first to occur in G_a, then and only then it is possible to obtain a $\mu \cdot p_1 \cdot p_2 \ldots p_\nu$-valued function belonging to G_ν from a μ-valued function belonging to G by the solution of a series of binomial equations. The latter are then of degree $p_1, p_2, p_3, \ldots p_\nu$, respectively.

§ 111. In the expression of a given function in terms of another belonging to the same family, we have met with rational fractional forms the denominators of which were factors of the discriminant of the given function. If we regard the elements $x_1, x_2, \ldots x_n$ as independent quantities, as we have thus far done, the discriminant of any function φ is different from zero, for the various conjugate values of φ have different *forms*. But if any relations exist among the elements x, it is no longer true that a difference in form necessarily involves a difference in *value*. It is therefore quite possible that if the coefficients in the equations

$$f(x) = x^n - c_1 x^{n-1} + c_2 x^{n-2} - \ldots \pm c_n = 0$$

are assigned special values, the discriminant

$$\Delta_\varphi = R(c_1, c_2, \ldots c_n)$$

may become zero. If this were the case, φ could not be employed in the expression of other functions of the same family. And it is conceivable that the discriminant of every function of a family might vanish. It is therefore necessary, in order to remove this uncertainty, to prove

Theorem XIII. *If only no two x's are equal, then whatever other relations may exist among the x's, there are in every family functions the discriminants of which do not vanish.*

A proof might be given similar to that of § 30. It is however more convenient to make use of the result there obtained, that under the given conditions there are still $n!$-valued functions of the form

$$\varphi = a_0 + a_1 x_1 + a_2 x_2 + \ldots + a_n x_n.$$

We suppose the a's and the x's to be free to assume imaginary (complex) as well as real values. This being the case, if the $n!$ values of φ are all different, we can select the coefficient a_0, so that the moduli of the values of φ are also all different. For if

$$\varphi_\lambda = m_\lambda + \mu_\lambda \sqrt{-1} \qquad (\lambda = 1, 2, \ldots n),$$

then we can take

$$a' = p + q \sqrt{-1}$$

in such a way that all the $n!$ quantities

$$\varphi'_\lambda = \varphi_\lambda + a' = (m_\lambda + p) + (\mu_\lambda + q)\sqrt{-1} \qquad (\lambda = 1, 2, \ldots n)$$

shall have different moduli. For from

$$(m_\lambda + p)^2 + (\mu_\lambda + q)^2 = (m_\kappa + p)^2 + (\mu_\kappa + q)^2$$

it would follow, if p and q are entirely arbitrary, that

$$m_\lambda = m_\kappa, \; \mu_\lambda = \mu_\kappa.$$

In fact we can, for example, take $p = q^2$ and q so large that even special values of q satisfy the conditions.

Suppose then that the φ'_λ's are arranged in order of the magnitudes of their moduli

$$\varphi'_1, \varphi'_2, \varphi'_3, \ldots \varphi'_{n!} \qquad (\text{mod. } \varphi'_\lambda > \text{mod. } \varphi'_{\lambda+1}).$$

We take then the integer e so great that

$$\varphi'^e_\lambda > (\varphi'^e_{\lambda+1} + \varphi'^e_{\lambda+2} + \ldots + \varphi'^e_{n!}) \qquad (\lambda = 1, 2, \ldots (n!-1)).$$

From every equation of the form

$$\Psi) \qquad \psi_a^r + \psi_b^r + \psi_c^r + \ldots = \psi_\alpha^r + \psi_\beta^r + \psi_\gamma^r + \ldots$$

it follows accordingly that $a = \alpha, b = \beta, c = \gamma, \ldots$ If now we apply the r substitutions of G to ψ_1^r, and add the results, the sum

$$\omega = \psi_1^r + \psi_{n_2}^r + \psi_{n_3}^r + \ldots + \psi_{n_r}^r$$

is a function of the required kind. For in the first place ω is evidently unchanged by G. And in the second place the properties of the equation Ψ) show that ω has ρ distinct values, and consequently Δ_ω is not zero.

CHAPTER VI.

THE NUMBER OF VALUES OF INTEGRAL FUNCTIONS.

§ 112. Thus far we have obtained only ocasional theorems in regard to the existence of classes of multiple-valued functions. We are familiar with the one- and two-valued functions on the one side and the $n!$-valued functions on the other. But the possible classes lying between these limits have not as yet been systematically examined. An important negative result was obtained in Chapter III, § 42, where it was shown that p cannot take any value which is not a divisor of $n!$. Otherwise no general theorems are as yet known to us. We can, however, easily obtain a great number of special results by the construction of intransitive and non-primitive groups. But these are all positive, while it is the negative results, those which assert the non-existence of classes of functions, that are precisely of the greatest interest.

The general theory of the construction of intransitive groups would require as we have seen in § 101, a systematic study of isomorphism in its broadest sense. We shall content ourselves therefore with noting some of the simplest constructions.

Thus, if there are $n = a + b + c + \ldots$ elements present, and if we form the symmetric or the alternating group of a of them, the symmetric or alternating group of b others, and so on, then on multiplying all these groups together, we obtain an intransitive group of degree n and of order

$$r = \varepsilon \, a!\, b!\, c! \ldots,$$

where $\varepsilon = 1, \tfrac{1}{2}, \tfrac{1}{4}, \tfrac{1}{8}, \ldots$, according as the number of alternating groups employed in the construction is $0, 1, 2, 3, \ldots$, the rest being all symmetric. For the number of values of the corresponding functions we have then

$$p = \frac{n!}{\varepsilon \, a!\, b!\, c! \ldots}.$$

THE NUMBER OF VALUES OF INTEGRAL FUNCTIONS. 129

By distributing n in different ways between $a, b, c, \ldots$, we can obtain a large number of classes of functions. For example, if $n = 5$, we may take

$a = 5$; $\epsilon = 1, \; \rho = 1; \;\; \varphi_1 = x_1 x_2 x_3 x_4 x_5$.
$a = 5$; $\epsilon = \frac{1}{2}, \; \rho = 2; \;\; \varphi_2 = (x_1 - x_2)(x_1 - x_3)\ldots(x_4 - x_5)$
$a = 4, \; b = 1$; $\epsilon = 1, \; \rho = 5; \;\; \varphi_3 = x_1 x_2 x_3 x_4$.
$a = 4, \; b = 1$; $\epsilon = \frac{1}{2}, \; \rho = 10; \;\; \varphi_4 = (x_1 - x_2)(x_1 - x_3)(x_1 - x_4)$
$\qquad\qquad\qquad\qquad\qquad\qquad (x_2 - x_3)(x_2 - x_4)(x_3 - x_4)$.
$a = 3, \; b = 2$; $\epsilon = 1, \; \rho = 10; \;\; \varphi_5 = x_1 x_2 x_3 + x_4 x_5$.
$a = 3, \; b = 2$; $\epsilon = \frac{1}{2}, \; \rho = 20; \;\; \varphi_6 = (x_1 - x_2)(x_1 - x_3)(x_2 - x_3)$
$\qquad\qquad\qquad\qquad\qquad\qquad + x_4 x_5$.
$\qquad\qquad\qquad\qquad\qquad\qquad \varphi_7 = x_1 x_2 x_3 + x_4 - x_5$.
$a = 3, \; b = 2$; $\epsilon = \frac{1}{4}, \; \rho = 40; \;\; \varphi_8 = (x_1 - x_2)(x_1 - x_3)(x_2 - x_3)$
$\qquad\qquad\qquad\qquad\qquad\qquad + x_4 - x_5$.
$a = 3, \; b = 1, \; c = 1$; $\epsilon = 1, \; \rho = 20; \;\; \varphi_9 = x_1 x_2 x_3$.

. . . .

The imprimitive groups give rise in a similar way to the construction of functions with certain values. For example, for $n = 6$, we may take any two systems of non-primitivity of three elements each, or any three systems of two elements each, and with these construct various groups, the theory of which depends only on that of groups of degrees two and three.

§ 113. General and fundamental results are not however to be obtained in this way. We approach the problem therefore from a different side, which permits us to give it a new form of statement.

Given a ρ-valued function φ_1 with a group G_1, we construct again the familiar table of § 41:

$\varphi_1; \quad s_1 = 1, \; s_2, \quad s_3, \quad \ldots s_r \quad ; \quad G_1$
$\varphi_2; \quad \sigma_2, \quad s_2 \sigma_2, \; s_3 \sigma_2, \; \ldots s_r \sigma_2; \quad G_1 \sigma_2$
$\varphi_3; \quad \sigma_3, \quad s_2 \sigma_3, \; s_3 \sigma_3, \; \ldots s_r \sigma_3; \quad G_1 \sigma_3$
. . . .
$\varphi_\rho; \quad \sigma_\rho, \quad s_2 \sigma_\rho, \; s_3 \sigma_\rho, \; \ldots s_r \sigma_\rho; \quad G_1 \sigma_\rho$.

We proceed then to examine the distribution of the substitutions of a given type among the lines of this table.

A) There are $n - 1$ transpositions $(x_1 x_a), (a = 2, 3, \ldots n)$. If

9

then $p < n$, and if the group G_1 of φ_1 does not contain any transposition of the form $(x_1 x_a)$, these $(n-1)$ transpositions are distributed among, at the most, $(n-2)$ lines of the table. Accordingly some line after the first must contain at least two of them. Suppose these two are

$$s_a \sigma_\lambda = (x_1 x_a), \quad s_\beta \sigma_\lambda = (x_1 x_b).$$

Then it appears that a combination of the two

$$(x_1 x_a x_b) = (x_1 x_a)(x_1 x_b) = s_a \sigma_\lambda (s_\beta \sigma_\lambda)^{-1} = s_a \sigma_\lambda \sigma_\lambda^{-1} s_\beta^{-1} = s_a s_\beta^{-1} = s_\gamma$$

occurs in G_1. Consequently, if $p < n$, G_1 contains either a transposition or a circular substitution of the third order, including in either case a prescribed element x_1. The same is obviously true of any prescribed element x_λ.

B) There are $\dfrac{n(n-1)}{2}$ transpositions of the form $(x_a x_\beta)$, $(a \neq \beta = 1, 2, \ldots n)$. If therefore $p \leq \dfrac{n(n-1)}{2}$, and if the first line of the table does not contain any transposition, then some other line contains at least two. If these have one element in common, as $(x_a x_\beta)$, $(x_a x_\gamma)$, then, as we have seen in A), their product $(x_a x_\beta x_\gamma)$ occurs in G_1. If they have no element in common, as $(x_a x_\beta)$, $(x_\gamma x_\delta)$, then their product $(x_a x_\beta)(x_\gamma x_\delta)$ also occurs in G_1. In either case G_1 therefore contains a substitution of not more than four elements.

C) There are $(n-1)(n-2)$ substitutions of the form $(x_1 x_a x_\beta)$, $(a \neq \beta = 2, 3, \ldots n)$. If therefore $p \leq (n-1)(n-2)$, and if G_1 contains no substitution of this form, some other line of the table contains at least two of them. A combination of these shows that G_1 contains substitutions which affect three, four, or five elements.

Proceeding in this way, we obtain a series of results, certain of which we present here in the following

Theorem I. 1) *If the number p of the values of a function is not greater than $n-1$, the group of the function contains a substitution of, at the most, three elements, including any prescribed element.* 2) *If p is not greater than $\dfrac{n(n-1)}{2}$, the group of the function contains a substitution of, at the most, four elements.* 3) *If p is not greater than $\dfrac{n(n-1)(n-2)}{3}$, the group of the function* contains a substitution of, at the most, six elements. 4) If ρ is not greater than $\dfrac{n(n-1)(n-2)\ldots(n-k+1)}{k}$, the group of the function contains a substitution of, at the most, $2k$ elements. 5) If ρ is not greater than $(n-1)(n-2)\ldots(n-k+1)$, the group of the function contains a substitution of, at the most, $2k-1$ elements, including any prescribed element, so that the group contains at least $\dfrac{n}{2k-1}$ such substitutions.

By the aid of these results the question of the number of values of functions is reduced to that of the existence of groups containing substitutions with a certain minimum number of elements.

§ 114. In combination with earlier theorems, the first of the results above leads to an important conclusion.

From Chapter IV, Theorem I, we know that the order of an intransitive group is at the most $(n-1)!$. Consequently, the number of values of a function with an intransitive group is at least $\dfrac{n!}{(n-1)!} = n$. For such a function therefore ρ cannot be less than n.

Again, the order of a non-primitive group is, at the most, $2!\left(\dfrac{n}{2}!\right)^2$, so that the number of values of a function with a non-primitive group is at least $\dfrac{n!}{2!\dfrac{n}{2}!\dfrac{n}{2}!}$. For $n=4$, this number is less than n; but for $n>4$, it is greater than n. For such a function then, if $n>4$, ρ cannot be less than n. Again for the primitive groups it follows from Chapter IV, Theorem XVIII, in combination with the first result of Theorem I, § 113, that if $\rho < n$, the corresponding group is either alternating or symmetric, that is, $\rho = 2$ or 1. The non-primitive group for which $n=4, \rho=4, r=8$ is already known to us, (§ 46). We have then

Theorem II. *If the number ρ of the values of a function is less than n, then either $\rho = 1$ or $\rho = 2$, and the group of the function is either symmetric or alternating. An exception occurs only for $n=4, \rho=3, r=8$, the corresponding group being that belonging to $x_1 x_2 + x_3 x_4$.*

132   THEORY OF SUBSTITUTIONS.

§ 115. On account of the importance of the last theorem we add another proof based on different grounds. *

Suppose φ to be a function with the $\rho < n$ values

$$\varphi_1, \varphi_2, \varphi_3, \ldots \varphi_\rho.$$

If we apply any substitution whatever to this series, the effect will be simply to interchange the ρ values among themselves. If in particular the substitutions applied belong to the group G_1 of φ_1, then the ρ values will be so interchanged that φ_1 retains its place. All the $r = \dfrac{n!}{\rho} > (n-1)!$ substitutions of G_1 therefore rearrange only the $\rho - 1$ values $\varphi_2, \varphi_3, \ldots \varphi_\rho$. Since $\rho < n$, there are at the most, only $(\rho - 1)! \leq (n-2)!$ such rearrangements. Consequently among the $r > (n-1)!$ substitutions of G_1 there must be at least two, σ and τ, which produce the same rearrangement of $\varphi_2, \varphi_3, \ldots \varphi_\rho$. Then $\sigma \tau^{-1}$ is a substitution different from identity, which leaves all the φ's unchanged, that is, which occurs in all the conjugate groups $G_1, G_2, \ldots G_\rho$. But if $n > 4$ there is no such substitution (Chapter III, Theorem XIII). Consequently $\rho \geq n$.

§ 116. Passing to the more general question of the determination of all functions whose number of values does not exceed a given limit dependent on n, we can dispose once for all of the less important cases of the intransitive and the non-primitive groups. For the purpose we have only to employ the results already obtained in Chapter IV.

In the case of intransitive groups we have found for the maximum orders:

1) $r = (n-1)!$. Symmetric group of $n-1$ elements. $\rho = n$.

2) $r = \dfrac{(n-1)!}{2}$. Alternating group of $n-1$ elements. $\rho = 2n$.

3) $r = 2!(n-2)!$. Combination of the symmetric group of $n-2$ elements with that of the two remaining elements. $\rho = \dfrac{n(n-1)}{2}$.

4) $r = (n-2)!$. Either the combination of the alternating group of $n-2$ elements with the symmetric group of the two remaining elements; or the symmetric group of $n-2$ elements, In both cases $\rho = n(n-1)$. Etc.

---
* L. Kronecker: Monatsber. d. Berl. Akad., 1889, p. 211.

For the non-primitive groups we have

1) $r = 2!\left(\dfrac{n}{2}!\right)^2$. Two systems of non-primitivity containing each $\dfrac{n}{2}$ elements. The group is a combination of the symmetric groups of both systems with the two substitutions of the systems themselves. $\rho = \dfrac{n!}{2\left(\dfrac{n}{2}!\right)^2}$. For $n = 4, 6, 8, \ldots$ we have $\rho = 3, 10, 35, \ldots$

2) $r = 3!\left(\dfrac{n}{3}!\right)^3$. Three systems of non-primitivity. The group is a combination of the symmetric groups of the three systems with the 3! substitutions of the systems themselves. $\rho = \dfrac{n!}{3!\left(\dfrac{n}{3}!\right)^3}$. For $n = 6, 9, 12, \ldots$ we have $\rho = 15, 280, 5770, \ldots$

3) $r = 3\left(\dfrac{n}{3}!\right)^3$. As in 2), except that only the alternating group of the three systems is employed. $\rho = \dfrac{n!}{3\left(\dfrac{n}{3}!\right)^3}$. For $n = 6, 9, 12, \ldots$ we have $\rho = 30, 560, 11540, \ldots$

The values of ρ increase, as is seen, with great rapidity.

§ 117. In extension of the results of § 113 we proceed now to examine the *primitive* groups which contain substitutions of four, but none of two or of three elements.

Such a group G must contain substitutions of one of the two types

$$s_1 = (x_a x_b)(x_c x_d), \quad s_2 = (x_a x_b x_c x_d).$$

The presence of s_2 requires that of $s_2^2 = (x_a x_c)(x_b x_d)$, which belongs to the former type. Disregarding the particular order in which the elements are numbered, we may therefore assume that the substitution

$$s_5 = (x_1 x_2)(x_3 x_4)$$

occurs in the group G.

We transform s_5 with respect to all the substitutions of G and obtain in this way a series of substitutions of the same type which connect x_1, x_2, x_3, x_4 with all the remaining elements (Chapter IV, Theorem XIX). The group G therefore includes substitutions similar to s_5 which contain besides some of the old elements x_1, x_2, x_3, x_4 other new elements $x_5, x_6, x_7, \ldots$

This can happen in three different ways, according as one, two, or three of the old elements are retained. Noting again that it is only the nature of the connection of the old elements with the new, not the order of designation of the elements that is of importance, we recognize that there are only five typical cases:

$$(x_1x_2)(x_3x_5), \quad (x_1x_3)(x_2x_5),$$
$$(x_1x_5)(x_2x_6), \quad (x_1x_5)(x_3x_6),$$
$$(x_1x_5)(x_6x_7).$$

In the first case, for example, it is indifferent whether we take

$$(x_1x_2)(x_3x_5), \quad (x_1x_2)(x_4x_5), \quad (x_3x_4)(x_1x_5), \quad (x_3x_4)(x_2x_5);$$

and in the last we may replace x_1 by x_2, x_3, or x_4, etc.

The first and fifth cases are to be rejected, since their presence is at once found to be inconsistent with the assumed character of the group. Thus we have

$$(x_1x_2)(x_3x_4) \cdot (x_1x_2)(x_3x_5) = (x_3x_4x_5),$$
$$[(x_1x_2)(x_3x_4) \cdot (x_1x_5)(x_6x_7)]^2 = (x_1x_5x_2),$$

the resulting substitutions in each case being inadmissible.

There remain therefore only three cases to be examined, according as G contains, beside s_1, one or the other of the substitutions

$A)$ $\qquad (x_1x_3)(x_2x_5),$
$B)$ $\qquad (x_1x_5)(x_2x_6),$
$C)$ $\qquad (x_1x_5)(x_3x_6),$

the first case involving one new element, the last two cases two new elements each.

§ 118. $A)$ The primitive group G contains the substitutions

$$s_5 = (x_1x_2)(x_3x_4), \quad s_4 = (x_1x_3)(x_2x_5),$$

and consequently also

$$t = s_5s_4 = (x_1x_5x_2x_3x_4), \quad s_1 = ts_4t^{-1} = (x_2x_3)(x_4x_5).$$

Since t is a circular substitution of prime order 5, it follows from § 83, Corollary I, that if $n \geq 7$, is at least three-fold transitive. Then G must contain a substitution u, which does not affect x_1 but replaces x_2 by x_6 and x_3 by x_7. If we transform s_5 with respect to this substitution, we obtain

$$s' = u^{-1}s_5 u = (x_1 x_6)(x_7 x_a).$$

If x_a is contained among x_2, x_3, x_4, x_5, then s' and s_1 have only one element in common and if x_a is contained among $x_8, x_9, \ldots$ then s' and s_5 have only one element in common. Both alternatives therefore lead to the rejected fifth case of the preceding Section.

If $n \geq 7$, G becomes either the alternating or the symmetric group. There is in this case no group of the required kind.

For $n = 4$ it is readily seen that there are two types of groups with substitutions of not less than four elements, both of which are however non-primitive.

Groups of the type A) *therefore occur only for* $n = 5$ *or* $n = 6$.

For $n = 5$ we have first the group of order 10,

$$G_1 = \{s_5, s_4\} = \{s_5, t\}.$$

If we add to G_1 the substitution $\sigma = (x_1 x_3 x_4 x_5)$, we obtain a second group of order 20

$$G_2 = \{s_5, t, \sigma\} = \{t, \sigma\}.$$

The latter group is that given on p. 39. G_1 and G_2 exhaust all the types for $n = 5$.

For $n = 6$ we obtain a group G_4 of the required type by adding to G_2 the substitution $(x_1 x_6)(x_2 x_3)$. Since G_1 is of order 10, the transitive group G_4 must be at least of order 60 (Theorem II, Chapter IV). And again, since $(x_1 x_6)(x_2 x_3)\sigma = \tau = (x_1 x_6 x_3 x_2 x_4 x_5)$, we may write $G_4 = \{t, \sigma, \tau\}$. We find then that

$$\tau t = \sigma^2 \tau^3, \quad \tau \sigma = t^3 \sigma \tau^3.$$

Consequently from § 37, it follows that G_4 is of order 120.

The 60 substitutions of G_4 which belong to the alternating group from another group of the required type

$$G_3 = \{G_1, (x_1 x_6)(x_2 x_3)\}.$$

§ 119. *B)* In this case G contains

$$s_5 = (x_1 x_2)(x_3 x_4) \quad \tau = (x_1 x_5)(x_2 x_6),$$

and consequently the combination

136   THEORY OF SUBSTITUTIONS.

$$v = s_5^{-1}\tau s_5 = (x_1 x_6)(x_2 x_5).$$

These three substitutions are not sufficient to connect the six elements $x_1, x_2, \ldots x_6$ transitively, there being no connection between x_3, x_4 and x_1, x_2, x_5, x_6. The group must therefore (§ 83) contain another substitution of the type $(x_\alpha x_\beta)(x_\gamma x_\delta)$ which connects x_1, x_2, x_5, x_6 with other elements. If this substitution should contain three of the elements x_1, x_2, x_5, x_6 and only one new one, it would have three elements in common with v. This would lead either to to the type $A)$ or to the rejected first case of § 117. If the new substitutions contained only one of the new elements x_1, x_2, x_5, x_6 and three new ones, then we should have the fifth case of § 117, and this is also to be rejected.

There remains only the case where the new substitution connects two of the elements x_1, x_2, x_5, x_6 with two others. It must then be of one of the forms

$$(x_1 x_a)(x_2 x_b), \quad (x_1 x_a)(x_5 x_b), \quad (x_1 x_a)(x_6 x_b),$$
$$(x_2 x_a)(x_5 x_b), \quad (x_2 x_a)(x_6 x_b), \quad (x_5 x_a)(x_6 x_b).$$

Of these the first, third, fourth and sixth stand in the relation defined by $C)$ to τ, while the first, second, fifth and sixth stand in the same relation to v.

All the groups $B)$ therefore occur under either $A)$ or $C)$, and we may pass at once to the last case.

§ 120. $C)$. In this case the required group contains

$$\sigma_1 = (x_1 x_2)(x_3 x_4) \quad \sigma_2 = (x_1 x_5)(x_3 x_6), \quad \sigma_3 = \sigma_1^{-1}\sigma_2\sigma_1 = (x_2 x_5)(x_4 x_6).$$

We consider first the case $n = 6$.

The elements x_1, x_2, x_5 are not yet connected with x_3, x_4, x_6. There must be a connecting substitution in the group of the type $(x_\alpha x_\beta)(x_\gamma x_\delta)$, where we may assume that x_α is contained among the the three elements x_1, x_2, x_5. If x_α were x_2 or x_5, then we should obtain, by transformation with respect to σ_1 or σ_2, a substitution $(x_1 x_b)(x_c x_d)$, so that we may assume $a = 1$. The possible cases are then

(α) $(x_1 x_2)(x_5 x_m)$, $(x_1 x_5)(x_2 x_m)$, $(x_2 x_5)(x_1 x_m)$ $\quad m = 3, 4, 6.$
(β) $(x_1 x_m)(x_n x_p)$, $\qquad\qquad\qquad\qquad\qquad\qquad m, n, p = 3, 4, 6.$
(γ) $(x_1 x_m)(x_2 x_n)$, $(x_1 x_m)(x_5 x_n)$, $\qquad\qquad m, n = 3, 4, 6.$

The substitutions of the first and second lines are to be rejected, since their products with $\sigma_1, \sigma_2, \sigma_3$ lead to the first case in § 117, or directly to substitutions with only three elements. There remain, for the different values of m and n, only the following cases:

$$(x_1 x_3)(x_2 x_4), \quad (x_1 x_4)(x_2 x_3),$$
$$(x_1 x_3)(x_2 x_6), \quad (x_1 x_6)(x_2 x_3),$$
$$(x_1 x_4)(x_2 x_6), \quad (x_1 x_6)(x_2 x_4),$$
$$(x_1 x_3)(x_5 x_4), \quad (x_1 x_4)(x_5 x_3),$$
$$(x_1 x_3)(x_5 x_6), \quad (x_1 x_6)(x_5 x_3),$$
$$(x_1 x_4)(x_5 x_6), \quad (x_1 x_6)(x_5 x_4).$$

The second and fourth lines and the third and sixth must be rejected, since their substitutions have three elements in common, the former with σ_1, the latter with σ_3. The first line stands in the same relation to σ_1 as the fifth to σ_2. We need therefore consider only the first line. The product of either of its substitutions by σ_1 gives the other. The required group therefore contains beside $\sigma_1, \sigma_2, \sigma_3$ also

$$\sigma_4 = (x_1 x_2)(x_2 x_4).$$

The group generated by these four substitutions

$$G' = \{\sigma_1, \sigma_2, \sigma_3, \sigma_4\} = \{\sigma_1, \sigma_2, \sigma_4\}$$

is of order 24. It is non-primitive, the systems of non-primitivity being x_1, x_3; x_2, x_4; and x_5, x_6. It can also be readily shown that there is no primitive group G_5 of the required type which contains G'.

For $n = 6$ there are only two groups of the required type. These are the groups G_3 $(r = 60)$ and G_4 $(r = 120)$ of § 118.

§ 121. If the degree of the required group is greater than 6, the indices m, n, p of the lines a), β), γ) in the preceding Section have a correspondingly larger range of values. It is again readily seen, however, that the three cases a) are inadmissible. But (β) (γ) both give rise to groups which satisfy the required conditions. The actual calculation shows that in every case a proper combination of the resulting substitutions gives a circular substitution of seven elements. Consequently the group G is at least $(n-6)$-fold transitive (§ 83).

If then $n \geq 9$, G is at least three-fold transitive, and therefore contains a substitution which does not affect x_1, but interchanges x_2 and x_3. If we transform σ_1 with respect to this substitution, we obtain
$$\sigma' = (x_1 x_3)(x_2 x_a),$$
and since σ_1 has three elements in common with σ, either we have the case A), or $x_a = x_4$ and σ_1 is equal to the σ_4 of the preceding Section.

In the latter case the subgroup which affects $x_1, x_2, \ldots x_6$ is itself at least simply transitive. Combining with this group the circular substitution of seven elements we obtain a two-fold transitive group. Consequently (§ 84) G is at least $(n-5)$-fold, and for $n \geq 9$, at least 4-fold transitive. G contains therefore the substitutions
$$\tau = (x_1)(x_2 x_3)(x_4 x_5 \ldots),$$
$$\tau^{-1} \sigma_1 \tau = (x_1 x_3)(x_2 x_5),$$
so that we return in every case to the type A). For $n \geq 9$ there is therefore no group of the required type.

Theorem III. *If the degree of a group, which contains substitutions of four, but none of three or of two elements, exceeds 8, the group is either intransitive or non-primitive.*

Combining this result with those of § 113 and § 116, we have

Theorem IV. *If the number ρ of the values of a function is not greater than $\frac{1}{2}n(n-1)$, then if $n > 8$, either 1) $\rho = \frac{1}{2}n(n-1)$, and the function is symmetric in $n-2$ elements on the one hand and in the two remaining elements on the other, or 2) $\rho = 2n$, and the function is alternating in $n-1$ elements, or 3) $\rho = n$, and the function is symmetric in $n-1$ elements, or 4) $\rho = 1$ or 2, and the function is symmetric or alternating in all the n elements.*[*]

§ 122. We insert here a lemma which we shall need in the proof of a more general theorem.[†]

From § 83, Corollary II, a primitive group, which does not include the alternating group, cannot contain a circular substitution

---

[*] Cauchy: Journ. de l'Ecole Polytech. X Cahier; Bertrand: *Ibid.* XXX Cahier; Abel: Oeuvres complètes I, pp. 13-21; J. A. Serret: Journ. de l'Ecole Polytech. XXXII Cahier; C. Jordan: Traité etc., pp. 67-75.

[†] C. Jordan: Traité etc., p. 664. Note C.

of a prime degree less than $\frac{2n}{3}$. If p is any prime number less than $\frac{2n}{3}$, and if p^f is the highest power of p which is contained in $n!$, then the order of a primitive group G is not divisible by p^f. For otherwise G would contain a subgroup which would be similar to the group K of degree n and order p^f (§ 39). But the latter group by construction contains a circular substitution of degree p, and the same must therefore be true of G. Consequently $\rho = \frac{n!}{r}$ must contain the factor p at least once.

What has been proven for p is true of any prime number less than $\frac{2n}{3}$ and consequently for their product. We have then

Theorem V. *If the group of a function with more than two values is primitive, the number of values of the function is a multiple of the product of all the prime numbers which are less than $\frac{2n}{3}$.*

§ 123. By the aid of this result we can prove the following

Theorem VI. *If k is any constant number, a function of n elements which is symmetric or alternating with respect to $n-k$ of them has fewer values than those functions which have not this property. For small values of n exceptions occur, but if n exceeds a certain limit dependent on k, the theorem is rigidly true.*[*]

If φ is an alternating function with respect to $n-k$ elements, the order of the corresponding group is a multiple of $\frac{1}{2}(n-k)!$, and the number of values of the function is therefore at the most

A) $\qquad 2n(n-1)(n-2)\ldots(n-k+1)$.

If ψ is a function which is neither symmetric nor alternating in $n-k$ elements, it may be transitive with respect to $n-k$ or more elements. But in the last case ψ must not be symmetric or alternating in the transitively connected elements.

We proceed to determine for both cases a minimum number of

---
*C. Jordan: Traité etc., p. 67.

values of ψ, and to show that if n is sufficiently large, this minimum is greater than the maximum number of values A) of φ.

§ 124. Suppose at first that ψ is transitive in less than $n-k$ elements. Then the order of the corresponding group is a divisor of
$$\lambda_1!\,\lambda_2!\,\lambda_3!\ldots \text{ where } \lambda_1+\lambda_2+\lambda_3+\ldots = n \quad (\lambda_a < n-k).$$
This product is a maximum when one of the λ's is as large as possible, i. e., equal to $n-k-1$, and a second λ is then also as large as possible, i. e., equal to $k+1$. It is further necessary that $k+1 > n-k$ i. e., $n > 2k+1$. The maximum order of the group is consequently
$$(n-k-1)!\,(k+1)!,$$
and the minimum number of values of ψ is

B) $$\frac{n!}{(n-k-1)!\,(k+1)!} = \frac{n(n-1)(n-2)\ldots(n-k)}{1\cdot 2\cdot 3\cdot\ldots(k+1)}.$$

It appears at once that the minimum B) exceeds the maximum A), as soon as
$$n > k + 2(k+1)!$$
This is therefore the limit above which, in the first case, the theorem admits of no exception.

§ 125. In the second case ψ is transitive in $n-\varkappa$ elements ($\varkappa \geq k$), but it is neither alternating nor symmetric in these elements. The group G of ψ is intransitive, and its substitutions are therefore products each of two others, of which the one set $\sigma_1, \sigma_2, \ldots$ connect transitively only the elements $x_1, x_2, \ldots x_{n-\varkappa}$, while the other set $\tau_1, \tau_2, \ldots$ connect only the remaining elements $x_{n-\varkappa+1}, \ldots x_n$.

The substitutions of the group G of ψ have, then, the product forms
$$\sigma_1\tau_1,\quad \sigma_2\tau_2,\quad \sigma_3\tau_3,\ldots \sigma_a\tau_a,\ldots \sigma_\beta\tau_\beta,\ldots$$
where, however, one and the same σ may occur in combination with different τ's. It is easily seen that all the σ's occur the same number of times, so that the order of the group G is a multiple of that of the group $\Sigma = [\sigma_1, \sigma_2, \ldots]$.

We will show that Σ is neither alternating nor symmetric; oth-

THE NUMBER OF VALUES OF INTEGRAL FUNCTIONS. 141

erwise G would be alternating or symmetric in $n-\varkappa$ elements, which is contrary to assumption. If the group Σ were alternating, it would be of order $\frac{1}{2}(n-\varkappa)!$. This exceeds the maximum number $\varkappa!$ of the order of the group $T=[\tau_1, \tau_2, \ldots]$ of $\varkappa$ elements, as soon as $n>2k$. Consequently G contains substitutions $\sigma_a\tau_\beta$, $\sigma_\beta\tau_\beta$, in which $\tau_a = \tau_\beta$ but $\sigma_a \neq \sigma_\beta$, and therefore substitutions $\sigma_a\tau_a (\sigma_\beta\tau_\beta)^{-1} = \sigma_a\sigma_\beta^{-1}$ which affect only the elements $x_1, x_2, \ldots x_{n-\varkappa}$ of the first set. The entire complex of these substitutions forms a self-conjugate subgroup H of G. This subgroup is unchanged by transformation with respect to either G or Σ, since $\tau_a, \tau_\beta, \ldots$ have no effect whatever on the substitutions of H. H is therefore a self-conjugate subgroup of the alternating group Σ, and must accordingly coincide with Σ (§ 92). $H = \Sigma$ is therefore a subgroup of G, and φ would, contrary to assumption, be alternating in $n-\varkappa$ elements.

§ 126. The maximum order of the group G is therefore equal to the product of $\varkappa!$ by the maximum order of a non-alternating transitive group of $n-\varkappa$ elements. We denote the latter order by $R(n-\varkappa)$. Then the minimum number of values of φ is

$$C) \qquad \frac{n!}{\varkappa! R(n-\varkappa)} = \frac{(n-\varkappa)!}{R(n-\varkappa)} \cdot \frac{n(n-1)\ldots(n-\varkappa+1)}{\varkappa!}.$$

We have now still to determine $R(n-\varkappa)$, the maximum order of a non-alternating transitive group of $n-\varkappa$ elements, or $\dfrac{(n-\varkappa)!}{R(n-\varkappa)}$, the minimum number of values of a non-alternating transitive function of $n-\varkappa$ elements.

If this function is *non-primitive* in the $n-\varkappa$ elements, it follows that the minimum number of values is

$$C_1) \qquad \frac{(n-\varkappa)!}{2\{[\frac{1}{2}(n-\varkappa)]!\}^2} = \frac{1}{2} \frac{(n-\varkappa)(n-\varkappa-1)\ldots\left(\frac{n-\varkappa}{2}+1\right)}{[\frac{1}{2}n-\varkappa)]!}.$$

Substituting this value in C) we obtain for the minimum number of values of φ

$$C_1') \qquad \frac{1}{2} \frac{n(n-1)\ldots(n-\varkappa+1)(n-\varkappa)\ldots\left(\frac{n-\varkappa}{2}+1\right)}{\varkappa! [\frac{1}{2}(n-\varkappa)]!}.$$

142 THEORY OF SUBSTITUTIONS.

We compare this number with the maximum number A) and examine whether, above a certain limit for n, C'_1) becomes greater than A), i. e., whether

$$n(n-1)\ldots(n-\varkappa+1)(n-\varkappa)\ldots\left(\frac{n-\varkappa}{2}+1\right)$$
$$> 4\varkappa!\,\frac{n-\varkappa}{2}!\,n(n-1)\ldots(n-\varkappa+1).$$

For sufficiently large n we have $\left(\frac{n-\varkappa}{2}+1\right) < n-k+1$. We have therefore to prove that

$$(n-k)(n-k-1)\ldots\left(\frac{n-\varkappa}{2}+1\right) > 4\varkappa!\,\frac{n-\varkappa}{2}!$$

This is shown at once, if we write the right hand member in the form

$$(4\varkappa!\,[k-\varkappa]!)\left(\left[\frac{n-\varkappa}{2}\right]\left[\frac{n-\varkappa}{2}-1\right]\ldots[k-\varkappa+1]\right).$$

For the first factor is constant as n increases, and the ratio of the left hand member to the second parenthesis has for its limit

$$2^{\frac{n+\varkappa}{2}-k}.$$

§ 127. Finally, if the function ψ of the $n-\varkappa$ elements is *primitive*, we recur to the lemma of § 122. From this it follows that the minimum number of values of ψ is the product of all the prime numbers less than $\frac{2}{3}(n-\varkappa)$. We will denote this product by

$$P\left[\frac{2(n-\varkappa)}{3}\right].$$

Introducing it in C), we have

$$C_2)\qquad P\left[\frac{2(n-\varkappa)}{3}\right]\frac{n(n-1)\ldots(n-\varkappa+1)}{\varkappa!}$$

We have then to show that, for sufficiently large values of n, the value A) is less than C_2), i. e., that

$$P\left[\frac{2(n-\varkappa)}{3}\right] > 2\varkappa!(n-\varkappa)(n-\varkappa-1)\ldots(n-k+1).$$

The right hand member of this inequality will be greatly increased if we replace every $n-\varkappa-a$ by the first factor $n-\varkappa$. There are $k-\varkappa$ factors of the form $n-\varkappa-a$. These will be replaced by

$(n-x)^{k-\kappa}$. If we write then $\nu = \dfrac{2(n-x)}{3}$, we have only to prove that for sufficiently large ν,

$$P(\nu) > [2(\tfrac{3}{2})^{k-\kappa} x!]\nu^{k-\kappa},$$

or

$$\frac{P(\nu)}{\nu^{k-\kappa}} > [2(\tfrac{3}{2})^{k-\kappa} x!].$$

This can be shown inductively by actual calculation, or by the employment of the theorem of Tchebichef, *that if $\nu > 3$, there is always a prime number between ν and $2\nu - 2$*.

For we have from this theorem

$$P(2\nu) > \nu P(\nu),$$
$$(\nu)^{k-\kappa} = 2^{k-\kappa} \nu^{k-\kappa}$$
$$\frac{P(2\nu)}{(2\nu)^{k-\kappa}} > \frac{P(\nu)}{\nu^{k-\kappa}} \frac{\nu}{2^{k-\kappa}}.$$

Now whatever value the first quotient on the right may have, we can always take t so great that the left hand member of

$$\frac{P(2^t\nu)}{(2^t\nu)^{k-\kappa}} > \frac{P(\nu)}{\nu^{k-\kappa}} \left(\frac{\nu}{2^{k-\kappa}}\right)^t$$

increases without limit, if only ν is taken greater than $2^{k-\kappa}$. The proof of the theorem is now complete.

The limits here obtained are obviously far too high. In every special case it is possible to diminish them. As we have, however, already treated the special cases as far as $\rho = \tfrac{1}{2}n(n-1)$, it does not seem necessary, from the present point of view, to carry these investigations further.

CHAPTER VII.

CERTAIN SPECIAL CLASSES OF GROUPS.

§ 128. We recur now to the results obtained in § 48, and deduce from these certain further important conclusions.*

Suppose that a group G is of order $r = p^a m$, where p is a prime number and m is prime to p. We have seen that G contains a subgroup H of order p^a. Let J be the greatest subgroup of G which is commutative with H. J contains H, and the order of J is therefore $p^a i$, where i is a divisor of m and is consequently prime to p.

Excepting the substitutions of H, J contains no substitution of an order p^β. For if such a substitution were present, its powers would form a group L of order p^β. But if in A) of § 48 we take for G_1, H_1, K_1 the present groups J, L, K, then since $\sigma_\gamma^{-1} H \sigma_\gamma = H$, we should have

$$\frac{p^a i}{p^a} = \frac{p^\beta}{d_1} + \frac{p^\beta}{d_2} + \cdots.$$

The left member of this equation is not divisible by p. Consequently we must have in at least one case $d_\gamma = p^\beta$, that is, L is contained in $\sigma_\gamma^{-1} H \sigma_\gamma = H$.

Again, every subgroup M of order p^a which occurs in G is obtained by transformation of H. For if we replace G_1, H_1, K_1 of A) § 48 by G, H, M, we obtain

$$\frac{p^a i}{p^a} = \frac{p^a}{d_1} + \frac{p^a}{d_2} + \cdots,$$

and for the same reason as before $d_\gamma = p^a$ in at least one case, and therefore $M = \sigma_\gamma^{-1} H \sigma_\gamma$. Since H, as a self-conjugate subgroup of J, is transformed into itself by the $p^a i$ substitutions of J, it follows that there are always exactly $p^a i$ substitutions of G which transform H into any one of its conjugates. There are therefore $\dfrac{m}{i}$ of the latter.

---

* L. Sylow: Math. Ann. V. 584-94.

Finally, if we replace G, H_1, K_1 of A) § 48 by G, J, H, we have

$$\frac{r}{p^a} = \frac{p^a m}{p^a} = \frac{p^a i}{d_1} + \frac{p^a i}{d_2} + \cdots$$

Since H_1 is contained in J_1, we must have $d_1 = p^a$, and since J contains no other substitutions of order p^β, no other d can be equal to p^a. It follows that

$$r = p^a i(kp+1), \quad m = i(kp+1).$$

The group H has therefore $kp+1$ conjugates with respect to G. We have then the following

Theorem I. *If the order r of a group G is divisible by p^a but by no higher power of the prime number p, and if H is one of the subgroups of order p^a contained in G, and J of order $p^a i$ the largest subgroup of G which is commutative with H, then the order of G is*

$$r = p^a i(kp+1).$$

Every subgroup of order p^a contained in G is conjugate to H. Of these conjugate groups there are $kp+1$, and every one of them can be obtained from H by $p^a i$ different transformations.

§ 129. In the discussion of isomorphism we have met with transitive groups whose degree and order are equal. In the following Sections we shall designate such groups as *the groups Ω*.

If we regard all simply isomorphic transitive groups, for which therefore the orders r are all equal, as forming a *class*, then every such class contains one and only one type of a group Ω (§ 98). The construction of all the groups Ω of degree and order r therefore furnishes representatives of all the classes belonging to r, together with the number of these classes. The construction of these typical groups is of especial importance, because isomorphic groups have the same factors of composition, and the latter play an important part in the algebraic solution of equations.

One type can be established at once, in its full generality. This type is formed by the powers of a circular substitution. A group Ω of this type is called a *cyclical group*, and every function of n elements which belongs to a cyclical group is called a *cyclical function*.

146   THEORY OF SUBSTITUTIONS.

We limit ourselves to the consideration of cyclical groups of prime degree p. If $s = (x_1 x_2 \ldots x_p)$, and if ω is any primitive p^{th} root of unity, then

$$\varphi = (x_1 + \omega x_2 + \omega^2 x_3 + \ldots + \omega^{p-1} x_p)^p$$

is a cyclical function belonging to the group $G = [1, s, s^2, \ldots s^{p-1}]$. For φ is converted by s^a into

$$(x_{a+1} + \omega x_{a+2} + \ldots + \omega^{p-1} x_a)^p = \omega^{ap}(x_{a+1} + \omega x_{a+2} + \ldots + \omega^{p-1} x_a)^p$$
$$= (x_1 + \omega x_2 + \ldots + \omega^{p-1} x_p)^p = \varphi,$$

so that φ is unchanged by the substitutions of G.

Moreover, if for any substitution t, which converts every x_a into x_{t_a}, we have $\varphi_t = \varphi_1$, and consequently

$$\sqrt[p]{\varphi_t} = \omega^{-\beta} \sqrt[p]{\varphi_1},$$

then, if the x's are independent elements, it follows that

$$\omega^{\gamma-1} x_{t_\gamma} = \omega^{-\beta} \omega^{\gamma-1} x_{t_\gamma}.$$

Consequently $t_\gamma = \gamma + \beta$, that is, the substitution t replaces $x_1, x_2 \ldots$, by $x_{1+\beta}, x_{2+\beta}, \ldots$; t is therefore contained among the powers of s, and φ belongs to the group G. It is obvious that for $r = p$, the group G furnishes the only possible type Ω.

§ 130. We proceed next to determine all types of groups Ω of degree and order pq, where p and q are prime numbers, which for the present we will assume to be unequal, p being the greater.

1) One type, that of the cyclical group, is already known to us. It is characterized by the occurrence of a substitution of order pq.

2) If there are any other types, none of them can contain a substitution of order pq, since this would lead at once to 1). The only possible orders are therefore $p, q, 1$. A substitution s of order p is certainly present. This and its powers form within Ω a subgroup H of order p. If Ω contains any further subgroups H' of order p, their number must be $\varkappa p + 1$, where $\varkappa > 0$. These subgroups would have only the identical substitution in common. They would therefore contain in all

$$(p-1)(p\varkappa + 1) + 1 = p[(p-1)\varkappa + 1] \geq pq$$

substitutions. This being impossible, we must have $\varkappa = 0$.

The subgroup H contains only p substitutions; the rest are all of order q. Their number is
$$pq - p = (q-1)p.$$
There are therefore p subgroups of order q, and consequently from Theorem I we must have
$$p = \lambda q + 1, \quad \lambda = \frac{p-1}{q},$$
that is, q must be a divisor of $p-1$. *Only in this case can there be any new type Ω.*

3) The group H is a self-conjugate subgroup of Ω. Consequently every substitution t of order q must transform the substitution s of H into s^a, where a might also be equal to 1. We write
$$s = (x_1^1 x_2^1 \ldots x_p^1)(x_1^2 x_2^2 \ldots x_p^2) \ldots (x_1^q x_2^q \ldots x_p^q),$$
(where the upper indices are merely indices, not exponents). Then no cycle of t can contain two elements with the same upper index. For otherwise in some power of t one of these elements would follow the other, and if this power of t were multiplied by a proper power of s, one of the elements would be removed.

With a proper choice of notation, we may therefore take for one cycle of t
$$(x_1^1 x_1^2 x_1^3 \ldots x_1^p).$$
It follows then from
$$t^{-1} s t = s^a$$
that t replaces x_2^b by x_{a+1}^{b+1}, x_3^b by $x_{2a+1}^{b+1}, \ldots x_{a+1}^b$ by $x_{aa+1}^{b+1}, \ldots$ so that we have
$$t = (x_1^1 x_1^2 \ldots x_1^q) \ldots (x_{a+1}^1 x_{aa+1}^2 x_{aa^2+1}^3 \ldots x_{aa^{q-1}+1}^q \ldots) \ldots$$
If now the latter cycle is to close exactly with the element $x_{aa^{q-1}+1}^q$, we must have
$$a a^q + 1 \equiv a + 1, \quad a^q \equiv 1 \quad (\text{mod. } p).$$
The solution $a = 1$ is to be rejected, for in this case we should have
$$t = (x_1^1 x_1^2 \ldots x_1^q)(x_2^1 x_2^2 \ldots x_2^q) \ldots (x_p^1 x_p^2 \ldots x_p^q),$$
$$st = (x_1^1 x_2^2 \ldots x_q^q x_{q+1}^1 x_{q+2}^2 \ldots) \ldots,$$
so that the latter substitution would contain a cycle of more than q elements, without being a power of s.

It follows then from the congruence $a^q \equiv 1$ (mod. p) that q is a divisor of $p-1$, as we have already shown; further that a, belonging to the exponent q, has $q-1$ values $a_1, a_2, \ldots a_{q-1}$; finally that all these values are congruent (mod. p) to the powers of any one among them. From $t^{-1}st = s^a$ follows

$$t^{-2}st^2 = s^{a^2},\ t^{-3}st^3 = s^{a^3}, \ldots$$

so that, if s is transformed by t into any one of the powers s^{a_λ}, there are also substitutions in Ω which transform s into $s^{a_1}, s^{a_2}, \ldots s^{a_{q-1}}$. Accordingly the particular choice of a_λ has no influence on the resulting group, so that if there is any type Ω generated by substitutions s and t, there is only one.

The group formed by the powers of t being commutative with that formed by the powers of s, the combination of these two substitutions gives rise to a group exactly of order pq. The remaining $pq - p - q + 1$ substitutions of the group are the first $q-1$ powers of the $p-1$ substitutions conjugate to t

$$s^{-\beta+1}ts^{\beta-1} = (x_\beta^1 x_\beta^2 x_\beta^3 \ldots x_\beta^q) \ldots \quad (\beta = 2, 3, \ldots p).$$

If p and q are unequal, we have therefore only one new type Ω.

§ 131. Finally we determine all types of groups Ω of degree and order p^2.

1) The cyclical type, characterized by the presence of a substitution of order p^2, is already known.

2) If there are other types, none of them can contain a substitution of order p^2. There are therefore in every case $p^2 - 1$ substitutions of order p and one of order 1. If s is any substitution of Ω, and t any other, not a power of s, then Ω is fully determined by s and t. For all the products

$$s^a t^b \quad (a, b = 0, 1, 2, \ldots p-1)$$

are different, and therefore

$$\Omega = [s^a t^b] \quad (a, b = 0, 1, 2, \ldots p-1).$$

We must have therefore

$$ts = s^{\varepsilon_1} t^{\epsilon_1},\ t^2 s = s^{\delta_2} t^{\epsilon_2}, \ldots t^{p-1}s = s^{\delta_{p-1}} t^{\epsilon_{p-1}}.$$

If now two of the exponents δ are equal, it follows from

that
$$t^a s = s^\delta t^\epsilon, \quad t^b s = s^\delta t^{\epsilon'} \quad (a \dotplus b, \quad \epsilon \dotplus \epsilon')$$
$$(t^a s)^{-1}(t^b s) = s^{-1} t^c s = (s^\delta t^\epsilon)^{-1}(s^\delta t^{\epsilon'}) = t^\gamma.$$

Since for t^c we may write t, it therefore appears that Ω contains a substitution t which is transformed by s into one of its powers t^η.

The same result holds, if all the exponents δ are different. For one of them is then equal to 1, since none of them can be 0, and from $t^a s = s t^\epsilon$ follows $s^{-1} t^a s = t^\epsilon$.

3) There is therefore always a substitution
$$t = (x_1^1 x_2^1 \ldots x_p^1)(x_1^2 x_2^2 \ldots x_p^2) \ldots (x_1^p x_2^p \ldots x_p^p)$$
which is transformed by s into a power of itself t^η. As in the preceding Section, we may take for one cycle of s
$$(x_1^1 x_1^2 \ldots x_1^p).$$
Then from $s^{-1} t s = t^a$ follows
$$s = (x_1^1 x_1^2 \ldots x_1^p)(x_2^1 x_{a+1}^2 x_{a^2+1}^3 \ldots x_{a^{p-1}+1}^p \ldots) \ldots$$
If the second cycle is to close after exactly p elements, we must have
$$a^p + 1 \equiv 2, \quad a^p \equiv 1 \pmod{p}.$$
This is possible only if $a = 1$. Accordingly
$$s = (x_1^1 x_1^2 \ldots x_1^p)(x_2^1 x_2^2 \ldots x_2^p) \ldots (x_p^1 x_p^2 \ldots {}_p^p).$$
The $p+1$ substitutions
$$s, t, st, st^2, \ldots st^{p-1}$$
are all different and no one of them is a power of any other one. Their first $p-1$ powers together with the identical substitution form the group Ω.

Summarizing the preceding results we have

Theorem II. *There are three types of groups Ω, for which the degree and order are equal to the product of two prime numbers: 1) The cyclical type, 2) one type of order pq $(p > q)$, 3) one type of order p^2. The first and third types are always present; the second occurs only when q is a divisor of $p-1$.*

§ 132. We consider now another category of groups, characterized by the property that their substitutions leave *no* element, or only *one* element, or *all* the elements unchanged. The degree of the groups we assume to be a prime number p.

Every substitution of such a group is *regular*, *i.e.*, is composed of equal cycles. For otherwise in a proper power of the substitution, different from the identity, two or more of the elements would be removed.

The substitutions which affect all the elements are cyclical, for p is a prime number. From this it follows that the groups are transitive, and again, from Theorem IX, Chapter IV, that the number of substitutions which affect all the elements is $p-1$. We may therefore assume that

$$s = (x_1 x_2 x_3 \ldots x_p)$$

and its first $p-1$ powers are the only substitutions of p elements which occur in the required group.

The problem then reduces to the determination of those substitutions which affect exactly $p-1$ elements. If t is any one of these, then $t^{-1}st$, being similar to s, and therefore affecting all the elements, must be a power of s

$$t^{-1}st = s^m = (x_1 x_{1+m} x_{1+2m} \ldots),$$

where every index is to be replaced by its least positive remainder (mod p). Since it is merely a matter of notation which element is not affected by t, we may assume that x_1 is the unaffected element. It follows that

$$t = (x_2 x_{m+1} x_{m^2+1} x_{m^3+1} \ldots) \ldots (x_{a+1} x_{am+1} x_{am^2+1} \ldots) \ldots$$

If now g is a primitive root (mod. p), then all the remainders (mod. p) of the first $p-1$ powers of g

(G) $\qquad g^1, g^2, g^3, \ldots g^{p-2}, g^{p-1} \equiv 1 \quad (\text{mod. } p)$

are different, and we may therefore put

$$m \equiv g^\mu \quad (\text{mod. } p).$$

We will denote the corresponding t by t_μ. It appears then that t_μ consists of μ cycles of $\dfrac{p-1}{\mu}$ elements each. For every cycle of t_μ closes as soon as

$$am^z+1 \equiv a+1,$$
$$m^z \equiv g^{\mu z} \equiv 1, \quad (\text{mod. } p)$$

and this first happens when $z = \dfrac{p-1}{\mu}$.

If there is any further substitution t_ν which leaves x_1 unchanged and which replaces every x_{a+1} by $x_{ag^\nu+1}$, then $t_\mu{}^a t_\nu{}^\beta$ replaces every x_{a+1} by $x_{ag^{a\mu+\beta\nu}+1}$. If now we take a and β so that $a\mu + \beta\nu$ is congruent (mod. p) to the smallest common divisor ω of μ and ν, we have in

$$t_\omega = t_\mu{}^a t_\nu{}^\beta$$

a substitution of the group, of which both t_μ and t_ν are powers. Proceeding in this way, we can express all the substitutions which leave x_1 unchanged as powers of a single one among them t_σ, where g^σ is the lowest power of g to which a substitution t of the group corresponds.

The group is determined by s and t_σ. Since t_σ is of order $\dfrac{p-1}{\sigma}$, it follows from Theorem II, Chapter IV, that the group contains in all $\dfrac{p(p-1)}{\sigma}$ substitutions. σ may be taken arbitrarily among the divisors of $p-1$.

§ 133. To obtain a function belonging to the group just considered, we start with the cyclical function belonging to s

$$\psi_1 = (x_1 + \omega x_2 + \omega^2 x_3 + \ldots + \omega^{p-1} x_p)^p,$$

where ω is any primitive p^{th} root of unity. Applying to ψ_1 the successive powers of t_σ, we obtain

$$\psi_2 = (x_1 + \omega x_{g^\sigma+1} + \omega^2 x_{2g^\sigma+1} + \ldots + \omega^{p-1} x_{(p-1)g^\sigma+1})^p,$$
$$\psi_3 = (x_1 + \omega x_{g^{2\sigma}+1} + \omega^2 x_{2g^{2\sigma}+1} + \ldots + \omega^{p-1} x_{(p-1)g^{2\sigma}+1})^p,$$
$$\cdots \cdots \cdots \cdots \cdots \cdots \cdots \cdots \cdots$$

The powers of s, forming a self-conjugate subgroup of the given group, leave all the ψ's unchanged. The powers of t_σ, and consequently all the substitutions $s^a t_\sigma{}^b$ of the group, merely permute the ψ's among themselves. Every symmetric function of

$$\psi_1, \psi_2, \ldots \psi_{\frac{p-1}{\sigma}}$$

is therefore unchanged by every substitution of the group. Accordingly if ψ' is any arbitrary quantity, the function

$$\Psi_\sigma = (\psi - \psi'_1)(\psi - \psi'_2) \ldots (\psi - \psi'_{p-1})_\sigma$$

is unchanged by all these substitutions and by no others. Ψ_σ therefore belongs to the given group.

§ 134. If, in particular, we take $\sigma = 1$, the order of the group becomes $p(p-1)$. The substitution t_σ is then of the form

$$t_1 = (x_2 x_{g+1} x_{g^2+1} \ldots x_{g^{p-2}+1}),$$

containing only one cycle. The group is therefore two-fold transitive (Theorem XIII, Chapter IV). A group of this type is called a *metacyclic group*, and the corresponding function Ψ_1 a *metacyclic function*.

If $\sigma = 2$, the order of the group is $\dfrac{p(p-1)}{2}$, and t_σ is of the form

$$t_2 = (x_2 x_{g^2+1} x_{g^4+1} \ldots)(x_{a+1} x_{ag^2+1} x_{ag^4+1} \ldots)$$

The indices in the first cycle, each diminished by 1, are the quadratic remainders (mod. p); a is any quadratic non-remainder (mod. p). The group is in this case called *half-metacyclic group* and the corresponding function Ψ_τ a *half-metacyclic function.*

§ 135. We can define all substitutions of the groups Ω of prime degree p, as well as those of the two preceding Sections, in a simple way by expressing merely the changes which occur in the indices of the elements $x_1, x_2, \ldots x_n$. Thus the substitutions of Ω are defined by

$$s_a = |z \ \ z+a| \quad (\text{mod. } p). \quad (a = 0, 1, 2, \ldots p-1),$$

The symbol here introduced is to be understood as indicating that in the substitution s_a every element x_z is replaced by x_{z+a}, and for $z + a$ its least not negative remainder (mod. p) is to be taken.

The groups of the preceding Section contain then, in the first place, all the substitutions s_a, and beside these, if we suppose every index to be diminished by one, also those substitutions for which every index is multiplied by the same factor, that is, for $z = 0, 1, 2, \ldots p-1$

---

* L. Kronecker; *cf.* F. Klein: Math. Ann. XV, 258.

CERTAIN SPECIAL CLASSES OF GROUPS. 153

$$\sigma_\beta = |z \ \beta z| \quad (\text{mod. } p) \quad (\beta = 1, 2, 3, \ldots p-1).$$

The symbol

$$t = |z \ \beta z + a| \quad (\text{mod. } p) \quad (a = 0, 1, \ldots p-1;\ \beta = 1, 2, \ldots p-1)$$

includes all the substitutions s_a, σ_β, and their combinations. Since

$$|z \ \beta z + a| \cdot |z \ \beta_1 z + a_1| = |z \ \beta\beta_1 z + a_1\beta + a|,$$

it follows that the substitutions t form a group of degree p and of order $p(p-1)$. This group therefore coincides with that of § 134.

If we prescribe that the β's shall take only the values

$$g^{1\sigma}, g^{2\sigma}, g^{3\sigma}, \ldots g^{\frac{p-1}{\sigma}\sigma},$$

the products $\beta\beta_1$ belong to the same series, and we obtain the group of degree p and of order $\dfrac{p(p-1)}{\sigma}$ considered above.

§ 136. The consideration of the fractional linear substitutions (mod. p) leads to groups of degree $p+1$ and of order $(p+1)p(p-1)$. These substitutions are of the form

$$u = \left| z \ \frac{az+\beta}{\gamma z+\delta} \right| \quad (\text{mod. } p),$$

where z is to take the values $0, 1, 2, \ldots p-1, \infty$, the elements of the group being accordingly $x_0, x_1, x_2, \ldots x_{p-1}, x_\infty$. The values a, β, γ, δ, determine a single substitution u, but it may happen that one and the same u results from different systems a, β, γ, δ. To avoid, or at least to limit this possibility, we make use of the *determinant* of u

D) $\qquad a\delta - \beta\gamma.$

According as this is a quadratic remainder or non-remainder, we divide numerator and denominator of $\dfrac{az+\beta}{\gamma z+\delta}$ by $\sqrt{a\delta - \beta\gamma}$ or by $\sqrt{\beta\gamma - a\delta}$. For the new coefficients we have then

D') $\qquad a\delta - \beta\gamma \equiv \pm 1 \quad (\text{mod. } p).$

If now for two different systems of coefficients a, β, γ, δ and $a_1, \beta_1, \gamma_1, \delta_1$ the relation

$$\frac{az+\beta}{\gamma z+\delta} \equiv \frac{a_1 z+\beta_1}{\gamma_1 z+\delta_1} \quad (\text{mod. } p)$$

were possible, it would follow from the comparison of the coefficients of z^2, z^1, and z^0, with the aid of D', that if $a, a', \beta, \beta', \ldots$ are real,

$$\frac{a}{a'} \equiv \frac{\beta}{\beta'} \equiv \frac{\gamma}{\gamma'} \equiv \frac{\delta}{\delta'} \equiv \sqrt{\frac{a\delta - \beta\gamma}{a'\delta' - \beta'\gamma'}} \equiv \pm 1 \quad (\text{mod. } p).$$

If, therefore, we restrict the range of the values of a, β, γ, δ to $0, 1, 2, \ldots p-1$, there are always two and only two different systems of coefficients which give the same substitution u.

With D') it is assumed that $a\delta - \beta\gamma$ is different from 0. This restriction is necessary, for the symbol u can represent a substitution only if different initial values of z give rise to different final values of z, i. e., if the congruence

$$\frac{az + \beta}{\gamma z + \delta} \equiv \frac{az_1 + \beta}{\gamma z_1 + \delta} \quad (\text{mod. } p)$$

is impossible. This is ensured by the assumption $a\delta - \beta\gamma \not\equiv 0$.

We determine now how many elements are unchanged by the substitution u. An index z can only remain unchanged by u if

E) $\qquad \gamma z^2 + (\delta - a)z - \beta \equiv 0 \quad (\text{mod. } p)$.

There are accordingly four distinct cases:

a) The two roots of E) are imaginary. This happens if

$$\left(\frac{a+\delta}{2}\right)^2 \mp 1 \qquad (a\delta - \beta\gamma = \pm 1)$$

is a quadratic non-remainder (mod. p). The corresponding substitutions affect all the elements $x_0, x_1, x_2, \ldots x_{p-1}, x_\infty$.

b) The two roots of E) coincide. This happens if

$$\left(\frac{a+\delta}{2}\right)^2 \mp 1 \equiv 0 \quad (\text{mod. } p) \quad (a\delta - \beta\gamma = \pm 1).$$

The corresponding substitutions leave one element unchanged.

c) The two roots of E) are real and distinct. This happens if

$$\left(\frac{a+\delta}{2}\right)^2 \mp 1 \qquad (a\delta - \beta\gamma = \pm 1)$$

is a quadratic remainder (mod. p). The corresponding substitutions leave two elements unchanged.

d) The equation E) may vanish identically. This happens if

$$\gamma \equiv 0, \ \beta \equiv 0, \ a \equiv \delta \quad (\text{mod. } p).$$

The corresponding substitution leaves all the elements unchanged. Finally we observe also that

M) $\left| z \, \dfrac{az+\beta}{\gamma z+\delta} \right| \cdot \left| z \, \dfrac{a_1 z+\beta_1}{\gamma_1 z+\delta_1} \right| = \left| z \, \dfrac{(aa_1+\beta\gamma_1)z+(a\beta_1+\beta\delta_1)}{(\gamma a_1+\delta\gamma_1)z+(\gamma\beta_1+\delta\delta_1)} \right|,$

N) $(a\delta-\beta\gamma)(a_1\delta_1-\beta_1\gamma_1) = (aa_1+\beta\gamma_1)(\gamma\beta_1+\delta\delta_1)$
$\qquad -(a\beta_1+\beta\delta_1)(\gamma a_1+\delta\gamma_1).$

We proceed now to collect our results. If we take a not $\equiv 0$ (mod. p), and β and γ arbitrarily, then for each of the $(p-1)p^2$ resulting systems we obtain two solutions of D'). Since however there are always two systems of coefficients which give the same substitutions u, we have in all, in the present case, p^3-p^2 substitutions. Again, if we take $a \equiv 0$ (mod. p) and δ arbitrarily, then restricting β to the values $1, 2, \ldots p-1$, we obtain from D') for every system a, δ, β two values of γ; but as two systems of coefficients give the same u, we have in this case $p(p-1)$ substitutions.

There are therefore in all $p^3-p = (p+1)p(p-1)$ fractional linear substitutions (mod. p). From M) it appears that these form a group. Among them there are $\dfrac{(p+1)p(p-1)}{2}$ substitutions which correspond to the upper sign in D'). From M) and N) it is clear that these also form a group. This latter group is called "*the group of the modular equations for p*".[*]

Both groups contain only substitutions which affect either $p+1$, or p, or $p-1$ elements, or no element. Those substitutions which leave the element x_∞ unchanged, for which accordingly $\gamma \equiv 0$, form the metacyclic group of § 134. As the latter is two-fold transitive, it follows (Theorem XIII, Chapter IV) that the group of order $(p+1)p(p-1)$ is three-fold transitive.

Theorem III. *The fractional linear substitutions* (mod. p) *form a group of degree $p+1$ and of order $(p+1)p(p-1)$. Those of which the determinants are quadratic remainders* (mod. p) *form a subgroup of order* $\dfrac{(p+1)p(p-1)}{2}$, *the group of the modular equations for p. If any substitution of these groups leaves more than two elements unchanged, it reduces to identity. The first of the two groups is three-fold transitive.*

---

[*] *Cf.* J. Gierster: Math. Ann. XVIII, p. 319.

To construct a function belonging to the group of the fractional linear substitutions, we form first as in § 133, a function Ψ_1 of the elements $x_0, x_1, x_2, \ldots x_{p-1}$ which belongs to the group of substitutions

$$t = |z \; \beta z + \alpha| \pmod{p}. \quad (\alpha = 0, 1, 2, \ldots p-1; \; \beta = 1, 2, \ldots p-1)$$

The substitutions u, applied to Ψ_1, produce $p+1$ values

$$\Psi_1, \Psi_2, \ldots \Psi_{p+1},$$

which these substitutions merely permute among themselves. Accordingly, if Ψ is any undetermined quantity, the function

$$\Sigma = (\Psi - \Psi_1)(\Psi - \Psi_2) \ldots (\Psi - \Psi_{p+1})$$

belongs to the given group.

§ 137. We have now finally to turn our attention to those groups all the substitutions of which are commutative.

We employ here a general method of treatment of very extensive application.*

Suppose that $\theta', \theta'', \theta''' \ldots$ are a series of elements of finite number, and of such a nature that from any two of them a third one can be obtained by means of a certain definite process. If the result of this process is indicated by f, there is to be, then, for every two elements θ', θ'', which may also coincide, a third element θ''', such that $f(\theta' \theta'') = \theta'''$. We will suppose further that

$$f(\theta', \theta'') = f(\theta'', \theta'),$$
$$f[\theta', f(\theta'', \theta''')] = f[f(\theta', \theta''), \theta'''],$$

but that, if θ'' and θ''' are different from each other, then

$$f(\theta', \theta'')] \neq f(\theta', \theta''').$$

These assumptions having been made, the operation indicated by f possesses the associative and commutative property of ordinary multiplication, and we may accordingly replace the symbol $f(\theta', \theta'')$ by the product $\theta' \theta''$, if in the place of complete equality we employ the idea of *equivalence*. Indicating the latter relation by the usual sign ∞, the equivalence

$$\theta' \theta'' \infty \theta'''$$

is, then, defined by the equation

$$f(\theta', \theta'') = \theta'''.$$

---

*L. Kronecker: Monatsber. d. Berl. Akad., 1870, p. 881. The following is taken for the most part *verbatim* from this article.

CERTAIN SPECIAL CLASSES OF GROUPS. 157

Since the number of the elements θ, which we will denote by n, is assumed to be finite, these elements have the following properties:

I) Among the various powers of an element θ there are always some which are equivalent to unity. The exponents of all these powers are integral multiples of one among them, to which θ may be said to *belong*.

II) If any θ belongs to an exponent ν, then there are elements belonging to every divisor of ν.

III) If the exponents ρ and σ, to which θ' and θ'' respectively belong, are prime to each other, then the product $\theta' \theta''$ belongs to the exponent $\rho\sigma$.

IV) If n_1 is the least common multiple of all the exponents to which the n elements θ belong, then there are also elements which belong to n_1.

The exponent n_1 is the greatest of all the exponents to which the various elements belong. Since, furthermore, n_1 is a multiple of every one of these exponents, we have for every θ the equivalence $\theta^{n_1} \infty 1$

§ 138. Given any element θ_1 belonging to the exponent n_1, we may extend the idea of equivalence, and regard any two elements θ' and θ'' as "*relatively equivalent*" when for any integral value of k
$$\theta' . \theta_1^k \infty \theta''$$
We retain the sign of equivalence to indicate the original more limited relation.

If now we select from the elements θ any complete system of elements which are not relatively equivalent to one another, this subordinate system satisfies all the conditions imposed on the entire system and therefore possesses all the properties enumerated above. In particular there will be a number n_2, corresponding to n_1, such that the n_2^{th} power of every θ of the new system is relatively equivalent to unity, *i. e.*, $\theta^{n_2} \infty \theta_1^k$. Again there are elements $\theta_{,,}$ in the new system of which no power lower than the n_2^{th} is relatively equivalent to unity. Since the equivalence $\theta^{n_1} \infty 1$ holds for every element, and consequently *à fortiori* every θ^{n_1} is *relatively* equiva-

158   THEORY OF SUBSTITUTIONS.

lent to unity, it follows from I) than n_1 is equal to n_2 or is a multiple of n_2.

If now
$$\theta_{\prime\prime}{}^{n_2} \backsim \theta_1{}^k,$$

and if both sides are raised to the power $\dfrac{n_1}{n_2}$, we obtain, writing $\dfrac{k}{n_2} = m$, the equivalence
$$\theta_1{}^{mn_1} \backsim 1.$$

From this it follows that, since θ_1 belongs to the exponent n_1, m is an integer and k is therefore a multiple of n_2. There is therefore an element θ_2, defined by the equivalence
$$\theta_2 \theta_1{}^m \backsim \theta_{\prime\prime}, \quad \text{or} \quad \theta_2 \backsim \theta_{\prime\prime} \theta_1{}^{n_1-m}$$

of which the $n_2{}^{\text{th}}$ power is not only relatively equivalent, but also absolutely equivalent to unity. This element belongs both relatively and absolutely, to the exponent n_2, for we have the relation
$$\theta_2{}^{n_2} \backsim \theta_{\prime\prime}{}^{n_2} \theta_1{}^{n_1 n_2 - m \, n_2} \backsim \theta_1{}^k \theta_1{}^{n_1 n_2 - m \, n_2} \backsim \theta_1{}^{n_1 n_2} \backsim 1.$$

Proceeding further, if we now regard any two elements θ' and θ'' as relatively equivalent when
$$\theta' \theta_1{}^h \theta_2{}^k \backsim \theta'',$$

we obtain, corresponding to θ_2, an element θ_3 belonging to the exponent n_3, where n_3 is equal to n_2 or a divisor of n_2; and so on. We obtain therefore in this way a fundamental system of elements $\theta_1, \theta_2, \theta_3, \ldots$ which has the property that the expressions
$$\theta_1{}^{h_1} \theta_2{}^{h_2} \theta_3{}^{h_3} \ldots \quad (h_i = 1, 2, \ldots n)$$

include in the sense of equivalence every element θ once and only once. The number $n_1, n_2, n_3, \ldots$, to which the elements $\theta_1, \theta_2, \theta_3, \ldots$ belong, are such that every one of them is equal to or is a multiple of the next following. The product $n_1 n_2 n_3 \ldots$ is equal to the entire number n of the elements θ, and this number n accordingly contains no other prime factors than those which occur in the first number n_1.

§ 139. In the present case the elements θ are to be replaced by substitutions every two of which are commutative. The number n of the elements θ becomes the order r of the group. We have then

CERTAIN SPECIAL CLASSES OF GROUPS.   159

Theorem IV. *If all the substitutions of a group are commutative, there is a fundamental system of substitutions $s_1, s_2, s_3, \ldots$ which possesses the property that the products*
$$s_1^{h_1} s_2^{h_2} s_3^{h_3} \ldots \quad (h_i = 1, 2, \ldots r_i)$$
include every substitution of the group once and only once. The numbers $r_1, r_2, r_3, \ldots$ are the orders of $s_1, s_2, s_3, \ldots$ and are such that every one is equal to or is divisible by the next following. The product of these orders $r_1, r_2, r_3, \ldots$ is equal to the order r of the group.

The number r_1 is determined as the maximum of the orders of the several substitutions. On the other hand the corresponding substitution s_1 is not fully determined, but may be replaced by any other substitution s_1' of order r_1. According then as we start from s_1 or s_1', the values of $r_2, r_3, \ldots$ might be different. We shall now show that this is not the case.

In the first place it is plain that if several successive s's belong to the same exponent r, these s's may be permuted among themselves, without any change in the r's. Moreover, every s_a can be replaced by $s_a{}^\mu s_{a+1}{}^\nu s_{a+2}{}^\tau \ldots$ without any change in the r's, provided only that μ is prime to r_a.

If now the given group can be expressed in the two different forms
$$s_1^{h_1} s_2^{h_2} s_3^{h_3} \ldots (h_i = 1, 2, \ldots r_i), \quad \sigma_1^{h_1} \sigma_2^{h_2} \sigma_3^{h_3} \ldots (h_i = 1, 2, \ldots \rho_i),$$
then $\rho_1 = r_1$, and
$$\sigma_1 = s_1{}^\mu s_2{}^\nu s_3{}^\tau \ldots$$
Since σ_1 belongs to r_1, at least one of the exponents $\mu, \nu, \ldots$ must be prime to r_1. From the first remark above, we may assume that this is μ, and from the second it follows that the group can also be expressed by
$$\sigma_1^{h_1} s_2^{h_2} s_3^{h_3} \ldots \quad (h_i = 1, 2, \ldots r_i; \; \rho_1 = r_1).$$
Consequently the groups
$$s_2^{h_2} s_3^{h_3} \ldots (h_i = 1, 2, \ldots r_i), \quad \sigma_2^{h_2} \sigma_3^{h_3} \ldots (h_i = 1, 2, \ldots \rho_i)$$
are identical. From this it follows, as before, that $\rho_2 = r_2$, and so on.

Theorem V. *The numbers $r_1, r_2, r_3, \ldots$ are invariant for a given group.**

---
*This theorem is due to Frobenius and Stickelberger. *cf.* their article: Uber Gruppen mit vertauschbaren Elementen; Crelle, 86, pp. 217-262.

CHAPTER VIII.

ANALYTICAL REPRESENTATION OF SUBSTITUTIONS.
THE LINEAR GROUP.

§ 140. In the preceding Chapter we have met with a fourth method of indicating substitutions, which consisted in assigning the analytic formula by which the final value of the index of every x is determined from its initial value. Thus, if the index z of every x_z is converted by a given substitution into $\varphi(z)$, so that x_z becomes $x_{\varphi(z)}$, the substitution is completely defined by the symbol

$$s = |z \ \varphi(z)|.$$

Obviously not every function can be taken for $\varphi(z)$, for it is an essential condition that the system of indices $\varphi(1), \varphi(2), \ldots \varphi(n)$ shall all be different and shall be identical, apart from their order, with the system $1, 2, 3, \ldots n$. On the other hand it is readily shown that every substitution can be expressed in this notation. For if it is required that

$$\varphi(1) = i_1, \ \varphi(2) = i_2, \ \ldots \varphi(n) = i_n,$$

we can construct, by the aid of Lagrange's interpolation formula, from

$$F(1) = (z-1)(z-2) \ldots (z-n)$$

a function $\varphi(z)$ which satisfies the conditions, viz:

$$\varphi(z) = i_1 \frac{F(z)}{F'(z)(z-1)} + i_2 \frac{F(z)}{F'(z)(z-2)} + \ldots + i_n \frac{F(z)}{F'(z)(z-n)}.$$

This function is of degree $n-1$ in z. It is evident that there are an infinite number of other functions $\varphi(z)$ which also satisfy the required conditions.

§ 141. If n is a prime number p, we can on the one hand diminish the restrictions imposed on φ by permitting the indices $1, 2, 3, \ldots p$ to be replaced by any complete system of remainders (mod. p), so that indices greater than p are also allowable. And on the other hand we can depress every form of $\varphi(z)$ to the degree $n-1$, since $z^p \equiv z$ (mod. p) for all values of z.

In particular, we have in this case $F(z) \equiv z^p - z$ and $F'(z) = pz^{p-1} - 1 \equiv -1$.

For $n = p$, the functions $\varphi(z)$ which are adapted for the expression of a substitution, for which therefore $\varphi(0), \varphi(1), \ldots \varphi(p-1)$ form a complete system of remainders (mod. p), are defined by the following theorem:

Theorem I. *In order that $z \varphi z |$ may express a substitution of p elements, it is necessary and sufficient that after $\varphi(z)$ and its first $p-2$ powers have been depressed to the degree $p-1$ by means of the congruence $z^p \equiv z$ (mod. p), and after all multiples of p have been removed, these $p-1$ powers of $\varphi(z)$ should all reduce to the degree $p-2$.* *

Let
$$\varphi(z) = A_0 + A_1 z + A_2 z^2 + \ldots + A_{p-1} z^{p-1}$$
be any integral function (mod. p), and suppose that
$$[\varphi(z)]^m \equiv A_0^{(m)} + A_1^{(m)} z + A_2^{(m)} z^2 + \ldots + A_{p-1}^{(m)} z^{p-1} \quad (\text{mod. } p).$$

Since for every $a < p-1$
$$0^a + 1^a + 2^a + \ldots + (p-1)^a \equiv 0 \quad (\text{mod. } p),$$
we have

S) $\quad [\varphi(0)]^m + [\varphi(1)]^m + \ldots + [\varphi(p-1)]^m \equiv (p-1) A_{p-1}^{(m)}$
$$\equiv - A_{p-1}^{(m)} \quad (\text{mod. } p).$$

If now $\varphi(z)$ is adapted for the expression of a substitution, then, as $\varphi(0), \varphi(1), \ldots \varphi(p-1)$ form a complete system of remainders (mod. p), we conclude that for $m < p-1$
$$A_{p-1}^{(m)} \equiv 0 \quad (\text{mod. } p).$$

This is therefore a necessary condition.

Conversely, if this condition is satisfied for a given function $\varphi(z)$, then since S) holds for $m = 1, 2, \ldots p-2$, it follows from formula B) of § 8 that
$$[z - \varphi(0)][z - \varphi(1)] \ldots [z - \varphi(p-1)] \equiv z^p - a z - \beta$$
$$\equiv (1-a) z - \beta \quad (\text{mod. } p).$$

---

* Hermite: Comptes rendus de l'Académie des Sciences, 57.

11

Accordingly, if $a \neq 0$, the linear congruence $(1-a)z - \beta \equiv (\text{mod. } p)$ is satisfied by the p integers $\varphi(0), \varphi(1), \ldots \varphi(p-1)$. These must therefore all be equal, and, as their sum is 0, every one of them is equal to 0. But in this case the congruence of degree $p-2$

$$\varphi(z) \equiv A_0 + A_1 z + A_2 z^2 + \ldots + A_{p-2} z^{p-2} \equiv 0 \quad (\text{mod. } p)$$

would have $p-1$ different roots $z = 0, 1, 2, \ldots p-1$, which is impossible. Consquently $a = 1$ and then $\beta = 0$, that is

$$[z - \varphi(0)][z - \varphi(1)] \ldots [z - \varphi(p-1)] \equiv z^p - z$$
$$\equiv z(z-1)(z-2) \ldots [z-(p-1)],$$

and the values $\varphi(0), \varphi(1), \ldots \varphi(p-1)$ coincide with $0, 1, \ldots p-1$, apart from their order. It appears therefore that the condition stated in Theorem I is both necessary and sufficient to insure that $|z \ \varphi z|$ defines a substitution.

§ 142. To distinguish the individual elements of any system we may also employ several indices in each case instead of a single index as heretofore. For example, in the case of p^2 elements the indices z and u of $x_{z,u}$ might each assume any value from 0 to $p-1$. Any substitution among these p^2 elements could then be denoted by

$$s = |z, u \quad \varphi(z, u), \psi(z, u)|$$

where $\varphi(z, u)$ and $\psi(z, u)$ must satisfy conditions similar to those of § 140.

If $n = p^k$ the elements could be denoted by

$$x_{z_1}, x_{z_2}, \ldots x_{z_k} \quad (z_i = 0, 1, 2, \ldots p-1),$$

and any substitution s by the symbol

$$s = |z_1, z_2, \ldots z_k \quad \varphi_1(z_1, z_2, \ldots z_k), \varphi_2(z_1, z_2, \ldots z_k), \ldots \varphi_k(z_1, z_2, \ldots z_k)|,$$

which is to be understood as indicating that every index z_i is to be converted into $\varphi_i(z_1, z_2, \ldots z_k)$. The functions $\varphi_1, \varphi_2, \ldots \varphi_k$ must then be so taken that the p^k different systems of indices $z_1, z_2, \ldots z_k$ give rise to p^k different systems $\varphi_1, \varphi_2, \ldots \varphi_k$.

These considerations could be further extended to include the case where n contains several distinct prime factors, but as the theory then becomes much more complicated, we do not enter upon it here.

§ 143. The simplest analytical expressions for substitutions of $n = m^k$ elements are those of the linear form

1) $\quad s_{a_1, a_2, \ldots a_k} = |z_1, z_2, \ldots z_k \quad z_1 + a_1, z_2 + a_2, \ldots z_k + a_k|.$

The a's are arbitrary integers (mod. m). They can therefore be selected in m^k different ways, so that there are m^k substitutions of this type. Again, since

2) $\quad s_{a_1, a_2, \ldots a_k} \cdot s_{\beta_1, \beta_2, \ldots \beta_k} = s_{a_1 + \beta_1, a_2 + \beta_2, \ldots a_k + \beta_k},$

these *arithmetic substitutions* [*] form a group of order and degree m^k. This group is transitive, since the a's can be so chosen that any given element $x_{z_1, z_2, \ldots z_k}$ is replaced by any other element $x_{\zeta_1, \zeta_2, \ldots \zeta_k}$. For this purpose we need only take

$$a_1 = \zeta_1 - z_1, \quad a_2 = \zeta_2 - z_2, \ldots a_k = \zeta_k - z_k.$$

There is only one substitution of the group which produces this result.

In order that an arithmetic substitution may leave any element x unchanged, it is necessary that

$$a_1 = 0, \quad a_2 = 0, \ldots a_k = 0 \quad (\text{mod. } m)$$

Such a substitution leaves all the elements unchanged, and therefore reduces to identity. The present group is therefore included in the groups Ω considered in the preceding Chapter.

By the continued application of the formula 2) we obtain

$$s_{a_1, a_2, \ldots a_k} = s_{1, 0, \ldots 0}^{a_1} \cdot s_{0, 1, \ldots 0}^{a_2} \ldots s_{0, 0, \ldots 1}^{a_k}$$

so that we may define the group by

3) $\quad G = \{ s_{1, 0, \ldots 0}, \; s_{0, 1, \ldots 0}, \; \ldots s_{0, 0, \ldots 1} \}.$

The substitutions contained in the parenthesis are all commutative, and the same property consequently holds for all the substitutions of G.

§ 144. We determine now the most general form of the substitutions

$$t = |z_1, z_2, \ldots z_k \quad \varphi_1(z_1, z_2, \ldots z_k), \varphi_2(z_1, z_2, \ldots z_k), \ldots \varphi_k(z_1, z_2, \ldots z_k)|,$$

which are commutative with G, for which therefore

$$t^{-1} s_{a_1, \ldots} t = s_{\beta_1, \ldots}$$

It is obviously sufficient to take for $s_{a_1, \ldots}$ the several generating substitutions given in 3). The substitution $t^{-1} s_{1, 0, \ldots 0} t$

Cauchy: Exercices III., p. 232.

replaces $\varphi_\lambda(z_1, z_2, \ldots z_k)$ by $\varphi_\lambda(z_1+1, z_2, \ldots z_k)$. Consequently, taking $\lambda = 1, 2, 3, \ldots k$ we have

$$\varphi_\lambda(z_1+1, z_2, \ldots z_k) \equiv \varphi_\lambda(z_1, z_2, \ldots z_k) + a_\lambda \pmod{m}.$$

Similarly in the case of the substitutions $s_{0,1,\ldots 0}, \ldots s_{0,0,\ldots 1}$ we obtain

$$\varphi_\lambda(z_1, z_2+1, \ldots z_k) \equiv \varphi_\lambda(z_1, z_2, \ldots z_k) + b_\lambda \pmod{m}$$

. . . .

$$\varphi_\lambda(z_1, z_2, \ldots z_k+1) \equiv \varphi_\lambda(z_1, z_2, \ldots z_k) + c_\lambda \pmod{m}.$$

From these congruences it appears at once that the φ_λ's are linear functions of the z's, having for their constant terms $\delta_\lambda = \varphi_\lambda(0, 0, 0, \ldots 0)$. The remaining coefficients are then readily found. In fact, we have

$$\varphi_\lambda(z_1, z_2, \ldots z_k) = a_\lambda z_1 + b_\lambda z_2 + \ldots + c_\lambda z_k + \delta_\lambda,$$

and therefore

$$t = | z_1, z_2, \ldots z_k \quad a_1 z_1 + b_1 z_2 + \ldots + c_1 z_k + \delta_1,$$
$$a_2 z_1 + b_2 z_2 + \ldots + c_2 z_k + \delta_2, \ldots |.$$

Conversely all substitutions of this type transform the group G into itself. Thus, for example, t transforms $s_{1,0,0,\ldots 0}$ into

$$. \ | a_1 z_1 + b_1 z_2 + \ldots + c_1 z_k + \delta_1, \ldots$$
$$a_1(z_1+1) + b_1 z_2 + \ldots + c_1 z_k + \delta_1, \ldots |,$$

i. e., into the arithmetic substitution

$$\zeta_1, \zeta_2, \ldots \zeta_k \quad \zeta_1 + a_1, \zeta_2 + a_2, \ldots \zeta_k + a_k | = s_{a_1, a_2, \ldots a_k}.$$

By left hand multiplication by

$$s_{\delta_1, \delta_2, \ldots \delta_k}^{-1} = s_{-\delta_1, -\delta_2, \ldots -\delta_k}$$

we can reduce t to the form

$$t' = |z_1, z_2, \ldots z_k \quad a_1 z_1 + b_1 z_2 + \ldots + c_1 z_k, a_2 z_1 + b_2 z_2 + \ldots + c_2 z_k,$$
$$\ldots a_k z_1 + b_k z_2 + \ldots + c_k z_k |.$$

Such a substitution is called a *geometric substitution*.[*]

We proceed to examine this type. We have already demonstrated

Theorem II. *All geometric substitutions and their combinations with the arithmetic substitutions, and no others, are commutative with the group of the arithmetic substitutions.*

---
[*] Cauchy: loc. cit.

ANALYTICAL REPRESENTATION OF SUBSTITUTIONS.

§ 145. We have first of all to determine whether the constants $a_\lambda, b_\lambda, \ldots c_\lambda$ can be taken arbitrarily. They must certainly be subjected to one condition, since two elements $x_{z_1, z_2, \ldots z_k}$ and $x_{\zeta_1, \zeta_2, \ldots \zeta_k}$ must not be converted into the same element unless the indices $z_1, z_2, \ldots z_k$ coincide in order with $\zeta_1, \zeta_2, \ldots \zeta_k$. More generally, given any system of indices $\zeta_1, \zeta_2, \ldots \zeta_k$, it is necessary that from

$$a_1 z_1 + b_1 z_2 + \ldots + c_1 z_k \equiv \zeta_1, \quad a_2 z_1 + b_2 z_2 + \ldots + c_2 z_k \equiv \zeta_2, \ldots \quad (\text{mod. } m)$$

the indices $z_1, z_2, \ldots z_k$ shall be determined without ambiguity. In other words, the m^k systems of values z must give rise to an equal number of systems of values ζ. The necessary and sufficient conditions for this is that the congruences

$$a_1 b_1 + b_1 z_2 + \ldots + c_1 z_k \equiv 0, \quad a_2 z_1 + b_2 z_2 + \ldots + c_2 z_k \equiv 0, \ldots \quad (\text{mod. } m)$$

shall admit only the one solution $z_1 = 0, z_2 = 0, \ldots z_k = 0$. If the determinant of the coefficients is denoted by Δ, these congruences are equivalent to

$$\Delta \cdot z_1 \equiv 0, \quad \Delta \cdot z_2 \equiv 0, \ldots \Delta \cdot z_k \equiv 0 \quad (\text{mod. } m).$$

The required condition is therefore satisfied if and only if Δ is prime to m. We have then

Theorem III. *In order that the symbol*

$$t = |z_1, z_2, \ldots z_k \quad a_1 z_1 + b_1 z_2 + \ldots + c_1 z_k, \quad a_2 z_1 + b_2 z_2 + \ldots + c_2 z_k, \ldots|$$

$$(\text{mod } m)$$

may denote a (geometric) substitution, it is necessary and sufficient that

$$\Delta = \begin{vmatrix} a_1, b_1, \ldots c_1 \\ a_2, b_2, \ldots c_2 \\ \cdot \quad \cdot \quad \cdot \quad \cdot \\ a_k, b_k, \ldots c_k \end{vmatrix}$$

should be prime to the modulus m.

§ 146. From this consideration it is now possible to determine the number r of the geometrical substitutions corresponding to a given modulus m.

We denote the number of distinct systems of ρ integers which are less than m and prime to m by $[m, \rho]$. It is to be understood that any number of the ρ integers of a system may coincide.

Suppose N to be the number of those geometric substitutions $1, t_2, t_3, \ldots$ which leave the first index z_1 unchanged. If then τ_2 is any substitution which replaces z_1 by $a_1 z_1 + b_1 z_2 + \ldots + c_1 z_k$, then $\tau_2, t_1 \tau_2, t_2 \tau_2, \ldots$ are all the substitutions which produce this effect, and these are all different from one another. Similarly, if τ_3 replaces z_1 by $a_1' z_1 + b_1' z_2 + \ldots + c_1' z_k$, then $\tau_3, t_2 \tau_3, t_3 \tau_3, \ldots$ are all the substitutions which produce this effect, and these are all different, and so on. We obtain therefore the number r of all the possible geometric substitutions by multiplying N by the number of substitutions $1, \tau_2, \tau_3, \ldots$

The choice of the systems $a_1, b_1, \ldots c_1;\ a_1', b_1', \ldots c_1';\ \ldots$ is limited by the condition that that the integers of a system cannot have a same common factor with m. There are therefore $[m, k]$ such systems, and an equal number of substitutions $1, \tau_2, \tau_3, \ldots$ Consequently
$$r = [m, k] N.$$

The substitutions t are of the form
$$|z_1, z_2, \ldots z_k \quad z_1, a_2 z_1 + b_2 z_2 + \ldots c_2 z_k, \ldots a_k z_1 + b_k z_2 + \ldots + c_k z_k|$$
(mod. m).

Since $a_2, a_3, \ldots a_k$ do not occur in the expression of the discriminant Δ, these integers can be chosen arbitrarily, i. e., in m^{k-1} different ways. The $b_\lambda, \ldots c_\lambda$ are subject to the condition that
$$\begin{vmatrix} b_2, \ldots c_2 \\ \cdot\ \cdot\ \cdot\ \cdot \\ b_k, \ldots c_k \end{vmatrix}$$
must be prime to m. If the number of systems here admissible is r', we have
$$r = [m, k] m^{k-1} r'.$$

The number r' has the same significance for a substitution of $k-1$ indices (mod. m) as r for k indices. Consequently
$$r = [m, k] m^{k-1} [m, k-1] m^{k-2} r'',$$
and so on. We obtain therefore finally
$$r = [m, k] m^{k-1} [m, k-1] m^{k-2} \ldots [m, 2] r^{(k-1)},$$
where $r^{(k-1)}$ corresponds to a single index, and therefore $r^{(k-1)} = [m, 1]$. Hence

ANALYTICAL REPRESENTATION OF SUBSTITUTIONS. 167

4) $\quad r = [m, k] m^{k-1} [m, k-1] m^{k-2} \ldots [m, 2] m [m, 1].$

The evaluation of $[m, k]$ presents little difficulty. We limit ourselves to the simple case where m is a prime number p, this being the only case which we shall hereafter have occasion to employ. We have then evidently

5) $\quad [p, \rho] = p^\rho - 1,$

since only the combination $0, 0, \ldots 0$ is to be excluded. By the aid of 5), we obtain from 4)

6) $\quad r = (p^k - 1) p^{k-1} (p^{k-1} - 1) p^{k-2} \ldots (p^2 - 1) p (p - 1)$
$\quad = (p^k - 1)(p^k - p)(p^k - p^2) \ldots (p^k - p^{k-1}).$ *

§ 147. The entire system of the geometric substitutions (mod. m) forms a group the order of which is determined from 4) or from 6). This group is known as the *linear group* (mod. m). If the degree is to be particularly noticed, we speak of the *linear group of degree* m^k.

It is however evident that all the substitutions of this group leave the element $x_{0, 0, \ldots 0}$ unchanged. For the congruences

$$a_1 z_1 + b_1 z_2 + \ldots + c_1 z_k \equiv z_1, \quad a_2 z_2 + b_2 z_2 + b_2 z_2 + \ldots + c_2 z_k \equiv z_2, \ldots$$
$$(\text{mod. } m)$$

have for every possible system of coefficients the solution

$$z_1 \equiv 0, \quad z_2 \equiv 0, \ldots z_k \equiv 0 \quad (\text{mod. } m).$$

We shall have occasion to employ the linear group in connection with the algebraic solution of equations.

Theorem IV. *The group of the geometric substitutions (mod. m), or the linear group of degree m^k is of the order given in* 4). *Its substitutions all leave the element* $x_{0, 0, \ldots 0}$ *unchanged. It is commutative with the group of the arithmetic substitutions.*

---
*Galois: Liouville Journal (1) XI, 1846, p. 410.

PART II.

APPLICATION OF THE THEORY OF SUBSTITUTIONS TO THE ALGEBRAIC EQUATIONS.

CHAPTER IX.

THE EQUATIONS OF THE SECOND, THIRD AND FOURTH DEGREES. GROUP OF AN EQUATION. RESOLVENTS.

§ 148. The problem of the algebraic solution of the equation of the second degree

1) $\qquad x^2 - c_1 x + c_2 = 0$

can be stated in the following terms: From the elementary symmetric functions c_1 and c_2 of the roots x_1 and x_2 of 1) it is required to determine the two-valued function x_1 by the extraction of roots.*
Now it is already known to us (Chapter I, § 13) that there is always a two-valued function, the square of which, $viz.$, the discriminant, is single-valued. In the present case we have

$$\Delta = (x_1 - x_2)^2 = (x_1 + x_2)^2 - 4 x_1 x_2 = c_1^2 - 4 c_2,$$
$$\sqrt{\Delta} = (x_1 - x_2) = \sqrt{c_1^2 - 4 c_2}.$$

Since there is only one family of two-valued functions, every such function can be rationally expressed in terms of $\sqrt{\Delta}$. For the linear two-valued functions we have

$$a_1 x_1 + a_2 x_2 = \frac{a_1 + a_2}{2}(x_1 + x_2) + \frac{a_1 - a_2}{2}(x_1 - x_2)$$
$$= \frac{a_1 + a_2}{2} c_1 + \frac{a_1 - a_2}{2} \sqrt{c_1^2 - 4 c_2},$$

and in particular, for $a_1 = 1, a_2 = 0$, and for $a_1 = 0, a_2 = 1$

$$x_1 = \frac{c_1}{2} + \tfrac{1}{2}\sqrt{c_1^2 - 4 c_2}, \quad x_2 = \frac{c_1}{2} - \tfrac{1}{2}\sqrt{c_1^2 - 4 c_2}.$$

---

*C. G. J. Jacobi: Observatiunculae ad theoriam aequationum pertinentes. Werke, Vol. III; p. 269. Also J. L. Lagrange: Réflexions sur la résolution algébrique des équations. Oeuvres. t. III, p. 205.

§ 149. In the case of the equation of the third degree
$$x^3 - c_1 x^2 + c_2 x - c_3 = 0$$
the solution requires not merely the determination of the three-valued function x_1, but that of the three three-valued functions x_1, x_2, x_3. With these the 3!-valued function
$$\xi = a_1 x_1 + a_2 x_2 + a_3 x_3$$
is also known, and conversely x_1, x_2, x_3 can be rationally expressed in terms of ξ. We have therefore to find a means of passing from the one-valued functions c_1, c_2, c_3 to a six-valued function by the extraction of roots.

In the first place the square root of the discriminant
$$\varDelta = (x_1-x_2)^2(x_1-x_3)^2(x_2-x_3)^2 = -27 c_3{}^2 + 18 c_3 c_2 c_1 - 4 c_3 c_1{}^3 \\ - 4 c_2{}^3 + c_2{}^2 c_1{}^2$$
furnishes the two-valued function
$$\pm (x_1 - x_2)(x_1 - x_3)(x_2 - x_3),$$
in terms of which all the two-valued functions are rationally expressible. The question then becomes whether there is any multiple-valued function of which a power is two-valued. This question has already been answered in Chapter III, § 59. The six-valued function
$$\varphi_1 = x_1 + \omega^2 x_2 + \omega x_3, \quad \omega = \left(\frac{-1 + \sqrt{-3}}{2} \right),$$
on being raised to the third power, gives
$$\varphi_1{}^3 = \tfrac{1}{2}(2 c_1{}^3 - 9 c_1 c_2 + 27 c_3 + 3 \sqrt{-3 \varDelta}) \\ = \tfrac{1}{2}(S_1 + 3 \sqrt{-3 \varDelta}).$$
again, if φ_2 is obtained from φ_1 by changing the sign of $\sqrt{-3}$, we have
$$\varphi_2{}^3 = (x_1 + \omega x_2 + \omega^2 x_3) = \tfrac{1}{2}(S_1 - 3\sqrt{-3 \varDelta}).$$
Accordingly
$$x_1 + \omega^2 x_2 + \omega x_3 = \sqrt[3]{\tfrac{1}{2}(S_1 + 3\sqrt{-3 \varDelta})},$$
$$x_1 + \omega x_2 + \omega^2 x_3 = \sqrt[3]{\tfrac{1}{2}(S_1 - 3\sqrt{-3 \varDelta})}.$$
Combining with these the equation
$$x_1 + x_2 + x_3 = c_1,$$

and observing that
$$1 + \omega + \omega^2 = 0,$$
we obtain the following results

$$x_1 = \tfrac{1}{3}\left[c_1 + \sqrt[3]{\tfrac{1}{2}(S_1 + 3\sqrt{-3\varDelta})} + \sqrt[3]{\tfrac{1}{2}(S_1 - 3\sqrt{-3\varDelta})}\right],$$

$$x_2 = \tfrac{1}{3}\left[c_1 + \omega\sqrt[3]{\tfrac{1}{2}(S_1 + 3\sqrt{-3\varDelta})} + \omega^2\sqrt[3]{\tfrac{1}{2}(S_1 - 3\sqrt{-3\varDelta})}\right],$$

$$x_3 = \tfrac{1}{3}\left[c_1 + \omega^2\sqrt[3]{\tfrac{1}{2}(S_1 + 3\sqrt{-3\varDelta})} + \omega\sqrt[3]{\tfrac{1}{2}(S_1 - 3\sqrt{-3\varDelta})}\right].$$

The solution of the equation of the third degree is then complete.

§ 150. In the case of the equation of the fourth degree
$$x^4 - c_1 x^3 + c_2 x^2 - c_3 x + c_4 = 0$$
it is again only the one-valued functions c_1, c_2, c_3, c_4 that are known. From these we have to obtain the four four-valued functions x_1, x_2, x_3, x_4, and with them the 24-valued function
$$\xi = a_1 x_1 + a_2 x_2 + a_3 x_3 + a_4 x_4$$
by the repeated extraction of roots.

In the first place the square root of a rational integral function of c_1, c_2, c_3, c_4 furnishes the two-valued function $\sqrt{\varDelta}$. Again, we have met in § 59 with a six-valued function
$$\varphi = (x_1 x_2 + x_3 x_4) + \omega(x_1 x_3 + x_2 x_4) + \omega^2(x_1 x_4 + x_2 x_3),$$
the third power of which is two-valued and therefore belongs to the family of $\sqrt{\varDelta}$. We can therefore obtain φ by the extraction of the cube root of a two-valued function. φ being determined, every function belonging to the same family is also known. The group of φ is
$$G = [1, (x_1 x_2)(x_3 x_4), (x_1 x_3)(x_2 x_4), (x_1 x_4)(x_2 x_3)]; \quad \rho = 6, \ r = 4.$$
To this same group belongs the function
$$\psi^2 = (x_1 x_2 - x_3 x_4)^2 (x_1 x_3 + x_2 x_4)^2,$$
which can therefore be rationally obtained from φ, while ψ, which can be obtained from ψ^2 by extraction of a square root, belongs to the group
$$H = [1, (x_1 x_2)(x_3 x_4)]; \quad \rho = 12, \ r = 2.$$
ψ may therefore be regarded as known. Finally
$$\chi^2 = [a_1(x_1 - x_2) + a_2(x_3 - x_4)]^2, \quad \tau = \beta_1(x_1 + x_2) + \beta_2(x_3 + x_4)$$

can be rationally expressed in terms of ψ, and χ is a 24-valued function. All rational functions of the roots, and in particular the roots themselves can then be rationally expressed in terms of χ. To determine the roots we may, for example, combine the four equations for χ and τ in which $a_1 = \pm a_2$ and $\beta_1 = \pm \beta_2$.

§ 151. In attempting the algebraic solution of the general equations of the fifth degree by the same method, we should not be able to proceed further than the construction of the two-valued functions. For we have seen in § 58 that for more than four independent quantities there is no multiple-valued function of which a power is two-valued. It is still a question, however, whether the solution of the equation fails merely through a defect in the method or whether the impossibility of an algebraic solution resides in the nature of the problem. It will hereafter be shown that the latter is the case. We shall demonstrate the full sufficiency of the method by the proof of the theorem that every irrational function of the coefficients which occurs in the algebraic solution of an equation is a rational function of the roots. All the steps leading from the given one-valued function to the required $n!$-valued functions can therefore be taken within the theory of the integral rational functions of the roots.

§ 152. We turn our attention next to the accurate formulation of the problem involved in the solution of algebraic equations.

Suppose that *all* the roots of an equation of the nth degree

1) $$f(x) = 0$$

are to be determined. If one of them x_1 is known, the problem is only partially solved. By the aid of the partial solution we can however reduce the problem and regard it now as requiring not the determination of the $n-1$ remaining roots of the equation of the nth degree $f(x) = 0$, but that of *all* the $n-1$ roots of the reduced equation

2) $$\frac{f(x)}{x - x_1} \equiv f_1(x) \equiv x^{n-1} - \gamma_1 x^{n-2} + \gamma_2 x^{n-3} - \ldots \pm \gamma_{n-1} = 0.$$

We have then still to accomplish the solution of 2). If one of its roots x_2 is known, we can reduce the problem still further to that of the determination of the $n-2$ roots of the equation

3) $\quad \dfrac{f_1(x)}{x-x_2} = f_2(x) = x^{n-2} - c_1 x^{n-3} + c_2 x^{n-4} \ldots \pm c_{n-2} = 0.$

Proceeding in this way we arrive finally at an equation of the first degree.

It appears therefore that solution of an equation of the nth degree involves a series of problems. All of these problems are however included in a single one, *that of the determination of a single root of a certain equation of degree $n!$*

Thus, if the entirely independent roots $x_1, x_2, \ldots x_n$ of the equation 1) are known, then the $n!$-valued function with arbitrary constants $a_1, a_2, \ldots a_n$

4) $\quad\quad\quad\quad \xi = a_1 x_1 + a_2 x_2 + \ldots + a_n x_n$

is also known. Conversely if ξ is known, every root of the equation 1) is known, for every rational integral function of $x_1, x_2, \ldots x_n$ can be rationally expressed in terms of the $n!$-valued function ξ.

The function ξ satisfies an equation of degree $n!$

5) $F(\xi) \equiv \xi^{n!} - A_1 \xi^{n!-1} + \ldots \pm A_{n!} \equiv (\xi - \xi_1)(\xi - \xi_2)\ldots(\xi - \xi_{n!}) = 0,$

the coefficients of which are rational integral functions of those of 1) and of $a_1, a_2, \ldots a_n$. This equation is, in distinction from 1), a very special one. For its roots are no longer independent, as was the case with 1), but every one of them is a rational function of every other, since all the values $\xi_1, \xi_2, \ldots \xi_{n!}$ belong to the group 1 with respect to $x_1, x_2, \ldots x_n$. The solution of 5), and consequently that of 1) is therefore complete, as soon as a single root of the former is known.

The equation $F(\xi) = 0$ is called the *resolvent equation*, and ξ the *Galois resolvent* of 1). We shall presently introduce the name "resolvent" in a more extended sense.

§ 153. If the coefficients $c_1, c_2, \ldots c_n$ of the equation 1) are entirely independent, the Galois resolvent cannot break up into rational factors. Conversely, if the Galois resolvent does not break up into factors, then, although relations may exist among the coefficients c, they are not of such a nature as to produce any simplification in the form of the solution. From this point of view the equation 1) is in this case a *general* equation, or, according to Kronecker, it has no *affect*.

On the other hand, if for particular values of the coefficients c the Galois resolvent $F(\bar{z})$ breaks up into irreducible factors with rational coefficients

6) $\qquad F(\bar{z}) = F_1(\bar{z})(F_2(\bar{z}) \ldots F_\nu(\bar{z}),$

then the unsymmetric functions $F_i(\bar{z})$, which in the case of fully independent coefficients are irrational, are now rationally known. The equation 1) is then a *special* equation, or according to Kronecker, an *affect* equation. The *affect* of an equation lies, then, in the manner in which the Galois resolvent breaks up or, again, in the fact that, as a result of particular relations among the coefficients $c_1, c_2, \ldots c_n$, or the roots $x_1, x_2, \ldots x_n$, certain unsymmetric irreducible functions $F_i(\bar{z})$ are rationally known. The determination of that which is to be regarded as rationally known in any case is obviously of the greatest importance. As a result of any change in this respect, an equation may gain or lose an affect.

If the group belonging to any one of the functions $F_i(\bar{z})$ is G_i, then every function belonging to G_i is rationally known, being a rational function of F_i. It is readily seen that the groups G_i all coincide. For if the subgroup common to them all is Γ, then the rationally known function $\alpha F_1 + \beta F_2 + \ldots + \lambda F_\nu$ belongs to Γ, and consequently every function belonging to Γ is rationally known. Accordingly the factor of F_i which proceeds from the application of the substitutions of Γ alone to any linear factor $\bar{z} \quad a_1 x_{i_1} \quad a_2 x_{i_2} \ldots - a_n x_{i_n}$ of F_i is itself rationally known. This is inconsistent with the assumed irreducibility of F_i, unless $\Gamma = G_1 = G_2 = \ldots = G_\nu.$

All the functions $F_1, F_2, \ldots F_\nu$ therefore belong to the same family, and this is characterized by a certain group G or by any function $\Phi(x_1, x_2, \ldots x_n)$ belonging to G. Every function belonging to G is rationally known and conversely every rationally known function belongs to G.

Theorem 1. *Every special or affect equation is characterized by a group G, or by a single relation between the roots*

$$\Phi(x_1, x_2, \ldots x_n) = 0.$$

The group G is called the Galois group of the equation. Every equation is accordingly completely defined by the system

$$\Sigma \cdot c_\lambda = c_1, \quad \Sigma x_\lambda x_\mu = c_2, \ldots; \quad \Phi(x_1, x_2, \ldots x_n) = 0.\ *$$

---

* Cf. Kronecker: Grundzüge einer arithmetischen Theorie der algebraischen Grössen. §§ 10, 11.

For example, given a quadratic equation
$$x^2 - 2c_1 x + c_2 = 0,$$
the corresponding Galois resolvent is
$$\xi^2 - 2(a_1 + a_2)c_1 \xi + (a_1 - a_2)^2 c_2 + 4a_1 a_2 c_1^2 = 0.$$
In general the latter equation is irreducible, and the quadratic equation has no affect. But, if we take
$$2c_1 = m + n, \quad c_2 = mn,$$
the equation in ξ becomes
$$(\xi - a_1 m - a_2 n)(\xi - a_1 n - a_2 m) \doteq 0 = (\xi - a_1 x_1 - a_2 x_2)(\xi - a_1 x_2 - a_2 x_1)$$
and the given quadratic equation has an affect.

Again, if $c_2 - c_1^2 = 0$, we have
$$(\xi - a_1 c_1 - a_2 c_2)^2 = 0 = (\xi - a_1 x_1 - a_2 x_2)(\xi - a_1 x_2 - a_2 x_1).$$
But if $c_2 - 2c_1^2 = 0$, we obtain
$$\xi^2 - 2(a_1 + a_2)c_1 \xi + 2(a_1^2 + a_2^2)c_1^2 = 0,$$
and this equation has no affect, so long as we deal only with real quantities. If however we regard $i = \sqrt{-1}$ as known, the equation has an affect, for the Galois resolvent then becomes
$$(\xi - (a_1 + a_2)c_1 + (a_1 - a_2)c_1 i)(\xi - (a_1 + a_2)c_1 - (a_1 - a_2)c_1 i) = 0.$$

§ 154. It is clear that every unsymmetric equation $\varphi(x_1, \ldots x_n) = 0$ between the roots produces an affect. On the other hand an equation between the coefficients $\Psi(c_1, c_2, \ldots c_n) = 0$ produces an affect only when, as a result, the Galois resolvent breaks up into rational factors. This is always the case if Ψ is the product of all the conjugate values of an unsymmetric function,
$$\Psi(c_1, c_2, \ldots c_n) = \overline{\prod} \Phi(x_{i_1}, x_{i_2}, \ldots x_{i_n});$$
for it follows then from § 2 that one of the factors of the product must vanish. Whether this condition is also necessary, is a question which we will not here consider.

A special equation might also be characterized by several relations between the roots or the coefficients or both. A direct consideration of the latter cases would again present serious difficulty. From the preceding Section we recognize, however, that whatever the number or the character of the relations may be, if they produce an affect, they can all be replaced by a single equation $\Phi(x_1, x_2, \ldots x_n) = 0$.

§ 155. We will illustrate the ideas and definitions introduced in the preceding Sections by an example.

Suppose that all the roots of an irreducible equation $f(x) = 0$ are rational functions of a single one among them

$$x_2 = \varphi_2(x_1),\ x_3 = \varphi_3(x_1),\ \ldots x_n = \varphi_n(x_1)$$

The two equations

$$f(x) = 0,\ f[\varphi_\lambda(x)] = 0$$

have then one root, x_1, in common, and since $f(x)$ is irreducible, all the other roots $x_2, x_3, \ldots x_n$ of the first equation are also roots of the second. Consequently $f(x) = 0$ is satisfied by

$$\varphi_\lambda(x_2),\ \varphi_\lambda(x_3),\ \ldots \varphi_\lambda(x_n),$$

and in general by every

$$\varphi_\alpha[\varphi_\beta(x_1)] \quad (\alpha, \beta = 2, 3, \ldots n).$$

Again

$$\varphi_\alpha[\varphi_\gamma(x_1)] \neq \varphi_\beta[\varphi_\gamma(x_1)],$$

for otherwise the two equations

$$f(x) = 0,\ \varphi_\alpha(x) - \varphi_\beta(x) = 0$$

would have the root $\varphi_\gamma(x_1)$ and consequently all the roots of $f(x) = 0$ in common. It would then follow that $\varphi_\alpha(x_1) = \varphi_\beta(x_1)$, i, e., two of the roots of $f(x)$ would be equal. This being contrary to assumption, it appears that the series

R$_1$) $x_\gamma,\ \varphi_2(x_\gamma),\ \varphi_3(x_\gamma) \ldots \varphi_n(x_\gamma)$

coincides, apart from the order of the elements, with the series

R$_2$) $x_1,\ \varphi_2(x_1),\ \varphi_3(x_1),\ \ldots \varphi_n(x_1).$

We determine now what substitutions can be performed among the roots $x_1, x_2, \ldots x_n$ without disturbing the relations existing among them. If such a substitution replaces x_1 by x_γ, then it must also replace every x_α by $\varphi_\alpha(x_\gamma)$. The substitution is therefore fully determined by the single sequence x_1, x_γ. There are accordingly only n substitutions which satisfy the conditions. These form a group Ω (§ 129); for the system is transitive and its degree and order are both equal to n.

This group Ω is the group of the given equation. For the relations which characterize the given equations are equivalent to the single relation

$$\Phi = \beta_1[x_2 - \varphi_2(x_1)] + \beta_3[x_3 - \varphi_3(x_1)] + \ldots + \beta_n[x_n - \varphi_n(x_1)] = 0,$$

and if this function Φ is to remain unaltered, then when x_1 is replaced by x_γ, every x_a must be replaced by $\varphi_a(x_\gamma) = \varphi_a[\varphi_\gamma(x_1)]$, exactly as under the application of the substitutions of Ω.

§ 156. Without entering further into the theory of the group of an equation we can still give here two of the most important theorems.

Theorem II. *The group of an irreducible equation is transitive. Conversely, if the group of an equation is transitive, the equation is irreducible.*

Thus if the group G of the equation

$$f(x) \equiv (x - x_1)(x - x_2) \ldots (x - x_n) = 0$$

is intransitive, suppose that it connects only the elements $x_1, x_2, \ldots x_a$ with one another. Then the function

$$\varphi = (x - x_1)(x - x_2) \ldots (x - x_a)$$

is unchanged by G, and consequently belongs either to the family of G or to one of its sub-families. In either case φ is rationally known, that is, the coefficients of φ are rational functions of known quantities, so that $\varphi(x)$ is a rational factor of $f(x)$.

Conversely, if $f(x)$ is reducible, a factor $\varphi(x)$ of the above form will be rationally known, and G can contain no substitution which replaces any element $x_1, x_2, \ldots x_a$ by x_{a+1}, for otherwise the rationally known function φ would not remain unchanged for all the substitutions of G. Consequently G is intransitive.

Theorem III. *If all the roots of an irreducible equation are rational functions of any one among them, the order of the group of the equation is n. Conversely, if the group of an equation is transitive, and if its order and degree are equal, then all the roots of the equation are rational functions of any one among them.*

The first part of the theorem follows at once from § 155. We proceed to prove the second part.

From the transitivity of the group follows the irreducibility of the equation.

If we specialize the given equation by adjoining to it the family belonging to x_1, the group will be correspondingly reduced. It will in fact then contain only substitutions which leave x_1 unchanged. But as the group is of the type Ω (§ 129), it contains only one substitution, identity, which leaves x_1 unchanged. Accordingly after the adjunction of x_1, all functions belonging to the group 1 or to any larger group are rationally known. In particular $x_1, x_2, \ldots x_n$ are rationally known, i. e., they are rational functions of x_1.

From this follows again the theorem which has already been proved in § 155:

Theorem VI. *If all the roots of an irreducible equation are rational functions of any one among them, they are rational functions of every one among them.*

§ 157. From Theorem III, the group of the Galois resolvent equation of a general equation is of order $n!$ To obtain it, we apply to the values
$$\xi_1, \xi_2, \ldots \xi_{n!}$$
all the substitutions of $x_1, x_2, \ldots x_n$ and regard the resulting rearrangements as substitutions of the ξ's. Since every substitution of the x's affects all of the ξ's, the group of the ξ's belongs to the groups Ω. The group of the x's and that of the ξ's are simply isomorphic (§ 72).

For an example we may take again the case of the equation of the third degree. The groups G of $f(x) = 0$ and Γ of $F(\xi) = 0$ are then

$$G = [1, (x_1 x_2), (x_1 x_3), (x_2 x_3), (x_1 x_2 x_3), (x_1 x_3 x_2)]$$
$$\Gamma = [1, (\xi_1 \xi_3)(\xi_2 \xi_4)(\xi_5 \xi_6), (\xi_1 \xi_6)(\xi_2 \xi_5)(\xi_3 \xi_4), (\xi_1 \xi_2)(\xi_3 \xi_6)(\xi_4 \xi_5),$$
$$(\xi_1 \xi_4 \xi_5)(\xi_2 \xi_3 \xi_6), (\xi_1 \xi_5 \xi_4)(\xi_2 \xi_6 \xi_3)].$$

If however, the given equation is an affect equation with a group G of order r, then of the $n!$ substitutions among the ξ's only those are to be retained which connect any ξ_i with those ξ's which together with ξ_i belong to one of the rational irreducible factors $F_i(\xi)$ of $F(\xi)$.

§ 158. We apply the name *resolvent* generally to every ρ-valued function $\varphi(x_1, x_2, \ldots x_n)$ of the roots of a given equation $f(x) = 0$. The equation of the ρth degree which is satisfied by φ and its conjugate values is called a *resolvent equation*. This designation is appropriate, in that the solution of a resolvent equation reduces the problem of the solution of the given equation. Thus, for example, in the case of the equation of the fourth degree, we employed the following system of resolvents (§ 150):

1) The 2-valued function $\sqrt{\Delta} = (x_1 - x_2)(x_1 - x_3) \ldots (x_3 - x_4)$,
2) The 6-valued function $\varphi = (x_1 x_2 + x_3 x_4) + \omega(x_1 x_3 + x_2 x_4) + \omega^2(x_1 x_4 + x_2 x_3)$,
3) The 12-valued function $\psi = (x_1 x_2 - x_3 x_4)(x_1 x_3 + x_2 x_4)$,
4) The 24-valued function $\chi = a_1(x_1 - x_2) + a_2(x_3 - x_4)$.

Originally the group of the equation was of order 24. After the solution of the quadratic equation of which the two-valued function $\sqrt{\Delta}$ was a root, the general, symmetric group reduced to the alternating group of 12 substitutions. The extraction of a cube root led then to the group G (§ 150) of order 4; another square root to the group H of order 2; and finally we arrived at the group 1, and the solution of the equation was complete, the function τ being superfluous.

The above reduction of the group of an equation to its ρth part by the solution of a resolvent equation of degree ρ is exceptional. For general equations and resolvents this reduction is not possible. We shall see later that it is possible for the biquadratic equation and the particular series of resolvents employed above only because the family of every resolvent was a self-conjugate subfamily of that of the succeeding one.

§ 159. Given any ρ-valued resolvent $\varphi(x_1, x_2, \ldots x_n)$, this satisfies an equation of the ρth degree

$$\varphi^\rho - A_1 \varphi^{\rho-1} + A_2 \varphi^{\rho-2} - \ldots \pm A_\rho = 0.$$

To determine the group of this equation we adopt the same method as in the preceding Section. Suppose that

$$\varphi_1, \varphi_2, \ldots \varphi_\rho$$

are the ρ-values of φ. Every substitution of the group G of the equation $f(x) = 0$ produces a corresponding substitution of these ρ values. The latter substitutions form the required group of the resolvent equation for φ. This group is isomorphic to G, and it is transitive, since G contains substitutions which replace φ_1 by every φ_λ. If, in particular, the group of φ_1 is a self-conjugate subgroup of G, then it is also the group of $\varphi_2, \varphi_3, \ldots \varphi_\rho$. The latter values are therefore all rational functions of φ_1, and the group of the resolvent equation is a group Ω, as in § 156.

§ 160. Following the example of Lagrange * we might attempt to accomplish the reduction and possibly the solution of the general Galois resolvent equation $F(\xi) = 0$ by employing particular resolvents. It will, however, appear later that this method cannot succeed in general for equations of a degree higher than the fourth.

---

* Mem. d. Berl. Akad. III; and Oeuvres III, p. 305 ff.

CHAPTER X.

THE CYCLOTOMIC EQUATIONS.

§ 161. The equation which is satisfied by a primitive p^{th} root of unity ω, (p, as usual, always denoting a prime number), is called the *cyclotomic equation*. It is of the form

1) $$\frac{x^p - 1}{x - 1} \equiv x^{p-1} + x^{p-2} + \ldots + x^2 + x + 1 = 0,$$

and its $p-1$ roots are

2) $$\omega, \omega^2, \omega^3 \ldots \omega^{p-1}.$$

We prove that the left member of 1) cannot be expressed as a product of two integral functions $\varphi(x)$ and $\psi(x)$ with integral cofficients. For if this were the case, we should have for $x = 1$

$$\varphi(1)\, \psi(1) = p,$$

and consequently one of the two integral factors, for example $\varphi(1)$, must be equal to ± 1. Moreover, since $\varphi(x) = 0$ has at least one root in common with 1), at least one of the expression $\varphi(\omega^a)$ must vanish. Consequently

$$\varphi(\omega_1)\, \varphi(\omega_1^2)\, \varphi(\omega_1^3) \ldots \varphi(\omega_1^{p-1}) = 0$$

where ω_1 may be any root of 1), since the series 2) is identical with the series $\omega_1, \omega_1^2, \omega_1^3, \ldots \omega_1^{p-1}$. The equation

$$\varphi(x)\, \varphi(x^2)\, \varphi(x^3) \ldots \varphi(x^{p-1}) = 0$$

has therefore all the quantities 2) as roots. Consequently the left member of this equation is divisible by 1). Suppose that

3) $$\varphi(x)\, \varphi(x^2) \ldots \varphi(x^{p-1}) = F(x)\,(x^{p-1} + x^{p-2} + \ldots + x + 1),$$

where $F(x)$ is an integral function with integral coefficients. From 3) we have for $x = 1$

$$[\varphi(1)]^{p-1} = p\, F(1),$$

and therefore $[\varphi(1)]^{p-1}$, which is equal to 1, must be divisible by p. Accordingly 1) is not reducible.

Theorem I. *The cyclotomic equation for the prime number* p

$$\frac{x^p - 1}{x - 1} \equiv x^{p-1} + x^{p-2} + \ldots + x + 1$$

is irreducible.

§ 162. If now g is a primitive root (mod. p), then the series 2) of the roots of the cyclotomic equation can be written

4) $\qquad \omega^g, \omega^{g^2}, \omega^{g^3} \ldots \omega^{g^{p-1}}.$

Since 1) is irreducible, the corresponding group is transitive. There is therefore a substitution present which replaces ω^g by ω^{g^2}. Then every ω^{g^a} is replaced by

$$(\omega^g)^{g^a} = \omega^{g^{a+1}},$$

and the substitution is therefore

$$s = \left(\omega^g \; \omega^{g^2} \; \omega^{g^3} \ldots \omega^{g^{p-1}} \right).$$

The $p-1$ powers of s form the group of 1). For they all occur in this group, and from § 156 the group contains only $p-1$ substitutions in all.

We form now the cyclical resolvent

$$(\omega + a\omega^g + a^2\omega^{g^2} + \ldots + a^{p-2}\omega^{g^{p-2}})^{p-1},$$

in which a denotes a primitive root of the equation

$$z^{p-1} - 1 = 0.$$

For brevity we write with Jacobi

$$\omega^{g^0} + a\omega^{g^1} + a^2\omega^{g^2} + \ldots + a^{p-2}\omega^{g^{p-2}} = (a, \omega).$$

From § 129 the resolvent $(a, \omega)^{p-1}$ is unchanged by s and its powers, that is, by the group of the equation. It can therefore be expressed as a rational function of a and the coefficients of 1).

If we denote a $(p-1)^{\text{th}}$ root of this rationally known quantity by τ_1 we have

5) $\qquad (a, \omega) = \tau_1.$

The quantity τ_1 is a $(p-1)$-valued function of the roots of 1). It is changed by every substitution of the group, for the substitution s converts it into

$$(a, \omega^g) = a^{-1}(a, \omega) = a^{-1}\tau_1.$$

182   THEORY OF SUBSTITUTIONS.

It follows from the general theory of groups that every function of the roots can be rationally expressed in terms of τ_1. We will however give a special investigation for this particular case. The group of the cyclotomic equation leaves the value of

6) $\qquad (a^\lambda, \omega)(a, \omega)^{p-1-\lambda}$

unchanged; for the effect of the substitution s is to convert this function into

$$(a^\lambda, \omega^g)(a, \omega^g)^{p-1-\lambda} = a^{-\lambda}(a^\lambda, \omega)a^{-p-1+\lambda}(a, \omega)^{p-1-\lambda}$$
$$= (a^\lambda, \omega)(a, \omega)^{p-1-\lambda}$$

i. e., into itself. If now we denote the rationally known value 6) by T_λ, where in particular $\tau_1^{p-1} = T_1$, we obtain for $\lambda = 1, 2, \ldots p-2$, the following series of equations:

$$(a, \omega) = \tau_1, \quad (a^2, \omega) = \frac{T_2}{T_1}\tau_1^2, \quad (a^3, \omega) = \frac{T_3}{T_1}\tau_1^3, \ldots (a^{p-2}, \omega) = \frac{T_{p-2}}{T_1}\tau_1^{p-2}.$$

Combining with these the obvious relation among the roots and the coefficients of 1)

$$(1, \omega) = -1,$$

we obtain by proper linear combinations

$$\omega = \frac{1}{p-1}\Big[-1 + \tau_1 + \frac{T_2}{T_1}\tau_1^2 + \ldots + \frac{T_{p-2}}{T_1}\tau_1^{p-2}\Big],$$

$$\omega^g = \frac{1}{p-1}\Big[-1 + a^{-1}\tau_1 + a^{-2}\frac{T_2}{T_1}\tau_1^2 + \ldots + a^{-p+2}\frac{T_{p-2}}{T_1}\tau_1^{p-2}\Big],$$

$$\omega^{g^2} = \frac{1}{p-1}\Big[-1 + a^{-2}\tau_1 + a^{-4}\frac{T_2}{T_1}\tau_1^2 + \ldots + a^{-2p+4}\frac{T_{p-2}}{T_1}\tau_1^{p-2}\Big],$$

. . . .

It is evident that a change in the choice of the particular root a or of the particular value of $\tau_1 = \sqrt[p-1]{T_1}$ only interchanges the value ω among themselves.

Theorem II. *The solution of the cyclotomic equation for the prime number p requires only the determination of a primitive root of the equation $z^{p-1} - 1 = 0$, and the extraction of the $(p-1)^{th}$ root of an expression which is then rationally known. The cyclotomic equation therefore reduces to two binomial equations of degree $p-1$.*

§ 163. The second of these operations can be still further simplified. The quantity T_1 is in general complex and of the form

$$T_1 = \rho(\cos\vartheta + i\sin\vartheta).$$

Since now $(a, \omega)^{p-1}$ and $(a^{-1}, \omega^{-1})^{p-1}$ are conjugate values, it follows that

$$(a, \omega)^{p-1}(a^{-1}, \omega^{-1})^{p-1} = \rho(\cos\vartheta + i\sin\vartheta)\rho(\cos\vartheta - i\sin\vartheta) = \rho^2.$$

Again it can be shown, exactly as in the preceding Section, that

$$(a, \omega)(a^{-1}, \omega^{-1})$$

belongs to the group of the cyclotomic equation and is consequently a rational function of a and of the coefficients of 1). If we denote its value by U we have

$$\sqrt[p-1]{\rho} = \sqrt{U}.$$

Accordingly for any integral value of k

$$(a, \omega) = \sqrt{U}\left(\cos\frac{\vartheta + 2k\pi}{p-1} + i\sin\frac{\vartheta + 2k\pi}{p-1}\right).$$

Since U and ϑ are both known, we have then

Theorem III. *The solution of the cyclotomic equation requires the determination of a primitive root of the equation $z^{p-1} - 1 = 0$, the division into $p-1$ equal parts of an angle which is then known, and the extraction of the square root of a known quantity.*

The latter quantity, U, is readily calculated. We have

$$U = (\omega + a\omega^g + a^2\omega^{g^2} + \ldots + a^{p-2}\omega^{g^{p-2}}).$$
$$(\omega^{-1} + a^{-1}\omega^{-g} + a^{-2}\omega^{-g^2} + \ldots + a^{-p+2}\omega^{-g^{p-2}}).$$

To reduce this product we begin by multiplying each pair of corresponding terms of the two parentheses together. The result is

$$1 + 1 + 1 + \ldots + 1 = p - 1.$$

Again, if we multiply every term of the first parenthesis by the kth term to the right of the corresponding term in the second parenthesis, we obtain

$$K) \qquad a^{-k}(\omega^{-g^k+1} + \omega^{-g^{k+1}+g} + \omega^{-g^{k+2}+g^2} + \ldots).$$

Now ω^{-g^k+1} is a pth root of unity ω_1 different from 1; for if

$$\omega^{-g^k+1} = 1,$$

then $-g^k + 1 \equiv 0$, $g^k \equiv 1$ (mod. p), i. e., $k = 0$ or $p-1$. The quantity K) is therefore equal to

$$a^{-k}(\omega_1 + \omega_1{}^g + \omega_1{}^{g^2} + \ldots \omega_1{}^{g^{p-2}}) = -a^{-k}$$

and consequently

$$U = p - 1 - (a^{-1} + a^{-2} + \ldots + {}^{-p+2}) = p - 1 - (-1)$$
$$= p$$

Theorem IV. *The quantity of Theorem III, of which the square root is to be extracted, has the value p.*

§ 104. The resolvent 5) was $(p-1)$-valued, and consequently the preceding method furnished at once the complete solution of the cyclotomic equation. By the aid of resolvents with smaller numbers of values, the solution of the equation can be divided into its simplest component operations.

Suppose that p_1 is a prime factor of $p-1$, and that $p-1 = p_1 q_1$. We form then the resolvent

$$(\omega + a_1 \omega^g + a_1{}^2 \omega^{g^2} + \ldots + a_1{}^{p-2} \omega^{g^{p-2}})^{p_1},$$

where a_1 is a primitive root of the equation

$$z^{p_1} - 1 = 0.$$

The values $a_1, a_1{}^2, \ldots a_1{}^{p_1}$ are all different, and the higher powers of a_1 take the same values again. It follows that, if

$$\varphi_0 = \omega \quad + \omega^{g^{p_1}} \quad + \omega^{g^{2p_1}} \quad + \ldots + \omega^{g^{(q_1-1)p_1}},$$
$$\varphi_1 = \omega^g \quad + \omega^{g^{p_1+1}} + \omega^{g^{2p_1+1}} + \ldots + \omega^{g^{(q_1-1)p_1+1}},$$
$$\cdot \quad \cdot \quad \cdot \quad \cdot \quad \cdot \quad \cdot \quad \cdot \quad \cdot \quad \cdot$$
$$\varphi_{p_1-1} = \omega^{g^{p_1-1}} + \omega^{g^{2p_1-1}} + \omega^{g^{3p_1-1}} + \ldots + \omega^{g^{p_1 q_1 - 1}},$$

the resolvent above can be written

$$(\varphi_0 + a_1 \varphi_1 + a_1{}^2 \varphi_2 + \ldots + a_1{}^{p_1-1} \varphi_{p_1-1})^{p_1},$$

or, again in Jacobi's notation,

$$(a_1, \varphi)^{p_1}.$$

By the same method as before we can show that this resolvent is unchanged if ω is replaced by ω^g, that is, that it belongs to the group of 1) and is consequently a rational function of a_1 and of the

THE CYCLOTOMIC EQUATIONS.   185

coefficients of 1). We denote its value by $N_1 = \nu_1^{p_1}$, and have accordingly

$$(a_1, \varphi) = \nu_1.$$

If then we write precisely as before,

$$(a_1^\lambda, \varphi)(a_1, \varphi)^{p_1-1-\lambda} = N_\lambda,$$

it appears that N_λ is rationally known, and that

$$\varphi_0 = \frac{1}{p_1}\left[-1 + \nu_1 + \frac{N_2}{N_1}\nu_1^2 + \ldots\right],$$

$$\varphi_1 = \frac{1}{p_1}\left[-1 + a_1^{-1}\nu_1 + a_1^{-2}\frac{N_2}{N_1}\nu_1^2 + \ldots\right],$$

. . . .

These several functions are all unchanged if ω is replaced by $\omega^{g^{p_1}}$, that is, they are unchanged by the subgroup

$$s^{p_1}, s^{2p_1}, s^{3p_1}, \ldots s^{q_1 p_1}.$$

We have therefore

Theorem V. *The p_1-valued resolvents $\varphi_0, \varphi_1, \ldots \varphi_{p_1-1}$ of the cyclotomic equation belong to the group formed by the powers of s^{p_1}. They can be obtained by determining a primitive root of $z^{p_1} - 1 = 0$ and extracting the p_1^{th} root of a quantity which is then rationally known.*

If p_2 is a second prime factor of $p-1$, and if $p-1 = p_1 p_2 q_2$, then the resolvent

$$(\omega + a_2 \omega^{g^{p_1}} + a_2^2 \omega^{g^{2p_1}} + \ldots + a_2^{q_1-1} \omega^{g(q_1-1)p_1})^{p_2},$$

in which a_2 is a primitive p_2^{th} root of unity, is reducible to the form

$$(\chi_0 + a_2 \chi_1 + a_2^2 \chi_2 + \ldots + a_2^{p_2-1} \chi_{p_2-1})^{p_2} = (a_2, \chi)^{p_2},$$

where

$$\chi_0 = \omega + \omega^{g^{p_1 p_2}} + \omega^{g^{2p_1 p_2}} + \ldots,$$
$$\chi_1 = \omega^{g^{p_1}} + \omega^{g^{p_1(p_2+1)}} + \omega^{g^{p_1(2p_2+1)}} + \ldots,$$

. . . . .

This resolvent is unchanged if ω is replaced by $\omega^{g^{p_1}}$, that is, it is unchanged by the substitutions $s^{p_1}, s^{2p_1} \ldots$ Consequently it can be rationally expressed in terms of φ_0, if a_2 is regarded as known. Again if we write

$$(a_2^\lambda, \chi)(a_2, \chi)^{p_2-1-\lambda} = M_\lambda, \quad M_1 = \mu_1^{p_2},$$

then M_λ is also a rational function of φ_0, and we have

$$\chi_0 = \frac{1}{p_2}\left[-\varphi_0 + \mu_1 + \frac{M_2}{M_1}\mu_1^2 + \ldots\right],$$

$$\chi_1 = \frac{1}{p_2}\left[-\varphi_0 + a_2^{-1}\mu_1 + a_2^{-2}\frac{M_2}{M_1}\mu_1^2 + \ldots\right],$$

. . . .

Theorem VI. *The $p_1 p_2$-valued resolvents $\chi_0, \chi_1, \ldots \chi_{p_1 p_2 - 1}$ of the cyclotomic equation belong to the group formed by the powers of $s^{p_1 p_2}$. They can be obtained by determining a primitive root of $z^{p_2} - 1 = 0$ and extracting a p_2^{th} root of a quantity which is rational in this primitive root and in φ_0.*

§ 165. By continuing the same process we have finally

Theorem VII. *If $p - 1 = p_1 p_2 p_3 \ldots$, the solution of the cyclotomic equation for the prime number p requires the determination of a primitive root of each of the equations*

$$z^{p_1} - 1 = 0, \quad z^{p_2} - 1 = 0, \quad z^{p_3} - 1 = 0, \ldots$$

and the extraction successively of the p_1^{th}, p_2^{th}, p_3^{th}, $\ldots$ roots of expressions each of which is rationally known in terms of the preceding known quantities.

If φ_0 is given, $\varphi_1, \varphi_2, \ldots \varphi_{p_1} - 1$ are rationally known, since they all belong to the same group. Similarly the coefficients of

7) $\begin{cases} (x-\omega)(x-\omega^{g^{p_1}})(x-\omega^{2p_1})\ldots(x-\omega^{g^{(q_1-1)p_1}}) = 0, \\ (x-\omega^g)(x-\omega^{g^{p_1+1}})(x-\omega^{g^{2p_1+1}})\ldots(x-\omega^{g^{(q_1-1)p_1+1}}) = 0, \\ \cdot \quad \cdot \quad \cdot \quad \cdot \quad \cdot \quad \cdot \quad \cdot \quad \cdot \end{cases}$

are all rationally known. Accordingly after the process of Theorem V has been carried out, the equation 1) breaks up into p_1 factors 7). Since the group belonging to each of these new equations is transitive in the corresponding roots, the factors 7) are again all irreducible, so long as only the coefficients of 1) and φ_0 are known.

THE CYCLOTOMIC EQUATIONS. 187

Again, after the process of Theorem VI has been carried out, χ_0 is known. All the values of this function belong to the same group and they are therefore all rational functions of χ_0. Similarly the coefficients of

$$8) \begin{cases} (x-\omega)(x-\omega^{g^{p_1 p_2}})(x-\omega^{g^{2p_1 p_2}})\ldots(x-\omega^{g^{(q_2-1)p_1 p_2}})=0 \\ (x-\omega^g)(x-\omega^{g^{p_1 p_2}+1})(x-\omega^{g^{2p_1 p_2}+1})\ldots(x-\omega^{g^{(q_2-1)p_1 p_2}+1})=0 \\ \cdot\quad\cdot\quad\cdot\qquad\cdot\qquad\cdot\qquad\cdot\qquad\cdot\qquad\cdot\qquad\cdot\qquad\cdot\qquad\cdot \end{cases}$$

are rationally known. The equations 7) are therefore now reducible, and each of them resolves into p_2 factors 8), which are again irreducible within the domain defined by χ_0. We can proceed in this way until we arrive at equations of the first degree.

§ 166. The particular case for which the prime factors of $p-1$ are all equal to 2 is of especial interest.

Theorem VIII. *If $2^m + 1$ is a prime number p, the cyclotomic equation belonging to p can be solved by means of a series of m quadratic equations. In this case the regular polygon of $p = 2^m + 1$ sides can be constructed by means of ruler and compass.*

In fact, for one root of the cyclotomic equation we have

$$\omega = \cos\frac{2\pi}{p} + i\sin\frac{2\pi}{p}, \quad \omega^{-1} = \cos\frac{2\pi}{p} - i\sin\frac{2\pi}{p},$$

$$\omega + \omega^{-1} = 2\cos\frac{2\pi}{p},$$

and consequently the angle $\dfrac{2\pi}{p}$ can be constructed with ruler and compass.

In order that $2^m + 1$ may be a prime number, it is necessary that $m = 2^\mu$. For if $m = 2^\mu m_1$, where m_1 is odd, then $2^m + 1 = (2^{2^\mu})^{m_1} + 1$ would be divisible by $2^{2^\mu} + 1$. If

$$\mu = 0, 1, 2, 3, 4,$$

the values of p are actually prime numbers

$$p = 3, 5, 17, 257, 65537,$$

and in these cases the corresponding regular polygons can therefore be constructed. But for $\mu = 5$ we have

$$2^{2^5} + 1 = 4294967297 = 641 \cdot 6700417,$$

so that it remains uncertain whether the form $2^{2^\mu}+1$ furnishes an infinite series of prime numbers.*

§ 167. We add the actual geometrical constructions for the cases $p=5$ and $p=17$.

For $p=5$, we take for a primitive root $g=2$, and obtain accordingly
$$g^0=1, \quad g^1=2, \quad g^2=4, \quad g^3=3 \quad (\text{mod. } 5).$$
Consequently
$$\varphi_0 = \omega + \omega^4, \quad \varphi_1 = \omega^2 + \omega^3$$
$$\varphi_0 + \varphi_1 = -1, \quad \varphi_0\varphi_1 = \varphi_0 + \varphi_1 = -1,$$
$$\varphi^2 + \varphi - 1 = 0,$$
$$\varphi_0 = \frac{-1+\sqrt{5}}{2}, \quad \varphi_1 = \frac{-1-\sqrt{5}}{2}.$$

If ω^2 is substituted for ω, the values φ_0 and φ_1 are interchanged. The algebraic sign of $\sqrt{5}$ is therefore undetermined unless a particular choice of ω is made. If we prescribe that
$$\omega = \cos\frac{2\pi}{5} + i\sin\frac{2\pi}{5},$$
then
$$\varphi_0 = \omega + \omega^4 = 2\cos\frac{2\pi}{5}, \quad \varphi_1 = \omega^2 + \omega^3 = 2\cos\frac{4\pi}{5},$$
consequently $\varphi_0 > 0$, $\varphi_1 < 0$, and the $\sqrt{5}$ in the expression above is to be taken positive. Furthermore
$$\chi_0 = \omega, \quad \chi_1 = \omega^4; \quad \chi_2 = \omega^2, \quad \chi_3 = \omega^3$$
$$\chi_0 + \chi_1 = \varphi_0, \quad \chi_0\chi_1 = 1,$$
$$\chi^2 - \varphi_0\chi + 1 = 0;$$
$$\chi_0 = \omega = \frac{-1+\sqrt{5}+i\sqrt{10+2\sqrt{5}}}{4},$$
$$\chi_1 = \omega^4 = \frac{-1+\sqrt{5}-i\sqrt{10+2\sqrt{5}}}{4},$$

---

*Cf. Gauss: Disquisit. arithm., § 362. The statement there made that Fermat supposed all the numbers $2^{2^\mu}+1$ to be prime, is corrected by Baltzer: Crelle 87, p. 172. At present the following exceptional cases are known:

$2^{2^5}+1$ divisible by 641......... (Landry),
$2^{2^{12}}+1$ divisible by 114689...... (J. Pervouchine),
$2^{2^{23}}+1$ divisible by 167772161... (J. Pervouchine; É. Lucas),
$2^{2^{36}}+1$ divisible by 274877906945 (P. Seelhoff).

Cf. P. Seelhoff: Schlömilch Zeitschrift, XXXI. pp. 172-4.

the sign of i being so taken that the imaginary part of ω is positive and that of ω^4 negative.

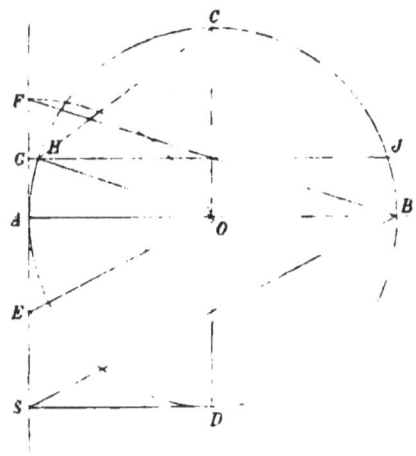

For the construction of the regular pentagon it is sufficient to know the resolvent $\varphi_0 = 2\cos\dfrac{2\pi}{5}$.

Suppose a circle of radius 1 to be described about O as a center. On the tangent at the extremity of the horizontal radius OA a distance $AE = \tfrac{1}{2}AO = \tfrac{1}{2}$ is laid off. Then

$$OE = \sqrt{1+\tfrac{1}{4}} = \dfrac{\sqrt{5}}{2}.$$

If now we take $EF = EO$, we have

$$AF = EO - EA = \dfrac{\sqrt{5}-1}{2} = \varphi_0,$$

$$AF = 2\cos\dfrac{2\pi}{5}.$$

Finally if AF is bisected in G and GHJ drawn to OA and OC _ to HJ, then $HOC = COJ = \dfrac{2\pi}{5}$, since $\cos HOC = AC$ $= \cos\dfrac{2\pi}{5}$. H, C, and J are therefore three successsive vertices of a regular pentagon.

§ 168. For $p = 17$ we take for a primitive root $g = 6$. Accordingly we have

$g^0, g^1, g^2, g^3, g^4, g^5, g^6, g^7, g^8, g^9, g^{10}, g^{11}, g^{12}, g^{13}, g^{14}, g^{15}, g^{16};$
1, 6, 2, 12, 4, 7, 8, 14, 16, 11, 15, 5, 13, 10, 9, 3, 1;

$$\varphi_0 = \omega + \omega^2 + \omega^4 + \omega^8 + \omega^{16} + \omega^{15} + \omega^{13} + \omega^9,$$
$$\varphi_1 = \omega^6 + \omega^{12} + \omega^7 + \omega^{14} + \omega^{11} + \omega^5 + \omega^{10} + \omega^3;$$
$$\varphi_0 + \varphi_1 = -1.$$

To find $\varphi_0\varphi_1$, we multiply every term of φ_0 into the kth term to the right of that immediately below it in φ_1, taking successively $k = 0, 1, \ldots$ We obtain then

$$\varphi_0\varphi_1 = \varphi_1 + \varphi_0 + \varphi_0 + \varphi_0 + \varphi_1 + \varphi_1 + \varphi_1 + \varphi_0 = 4(\varphi_0 + \varphi_1) = -4.$$

Consequently

$$\varphi_0 + \varphi_1 = -1, \quad \varphi_0\varphi_1 = -4$$
$$\varphi^2 + \varphi - 4 = 0,$$
$$\varphi_0 = \frac{-1 + \sqrt{17}}{2}, \quad \varphi_1 = \frac{-1 - \sqrt{17}}{2},$$

where the sign of $\sqrt{17}$ is undetermined until the particular root ω is specified. If we take

$$\omega = \cos\frac{2\pi}{17} + i\sin\frac{2\pi}{17},$$

we have for the determination of the sign

$$\varphi_1 = (\omega^3 + \omega^{14}) + (\omega^5 + \omega^{12}) + (\omega^6 + \omega^{11}) + (\omega^7 + \omega^{10})$$
$$= 2\left[\cos\frac{6\pi}{17} + \cos\frac{10\pi}{17} + \cos\frac{12\pi}{17} + \cos\frac{14\pi}{17}\right]$$
$$= 2\left[\cos\frac{6\pi}{17} - \cos\frac{7\pi}{17} - \cos\frac{5\pi}{17} - \cos\frac{3\pi}{17}\right],$$
$$\therefore \varphi_1 < 0,$$

and $\sqrt{17}$ above must be taken positive.

We have further

$$\chi_0 = \omega + \omega^4 + \omega^{16} + \omega^{13}, \quad \chi_1 = \omega^2 + \omega^8 + \omega^{15} + \omega^9;$$
$$\chi_2 = \omega^6 + \omega^7 + \omega^{11} + \omega^{10}, \quad \chi_3 = \omega^{12} + \omega^{14} + \omega^5 + \omega^3;$$
$$\chi_0 + \chi_1 = \varphi_0, \quad \chi_2 + \chi_3 = \varphi_1,$$
$$\chi_0\chi_1 = \chi_3 + \chi_1 + \chi_0 + \chi_2 = -1, \quad \chi_2\chi_3 = \chi_0 + \chi_3 + \chi_2 + \chi_1 = -1;$$
$$\chi^2 - \varphi_0\chi - 1 = 0, \quad \chi^2 - \varphi_1\chi - 1 = 0;$$
$$\chi_0, \chi_1 = \frac{\varphi_0}{2} \pm \sqrt{\frac{\varphi_0^2 + 4}{4}}, \quad \chi_2, \chi_3 = \frac{\varphi_1}{2} \pm \sqrt{\frac{\varphi_1^2 + 4}{4}}.$$

The algebraic signs are again easily determined. We have

THE CYCLOTOMIC EQUATIONS.

$$\chi_0 = (\omega + \omega^{16}) + (\omega^4 + \omega^{13}) = 2\left(\cos\frac{2\pi}{17} + \cos\frac{8\pi}{17}\right) > 0,$$

$$\chi_2 = (\omega^6 + \omega^{11}) + (\omega^7 + \omega^{10}) = 2\left(\cos\frac{12\pi}{17} + \cos\frac{14\pi}{17}\right) < 0,$$

Consequently

$$\chi_0 = \frac{\varphi_0}{2} + \sqrt{\frac{\varphi_0^2}{4} + 1}, \quad \chi_2 = \frac{\varphi_0}{2} - \sqrt{\frac{\varphi_0^2}{4} + 1};$$

$$\chi_1 = \frac{\varphi_1}{2} - \sqrt{\frac{\varphi_1^2}{4} + 1}, \quad \chi_3 = \frac{\varphi_1}{2} + \sqrt{\frac{\varphi_1^2}{4} + 1}.$$

With χ_0 as a basis we proceed further:

$$\psi_0 = \omega + \omega^{16}, \quad \psi_1 = \omega^4 + \omega^{13},$$

$$\psi_0 + \psi_1 = \chi_0, \quad \psi_0\psi_1 = \chi_3 = \frac{\varphi_1}{2} + \sqrt{\frac{\varphi_1^2}{4} + 1},$$

$$\psi_0^2 - \chi_0\psi + \chi_3 = 0,$$

$$\psi_0, \psi_1 = \frac{\chi_0}{2} \pm \sqrt{\frac{\chi_0^2}{4} - \chi_3}.$$

Since now $\psi_0 = 2\cos\frac{2\pi}{17}$, $\psi_1 = 2\cos\frac{8\pi}{17}$, we have $\psi_0 > \psi_1$, and therefore

$$\psi_0 = \frac{\chi_0}{2} + \sqrt{\frac{\chi_0^2}{4} - \chi_3}, \quad \psi_1 = \frac{\chi_0}{2} - \sqrt{\frac{\chi_0^2}{2} - \chi_3}.$$

These results suffice for the construction of the regular polygon of 17 sides. Suppose a circle of radius 1 to be described about O as a center, and a tangent to be drawn at the extremity of the horizontal radius OA. On the tangent take a length $AE = \frac{1}{4}OA = \frac{1}{4}$; then

$$OE = \sqrt{1 + \tfrac{1}{16}} = \frac{\sqrt{17}}{4}.$$

Further, if $EF = EF' = EO$, we have

$$AF = \frac{\sqrt{17} - 1}{4} = \frac{\varphi_0}{2}, \quad AF' = \frac{\sqrt{17} + 1}{4} = -\frac{\varphi_1}{2};$$

$$OF = \sqrt{\frac{\varphi_0^2}{4}+1}, \qquad OF'' = \sqrt{\frac{\varphi_1^2}{4}+1}.$$

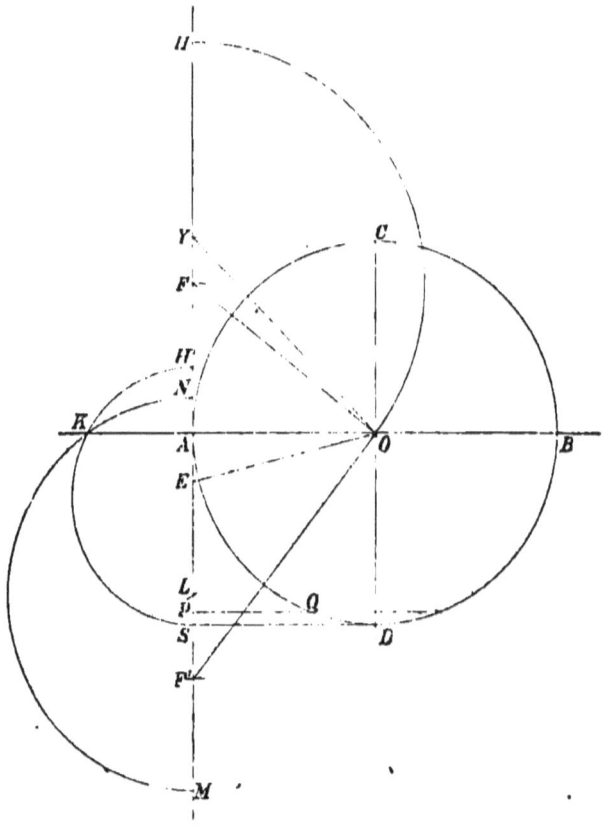

Taking then
$$FH = FO, \quad F'H' = F''O,$$
we have
$$AH = AF + FO = \frac{\varphi_0}{2} + \sqrt{\frac{\varphi_0^2}{4}+1} = \chi_0,$$
$$AH' = -AF' + F'O = \frac{\varphi_1}{2} + \sqrt{\frac{\varphi_1^2}{4}+1} = \chi_3.$$

We bisect AH in Y; then
$$AY = \tfrac{1}{2}\chi_0.$$

We take now $AS = 1$, and describe a semicircle on $H'S$ as a diameter: if this meets the continuation of OA in K, then

$$AK^2 = AS \cdot AH' = \chi_3.$$

Again if we take $LK = AY$ and $KL = LM = LN$, and describe a circle of radius LK about L, we obtain

$$AN + AM = 2KL = 2AY = \chi_0 = \psi_0 + \psi_1$$
$$AN \cdot AM = AK^2 = AS \cdot AH' = \chi_3 = \psi_0 \psi_1.$$

The greater of the two lengths AN, AM is equal to ψ_0; we may write therefore

$$AM = \psi_0 = 2\cos\frac{2\pi}{17}.$$

If P is the middle point of AM, and if $QP \quad AO$, and OD AO, then Q and D are two successive vertices of the required regular polygon of 17 sides.

§ 169. We consider now the case $p_1 = 2$, under the assumption $p > 2$. If g is a primitive root, then

$$\varphi_0 = \omega + \omega^{g^2} + \omega^{g^4} + \ldots + \omega^{g^{p-3}},$$
$$\varphi_1 = \omega^g + \omega^{g^3} + \omega^{g^5} + \ldots + \omega^{g^{p-2}},$$
$$\varphi_0 + \varphi_1 = -1.$$

To determine $\varphi_0 \varphi_1$ we use the same method as in the preceding Sections, and obtain

$$\varphi_0 \varphi_1 = [\omega^{g+1} + \omega^{g^3 + g^2} + \ldots + \omega^{g^{p-2} + g^{p-3}}]$$
$$+ [\omega^{g^3 + 1} + \omega^{g^5 + g^2} + \ldots + \omega^{g + g^{p-3}}]$$
$$+ \ldots$$

The exponents which occur here in any bracket

$$g^{2a+1} + 1, \quad g^2(g^{2a+1} + 1), \quad g^4(g^{2a+1} + 1), \ldots$$

are plainly either all quadratic remainders, or all quadratic non-remainders, or all equal to 0. In the first case the value of the corresponding bracket is φ_0, in the second case φ_1, and in the third case $\dfrac{p-1}{2}$.

13

Consequently

S)
$$\begin{cases} \varphi_0\varphi_1 = m_1\varphi_0 + m_2\varphi_1 + m_3\dfrac{p-1}{2}, \\ m_1 + m_2 + m_3 = \dfrac{p-1}{2}, \end{cases}$$

where m_1, m_2, m_3 represent the number of brackets of the several kinds.

If $g^{2a+1} + 1 \equiv 0 \pmod{p}$, then $2(2a+1) \equiv p-1$ and accordingly $\dfrac{p-1}{2}$ is an odd number. The third case occurs therefore only when $p = 4k+3$, and then only once, viz. for $a = \dfrac{p-3}{4}$. Consequently $m_3 = 0$ or $m_3 = 1$ according as $\dfrac{p-1}{2}$ is even or odd.

Since $\varphi_0\varphi_1$ is rational and integral in the coefficients of the cyclotomic equation 1), this product is an integer, and we may therefore take

$$\varphi_0\varphi_1 - m_3\frac{p-1}{2} = n = -n(\varphi_0 + \varphi_1)$$

where n is an integer. It follows then from S) that

$$(m_1 + n)\varphi_0 + (m_2 + n)\varphi_1 = 0$$

In this equation all the powers of ω can be reduced to powers lower than the pth, and the equation can then be divided by ω. The resulting equation is then of degree $p-2$ at the highest, and still has the root ω in common with the equation 1) of degree $p-1$. It is therefore an identity and

$$m_1 = m_2 = -n.$$

Consequently we have for the values of m_1, m_2 and φ_0, φ_1

$$m_1 = m_2 = \frac{p-1}{4}, \quad \varphi_0\varphi_1 = -\frac{p-1}{4}, \qquad \left(\frac{p-1}{2}\text{even}\right)$$

$$m_1 = m_2 = \frac{p-3}{4}, \quad \varphi_0\varphi_1 = \frac{p-1}{2} - \frac{p-3}{4} = \frac{p+1}{4}, \quad \left(\frac{p-1}{2}\text{odd}\right)$$

$$\varphi_0\varphi_1 = \frac{1-(-1)^{\frac{p-1}{2}}p}{4},$$

$$(\varphi - \varphi_0)(\varphi - \varphi_1) = \varphi^2 + \varphi + \frac{1 - (-1)^{\frac{p-1}{2}} p}{4},$$

$$\varphi_0 = \frac{-1 + \sqrt{(-1)^{\frac{p-1}{2}} p}}{2}, \quad \varphi_1 = \frac{-1 - \sqrt{(-1)^{\frac{p-1}{2}} p}}{2},$$

where the algebraic sign of the square root is necessarily undetermined so long as the particular value of ω is not specified.

§ 170. We consider now the two equations

$$z_0 \equiv (x - \omega)(x - \omega^{g^2})(x - \omega^{g^4}) \ldots (x - \omega^{g^{p-3}}) = 0,$$
$$z_1 \equiv (x - \omega^g)(x - \omega^{g^3})(x - \omega^{g^5}) \ldots (x - \omega^{g^{p-2}}) = 0.$$

The roots and consequently the coefficients of these equations are unchanged if ω is replaced by ω^{g^2}. If therefore, when the several factors are multiplied together, any coefficient contains a term $m\omega^a$, it must also contain terms $m\omega^{ag^2}$, $m\omega^{ag^4}$, ..., that is, it contains φ_0 or φ_1, according as a is a quadratic remainder or non-remainder (mod. p). Accordingly every coefficient will be of the form

$$m'\varphi_0 + m''\varphi_1 = \frac{a + b\sqrt{(-1)^{\frac{p-1}{2}} p}}{2},$$

and on introducing these values we have

$$z_0 = \frac{X + Y\sqrt{(-1)^{\frac{p-1}{2}} p}}{2},$$

where X and Y are integral functions of x with integral coefficients. Again since z_1 is obtained from z_0 by replacing ω by ω^g, or φ_0 by φ_1, that is, by a change of the sign of the square root, we have

$$z_1 = \frac{X - Y\sqrt{(-1)^{\frac{p-1}{2}} p}}{2}.$$

Hence

$$z_0 z_1 = \frac{x^p - 1}{x - 1} = \frac{X^2 - (-1)^{\frac{p-1}{2}} p Y^2}{2},$$

and we have

Theorem IX. *For every prime number p*

$$4\left(\frac{x^p-1}{x-1}\right) = 4(x^{p-1}+x^{p-2}+\ldots+x+1) = X^2 - (-1)^{\frac{p-1}{2}} p\, Y^2,$$

where X and Y are integral functions of x with integral coefficients. *

---

* The extensive literature belonging to this Chapter is found in Bachmann: Lehre von der Kreistheilung, Leipzig, Teubner, 1872. The present treatment follows in part the method there employed. The two figures are taken from Bachmann's work.

CHAPTER XI.

THE ABELIAN EQUATIONS.

§ 171. The cyclotomic equation has the property that it is irreducible and that every one of its roots is a rational function of every other one. We turn now to the treatment of the more general problem of those irreducible equations of which one root x'_1 is a rational function of another root x_1. Among these equations the cyclotomic equation is evidently included as a special case.

Suppose that

1) $$f(x) = 0$$

is the given irreducible equation, and that two of the roots x'_1 and x_1 are connected by the relation

2) $$x'_1 = \theta(x_1),$$

where θ is a rational function. Then

$$f(x_1) = 0, \quad f[\theta(x_1)] = 0,$$

so that the irreducible equation 1) has the root x_1, and consequently all its roots, in common with the equation

3) $$f[\theta(x)] = 0.$$

In particular $x'_1 = \theta(x_1)$ must be a root of 3), so that

$$f\{\theta[\theta(x_1)]\} = 0.$$

Consequently $\theta[\theta(x_1)]$ is a root of 1) and therefore of 3). Then $\theta\{\theta[\theta(x_1)]\}$ is a root of 1), and so on. With the notation

$$\theta[\theta(x)] = \theta^2(x)], \quad \theta[\theta^2(x)] = \theta^2[\theta(x)] = \theta^3(x), \ldots,$$

it appears that all the members of the infinite series

$$x_1, \ \theta(x_1), \ \theta^2(x_1), \ \theta^3(x_1), \ \ldots \ \theta^k(x_1), \ \ldots$$

are roots of the equation 1). Since however the latter has only a finite number of roots it follows from a familiar process of reasoning that there must be in the series a function $\theta^m(x_1)$ which is equal to the initial value x_1, while all the preceding functions

$$x_1, \theta(x_1), \theta^2(x_1), \ldots \theta^{m-1}(x_1)$$

are different from one another. The continuation of the series then reproduces these same values in the same order, so that only

$$\theta^m(x_1) = \theta^{2m}(x_1) = \ldots$$

take the initial value x_1, and that for $k < m$ only

$$\theta^{m+k}(x_1) = \theta^{2m+k}(x_1) = \ldots$$

are equal to $\theta^k(x_1)$.

If the system of m roots thus obtained does not include all the roots of the equation 1), suppose that x_2 is any remaining root. Then x_2 also satisfies 3), and therefore $\theta(x_2)$ is a root of 1), and so on. Accordingly we have now a new system of μ different roots

$$x_2, \theta(x_2), \theta^2(x_2), \ldots \theta^{\mu-1}(x_2).$$

But since the equations

4) $\qquad \theta^m(y) - y = 0 \quad \theta^\mu(z) - z = 0$

have each one root $y = x_1$, $z = x_2$ in common with the irreducible equation 1), they are satisfied by all the roots of the latter. The former equation of 4) is therefore satisfied by x_2, the latter by x_1, consequently m is a multiple of μ and μ is a multiple of m, i. e. $m = \mu$.

Again all the roots of the second series are different from those of the first. For if

$$\theta^b(x_2) = \theta^a(x_1) \quad (a, b < m),$$

then, on applying the operation θ^{m-b}, we should have

$$\theta^m(x_2) = x_2 = \theta^{m+a-b}(x_1),$$

and x_2 would be contained in the first series, which is contrary to assumption.

If there is another root x_3, not included among the $2m$ already found, the same reasoning applies again. We have therefore

Theorem I. *If one root of an irreducible equation $f(x) = 0$ is a rational function of another, then the roots can be divided into ν systems of m roots each, as in the following table:*

$$5) \quad \begin{cases} x_1, & \theta(x_1), & \theta^2(x_1), & \ldots & \theta^{m-1}(x_1), \\ x_2, & \theta(x_2), & \theta^2(x_2), & \ldots & \theta^{m-1}(x_2), \\ \cdot & \cdot & \cdot & \cdot & \cdot \\ x_\nu, & \theta(x_\nu), & \theta^2(x_\nu), & \ldots & \theta^{m-1}(x_\nu). \end{cases}$$

The function θ is such that for every root x_a

$$\theta^m(x_a) = x_a,$$

and the equation $f(x) = 0$ is of degree $m\nu$.

§ 172. We can now determine the group of the equation 1). Since the equation is irreducible, its group is transitive (§ 156); it therefore contains at least one substitution which replaces x_1 by any arbitrary x_a. It follows then that all the roots of the first line of 5) are replaced by those of the ath line. The group of 1) is therefore non-primitive and has ν systems of non-primitivity of m elements each. The number of admissible permutations of the ν systems is not as yet determinate; in the most general case there are $\nu!$ of them. If any x_a is replaced by $\theta^\lambda(x_a)$, then every $\theta^k(x_a)$ is replaced by $\theta^{k+\lambda}(x_a)$; there are therefore m possible substitutions within the single system. The order of the group of 1) is therefore a multiple of m^ν and a divisor of $\nu!m^\nu$.

Theorem II. *The group of the equation* 1) *is non-primitive. It contains ν systems of non-primitivity, which correspond to the several lines of* 5). *The order of the group is*

$$r = r_1 m^\nu,$$

where r_1 is a divisor of $\nu!$.

§ 173. In the following treatment we employ again the notation of Jacobi

$$z_0 + \omega z_1 + \omega^2 z_2 + \ldots + \omega^{m-1} z_{m-1} = (\omega, z),$$

where ω is a root of the equation $\omega^m - 1 = 0$. Similarly we write

$$x_a + \omega \theta(x_a) + \omega^2 \theta^2(x_a) + \ldots + \omega^{m-1} \theta^{m-1}(x_a) = (\omega, \theta(x_a)).$$

We form then the following resolvents:

$$\varphi_1 = (1, \theta(x_1)), \quad \varphi_2 = (1, \theta(x_2)), \ldots \varphi_\nu = (1, \theta(x_\nu)).$$

φ_1 is symmetric in the elements of the first system and is changed in value only when the first system is replaced by another; it is therefore a ν-valued function, its values being $\varphi_1, \varphi_2, \ldots \varphi_\nu$.

Every symmetric function of the φ's is a rational function of the coefficients of 1). The quantities

$$S_1(\varphi_1) = \varphi_1 + \varphi_2 + \ldots + \varphi_\nu,$$
$$S_2(\varphi_1\varphi_2) = \varphi_1\varphi_2 + \varphi_1\varphi_3 + \ldots + \varphi_{\nu-1}\varphi_\nu,$$
$$\cdots \cdots \cdots \cdots \cdots$$

that is, the coefficients of the equation of the νth degree

6) $\qquad \varphi^\nu - S_1\varphi^{\nu-1} + S_2\varphi^{\nu-2} - \ldots \pm S_\nu = 0,$

of which $\varphi_1, \varphi_2, \ldots \varphi_\nu$ are the roots, are therefore rationally known.

Theorem III. *The resolvent*

$$(1, \theta(x_1)) = x_1 + \theta(x_1) + \theta^2(x_1) + \ldots + \theta^{m-1}(x_1)$$

is a root of an equation of degree ν, the coefficients of which are rationally known in terms of the coefficients of the equation $f(x) = 0$.

§ 174. The equation 6) has no *affect* (§ 153), unless further relations are explicitly assigned among the roots $x_1, x_2, \ldots x_n$. If however any root φ_a of 6) has in any way been determined, the values of the corresponding $x_a, \theta(x_a), \ldots$ can be obtained algebraically by exactly the same method as that employed in the preceding Chapter.

Thus, the equation θ) of which the roots are $x, \theta(x), \theta^2(x), \ldots$ $\theta^{m-1}(x)$ is irreducible. Its group consists of the powers of the substitution $(1\,\theta\,\theta^2 \ldots \theta^{m-1})$. And if we write

$$(\omega, \theta(x))^m = T_1,$$

where ω is now assumed to be a primitive m^{th} root of unity, we have

$$[\theta(x) + \omega\,\theta^2(x) + \ldots + \omega^{m-1}\theta^m(x)]^m = \omega^m[\theta(x) + \omega\theta^2(x) + \ldots \omega^{m-1}\theta^m(x)]$$
$$= (\omega, \theta(x))^m,$$

that is T_1 is unchanged by the substitution $(1\,\theta\,\theta^2 \ldots \theta^{m-1})$. Consequently T_1 is a rational function of the coefficients of θ) and of the primitive m^{th} root of unity ω. The mth root of this known quantity T_1 we denote by τ_1. . Again if we write

$$(\omega^\lambda, \theta(x))\,(\omega, \theta(x))^{m-\lambda} = T_\lambda,$$

it can be shown by the same method that we have already frequently employed that T_1 is also rational in ω and the coefficients of θ). Taking successively $\lambda = 0, 1, 2, \ldots m-1$, and combining the resulting equations, we have then

$$mx = \varphi + \tau_1 + \frac{T_2}{T_1}\tau_1^2 + \frac{T_3}{T_1}\tau_1^3 + \ldots + \frac{T_{m-1}}{T_1}\tau_1^{m-1},$$

$$m\,\theta(x) = \varphi + \omega\tau_1 + \omega^2\frac{T_2}{T_1}\tau_1^2 + \omega^3\frac{T_3}{T_1}\tau_1^3 + \ldots + \omega^{m-1}\frac{T_{m-1}}{T_1}\tau_1^{m-1},$$

$$m\,\theta^2(x) = \varphi + \omega^2\tau_1 + \omega^4\frac{T_2}{T_1}\tau_1^2 + \omega^6\frac{T_3}{T_1}\tau_1^3 + \ldots + \omega^{2(m-1)}\frac{T_{m-1}}{T_1}\tau_1^{m-1},$$

. . . .

The function τ_1 being m-valued, x also admits only m values, and these coincide with $x, \theta(x), \theta^2(x), \ldots$. If any other m^{th} root of T_1 is substituted for τ_1, the m values of x are permuted cyclically.

Theorem IV. *If the m roots of an equation of degree m are*

$$x_1, \theta(x_1), \theta^2(x_1), \ldots \theta^{m-1}(x_1),$$

where $\theta(x)$ is a rational function for which $\theta^m(x_1) = x_1$, then the solution of the equation requires only the determination of a primitive root of $z^m - 1 = 0$ and the extraction of the mth root of a known quantity.

Theorem V. *If one root of an equation of prime degree is a rational function of another root, the equation can be solved algebraically.*

For in this case we have $m\nu = p$ and $m > 1$; consequently $m = p$ and $\nu = 1$.

§ 175. If all the coefficients of $f(x)$ are real, the process of the preceding Section admits of further reduction. The quantity

$$T_1 = (\omega, \theta(x))^m = \rho(\cos\vartheta + i\sin\vartheta)$$

can be rationally expressed in terms of ω and the coefficients of f. The latter being real, the occurrence of $i = \sqrt{-1}$ in T_1 is due entirely to the presence of ω. Consequently,

$$T_1' = (\omega^{-1}, \theta(x))^m = \rho(\cos\vartheta - i\sin\vartheta), \quad T_1 T_1' = \rho^2,$$
$$\sqrt{T_1 T_1'} = \sqrt{\rho^2} = (\omega, \theta(x))(\omega^{-1}, \theta(x)) = U,$$

where U is again rationally known, since it is unchanged by the group of φ. We have then.

$$\tau_1 = \sqrt[m]{U}\left(\cos\frac{\vartheta + 2k\pi}{m} + i\sin\frac{\vartheta + 2k\pi}{m}\right).$$

Theorem VI. *If all the coefficients of $f(x)$ are real, the second operation of Theorem IV can be replaced by the extraction of a square root of a known quantity and the division of a known angle into m equal parts.*

§ 176. If the m of the Theorem IV is a compound number, the solution can be divided into a series of steps by the aid of special resolvents. If $m = m_1 m_1'$, where m_1 is any arbitrary factor of m, we take

$$\psi_1 = x_1 + \theta^{m_1}(x_1) + \theta^{2m_1}(x_1) + \ldots + \theta^{(m_1'-1)m_1}(x_1),$$
$$\psi_2 = \theta(x_1) + \theta^{m_1+1}(x_1) + \theta^{2m_1+1}(x_1) + \ldots + \theta^{(m_1'-1)m_1+1}(x_1),$$
$$\cdot \cdot \cdot \cdot \cdot \cdot \cdot \cdot \cdot \cdot \cdot \cdot \cdot \cdot \cdot \cdot \cdot \cdot \cdot \cdot$$
$$\psi_{m_1} = \theta^{m_1-1}(x_1) + \theta^{2m_1-1}(x_1) + \theta^{3m_1-1}(x_1) + \ldots + \theta^{m_1'm_1-1}(x_1),$$

and consider the resolvent

$$[x_1 + a_1\theta(x_1) + a_1^2\theta^2(x_1) + \ldots + a_1^{m-1}\theta^{m-1}(x_1)]^{m_1}$$

in which a_1 is a primitive m_1^{th} root of unity. This resolvent is equal to

$$(\psi_1 + a_1\psi_2 + a_1^2\psi_3 + \ldots + a_1^{m_1-1}\psi_{m_1})^{m_1}.$$

If x_1 is replaced by $\theta(x_1)$, then $\psi_1, \psi_2, \ldots \psi_{m_1}$ are cyclically permuted, and the resolvent is unchanged. It can therefore be rationally expressed in terms of a_1 and known quantities. We denote this expression by N_1 and its m_1^{th} root by ν_1, so that

$$\psi_1 + a_1\psi_2^2 + a_1^2\psi_3 + \ldots + a_1^{m_1-1}\psi_{m_1} = \nu_1.$$

If then, as before, we write

$$(\psi_1 + a_1^\lambda \psi_2 + a_1^{2\lambda}\psi_3 + \ldots)(\psi_1 + a_1\psi_2 + a_1^2\psi_3 + \ldots)^{m_1-\lambda} = N_\lambda,$$

it appears that N_λ is rationally known, and that

$$m_1\psi_1 = \varphi_1 + \nu_1 + \frac{N_2}{N_1}\nu_1^2 + \frac{N_3}{N_1}\nu_1^3 + \ldots,$$

$$m_1\psi_2 = \varphi_1' + a_1^{-1}\nu_1 + a_1^{-2}\frac{N_2}{N_1}\nu_1^2 + a_1^{-3}\frac{N_3}{N_1}\nu_1^3 + \ldots,$$

THE ABELIAN EQUATIONS.   203

Theorem VII. *The m_1-valued resolvent ψ_1 can be obtained by determining a primitive root of $z^{m_1} - 1 = 0$ and extracting the m_1^{th} root of a known quantity.*

By continuing the same process, we obtain

Theorem VIII. *If the roots of an equation of the mth degree are*

$$x_1, \theta(x_1), \theta^2(x_1), \ldots \theta^{m-1}(x_1), \quad [\theta^m(x_1) = 1]$$

and if $m = m_1 m_2 m_3 \ldots$, the solution of the equation requires only the determination of a primitive root of each of the equations

$$z^{m_1} - 1 = 0, \ z^{m_2} - 1 = 0, \ z^{m_3} - 1 = 0, \ldots$$

and the successive extraction of the $m_1^{th}, m_2^{th}, m_3^{th}, \ldots$ roots of expressions, each of which is rationally known in terms of the preceding results.

§ 177. The solution can also be accomplished by a still different method.

Suppose that $m = m_1 m_2 \ldots m_\omega = m_1 n_1 = m_2 n_2 = \ldots = m_\omega n_\omega$. Then we can form the following equations:

A_1) $\begin{cases} g_1(x) = 0, \text{ with the roots } x_1, \theta^{n_1}(x_1), \theta^{2n_1}(x_1), \ldots \theta^{(m_1-1)n_1}(x_1), \\ \text{and with coefficients which are rational functions of a resolvent } \chi_1 = x_1 + \theta^{n_1}(x_1) + \ldots \quad \chi_1 \text{ is a root of an equation} \\ h_1(\chi) = 0 \text{ of degree } m_1. \end{cases}$

A_2) $\begin{cases} g_2(x) = 0, \text{ with the roots } x_1, \theta^{n_2}(x_1), \theta^{2n_2}(x_1), \ldots \theta^{(n_2-1)m_2}(x_1), \\ \text{and with coefficients which are rational functions of a resolvent } \chi_2 = x_1 + \theta^{n_2}(x_1) + \ldots \quad \chi_2 \text{ is a root of an equation} \\ h_2(\chi) = 0 \text{ of degree } m_2. \end{cases}$

. . . .

A_ω) $\begin{cases} g_\omega(x) = 0, \text{ with the roots } x_1, \theta^{m_\omega}(x_1), \theta^{2m_\omega}(x_1), \ldots \theta^{(n_\omega-1)m_\omega}(x_1), \\ \text{and with cofficients which are rational functions of a resolvent } \chi_\omega = x_1 + \theta^{m_\omega}(x_1) + \ldots \quad \chi_\omega \text{ is a root of an equation} \\ h_\omega(\chi) = 0 \text{ of degree } m_\omega. \end{cases}$

If now we select $m_1, m_2, \ldots m_\omega$ so that they are prime to each other, then the equations

$$g_1(x) = 0, \ g_2(x) = 0, \ldots g_\omega(x) = 0$$

have only the one root x_1 common to any two of them. By the method of the greatest common divisor this root can be rationally expressed in the coefficients of $g_1, g_2, \ldots g_\omega$, that is, in the coefficients of $f(x)$ and in $\chi_1, \chi_2, \ldots \chi_\omega$.

The solution of $f(x) = 0$ depends therefore on the determination of one root of each of the equations

$$h_1(\chi) = 0, \quad h_2(\chi) = 0, \ldots h_\omega(\chi) = 0$$

of degrees $m_1, m_2, \ldots m_\omega$, respectively. If

$$m = p_1^{a_1} p_2^{a_2} \ldots p_\omega^{a_\omega}$$

where $p_1, p_2, \ldots p_\omega$ are the different prime factors of m, then we are to take

$$m_1 = p^{a_1}, \quad m_2 = p^{a_2}, \ldots m_\omega = p_\omega^{a_\omega}.$$

If for any one of the equations $h_\lambda(\chi) = 0$ the exponent a_λ is greater than 1, then recourse must be had to the earlier method of solution to determine a χ_λ.

§ 178. In illustration of this type of equations we add the two following examples. In the one case we take

$$\theta(x) = \frac{a_1 x + \beta_1}{\gamma_1 x + \delta_1} = \frac{ax + \beta}{\gamma x + \delta},$$

where $a_1 = a, \beta_1 = \beta, \gamma_1 = \gamma, \delta_1 = \delta$ are real quantities. We assume that $a\delta - \beta\gamma \neq 0$; otherwise we should have

$$\theta(x) = \frac{\beta}{\delta} = \frac{a}{\gamma}.$$

The functions $\theta^2(x), \theta^3(x), \ldots$ will also be linear in x with real coefficients. We may write

$$\theta^2(x) = \frac{a_2 x + \beta_2}{\gamma_2 x + \delta_2}, \quad \theta^3(x) = \frac{a_3 x + \beta_3}{\gamma_3 x + \delta_3}, \ldots \theta^m(x) = \frac{a_m x + \beta_m}{\gamma_m x + \delta_m}.$$

Again

$$\theta^m(x) = \theta^{m-1}[\theta(x)] = \frac{a_{m-1}(a_1 x + \beta_1) + \beta_{m-1}(\gamma_1 x + \delta_1)}{\gamma_{m-1}(a_1 x + \beta_1) + \delta_{m-1}(\gamma_1 x + \delta_1)},$$

and a comparison of the two expressions for $\theta^m(x)$ shows that

$$a_m = a_1 a_{m-1} + \gamma_1 \beta_{m-1}, \quad \beta_m = \beta_1 a_{m-1} + \delta_1 \beta_{m-1},$$
$$\gamma_m = a_1 \gamma_{m-1} + \gamma_1 \delta_{m-1}, \quad \delta_m = \beta_1 \gamma_{m-1} + \delta_1 \delta_{m-1}.$$

From these equations we obtain at once the characteristic relation

7) $\quad a_m \delta_m - \beta_m \gamma_m = (a_1 \delta_1 - \beta_1 \gamma_1)(a_{m-1}\delta_{m-1} - \beta_{m-1}\gamma_{m-1})$
$\qquad = (a_1 \delta_1 - \beta_1 \gamma_1)^2 (a_{m-2}\delta_{m-2} - \beta_{m-2}\gamma_{m-2})$

$$\cdots\cdots\cdots\cdots\cdots\cdots\cdots$$

$$= (a\delta - \beta\gamma)^m.$$

By dividing every coefficient of $\theta(x)$ by $\sqrt{a\delta - \beta\gamma}$ or by $\sqrt{\beta\gamma - a\delta}$, according as $a\delta - \beta\gamma$ is positive or negative, we can arrange that for the new coefficients of $\theta(x)$, and consequently for those of every $\theta^k(x)$, the relation shall hold

8) $\qquad\qquad a\delta - \beta\gamma = \pm 1.$

We determine now under what conditions it can happen that

$$\theta^m(x) = x.$$

The values of x which are unchanged by the operation θ satisfy the equation

$$x = \frac{ax + \beta}{\gamma x + \delta},$$

$$\gamma x^2 + (\delta - a)x - \beta = 0.$$

For these fixed values we have therefore, according as $a\delta - \beta\gamma = \pm 1$,

$$\gamma x' = \frac{a-\delta}{2} + \sqrt{\left(\frac{a+\delta}{2}\right)^2 \mp 1}, \quad \gamma x'' = \frac{a-\delta}{2} - \sqrt{\left(\frac{a+\delta}{2}\right)^2 \mp 1},$$

and consequently

$$\frac{\gamma x' - a}{\gamma x'' - a} = \pm \left[\frac{a+\delta}{2} - \sqrt{\left(\frac{a+\delta}{2}\right)^2 \mp 1}\right]^2 = N.$$

A) We assume in the first instance that x' and x'' are distinct, that is that $N \neq 1$. We have then

$$\frac{\theta(x) - x'}{\theta(x) - x''} = N \frac{x - x'}{x - x''}, \quad \frac{\theta^2(x) - x'}{\theta^2(x) - x''} = N^2 \frac{x - x'}{x - x''}, \cdots$$

$$\frac{\theta^m(x) - x'}{\theta^m(x) - x''} = N^m \frac{x - x'}{x - x''}.$$

The necessary and sufficient condition that $\theta^m(x) = x$ is therefore that

$$N^m = (\pm 1)^m\left[\frac{a+\delta}{2} - \sqrt{\left(\frac{a+\delta}{2}\right)^2 \mp 1}\,\right]^{2m} = 1.$$

This condition can be satisfied by complex or by real values of the quantity in the bracket. In the former case the upper algebraic sign must be taken, and further

$$\left(\frac{a+\delta}{2}\right)^2 < 1,$$

so that we may write

$$\frac{a+\delta}{2} = \cos\varphi$$

$$N^m = (\cos\varphi - i\sin\varphi)^{2m} = \cos 2m\varphi - i\sin 2m\varphi.$$

Accordingly, we must have $2m\varphi = 2k\pi$, $\varphi = \dfrac{k\pi}{m}$, and

9) $$\frac{a+\delta}{2} = \cos\frac{k\pi}{m},$$

where k is any integer prime to m. If the condition 9) is fulfilled, the function $\theta^m(x)$ will be the first of the series $\theta(x), \theta^2(x), \ldots$ to take the initial value x.

If the quantity in the bracket is real, it must be either $+1$ or -1, since one of its powers is to be equal to 1. The case $N = +1$ is to be rejected, since then $x' = x''$. The case $N = -1$ gives

$$a + \delta = 0, \quad a^2 + \beta\gamma = 0,$$

and $\theta^2(x) = x$, which agrees with the condition 9), since $m = 2$.

B) It remains to consider the case $x' = x''$. We have then

$$\left(\frac{a+\delta}{2}\right)^2 \pm 1 = 0.$$

The lower sign must be taken, and accordingly

$$a + \delta = \pm 2, \qquad a\delta - \beta\gamma = +1.$$

It follows that

$$\theta^2(x) = \frac{(2a \mp 1)x + 2\beta}{2\gamma x + (2\delta \mp 1)},$$

$$\theta^2(x) = \frac{(3a \mp 2)x + 3\beta}{3\gamma x + (3\delta \mp 2)},$$

$$\cdots\cdots\cdots$$

$$\theta^m(x) = \frac{[ma \mp (m-1)]x + m\beta}{m\gamma x + [m\delta \mp (m-1)]}.$$

If now $\theta^m(x) = x$, we have

THE ABELIAN EQUATIONS. 207

$$\gamma x^2 + (\delta - a)x - \beta = 0,$$

that is, we must have already had $\theta(x) = x$. And, again, it is clear that, as m increases, $\theta^m(x)$ approaches the limiting value

$$\theta^\infty(x) = \frac{(a \mp 1)x + \beta}{\gamma x + (\delta \mp 1)}.$$

We have shown therefore that

$$a + \delta = 2\cos\frac{k\pi}{m}, \quad a\delta - \beta\gamma = +1,$$

where k is prime to m, are the sufficient and necessary conditions that $\theta^m(x)$ shall be the first of the functions $\theta^\lambda(x)$ which takes the initial value x. For $m = 2$, the second condition is not required.

§ 179. For the second example we take for $\theta(x)$ any integral rational function of x with constant coefficients.

For every integral m the difference $\theta^m(x) - x$ is divisible by $\theta(x) - x$. For if

$$\theta(x) - x = (x - z_1)(x - z_2) \ldots (x - z_\nu),$$

then for every z_a $\theta(z_a) = z_a$, and consequently $\theta^2(z_a) = \theta^3(z_a) = \ldots = \theta^k(z_a) = z_a$. Moreover

$$\theta^{k+1}(x) - \theta^k(x) = [\theta^k(x) - z_1][\theta^k(x) - z_2]\ldots[\theta^k(x) - z_\nu],$$

and consequently

$$\frac{\theta^{k+1}(x) - \theta^k(x)}{\theta(x) - x} = \frac{\theta^k(x) - z_1}{x - z_1} \frac{\theta^k(x) - z_2}{x - z_2} \cdots \frac{\theta^k(x_\nu) - z_\nu}{x - z_\nu} = P_k(x),$$

where P is a rational integral function of x and of the coefficients of θ, since it is symmetric in the roots $z_1, z_2, \ldots z_\nu$. If now we take k successively equal to $0, 1, 2, \ldots m-1$, and add the resulting equations, we have as asserted

$$\theta^m(x) - x = [\theta(x) - x] Q(x),$$

where Q is a rational integral function of x. From this equation it follows that for every root of

$$Q(x) = 0$$

we have $\theta^m(x) = x$, and conversely that every root of

$$\theta^m(x) - x = 0,$$

which is not contained among $z_1, z_2, \ldots z_\nu$ also makes $Q(x)$ vanish. Every root ξ of $Q(x) = 0$ therefore gives

$$\theta^m(\xi) = \xi,$$

and consequently also

$$\theta^{m+1}(\xi) = \theta(\xi), \quad \theta^m[\theta(\xi)] = \theta(\xi),$$

so that $\theta(\xi)$, and likewise $\theta^2(\xi), \theta^3(\xi), \ldots$ are all roots of $Q(x) = 0$. Again since ξ is different from the z's, $\theta(\xi) \not\equiv \xi$, and $\theta(\xi) \not\equiv z_a$.

Theorem IX. *If $\theta(x)$ is a rational integral function of x of degree ν, then the roots of the equation of degree $(\nu-1)m$*

$$Q(x) \equiv \frac{\theta^m(x) - x}{\theta(x) - x} = 0$$

can be arranged as in Theorem I, 5). If m is a prime number then each of the $\nu-1$ rows of 5) contains m roots

$$\xi, \theta(\xi), \theta^2(\xi), \ldots \theta^{m-1}(\xi) \quad [\theta^m(\xi) = \xi].$$

§ 180. Conversely if the equation $f(x) = 0$ has the roots

$$x_0, x_1 = \theta(x_0), x_2 = \theta^2(x_0), \ldots x_{m-1} = \theta^{m-1}(x_0); \quad [\theta^m(x_0) = x_0],$$

every one of these roots will also satisfy the equation

$${}^m(x) - x = 0,$$

but no one of them will satisfy

$$\theta(x) - x = 0;$$

consequently $f(x)$ is a divisor of the quotient

$$\frac{\theta^m(x) - x}{\theta(x) - x} = Q(x).$$

The restriction that $\theta(x)$ shall be an integral function is unessential. For if (x) is fractional

$$\theta(x) = \frac{g_1(x)}{g_2(x)},$$

where g_1 and g_2 are integral functions, then in

$$\theta(x_0) = \frac{g_1(x_0)[g_2(x_1) g_2(x_2) \ldots g_2(x_{m-1})]}{g_2(x_0) \ g_2(x_1) g_2(x_2) \ldots g_2(x_{m-1})}$$

the denominator, being a symmetric function of the roots of $f(x) = 0$, is a rational function of the coefficients of $f(x)$; and the second factor of the numerator, being symmetric in $x_1, x_2, \ldots x_{m-1}$ is a rational integral function of x_0. Consequently $\theta(x_0)$ is a rational integral function of x_0, which can be reduced to the $(m-1)^{\text{th}}$ degree by the aid of $f(x_0) = 0$.

We have therefore

Theorem X. *Every polynomial of the equations treated in § 174 is a factor of an expression*

$$\frac{\theta_1^m(x) - x}{\theta_1(x) - x},$$

where $\theta_1(x)$ is an integral function of the $(m-1)^{th}$ degree.

For example, if we take $\theta_1 = x^2 + bx + c$, we may reduce this by the linear transformation $y = x + a$ to the form $\theta_1 = x^2 + a$. Then

$$\theta_1^3(x) - x = (\theta_1(x) - x)[x^6 + x^5 + (3a+1)x^4 + (2a+1)x^3 \\ + (3a^2 + 3a + 1)x^2 + (a^2 + 2a + 1)x + (a^3 + 2a^2 + a + 1)].$$

The discriminant of the second factor on the right is

$$\Delta = -(4a + 7)(16a^2 + 4a + 7)^2.$$

If now we take

$$4a + 7 = -k^2, \quad a = -\frac{k^2 + 7}{4},$$

the second factor breaks up into two, and this is the only way in which such a reduction can be effected. We have then

$$64 \frac{\theta_1^3(x) - x}{\theta_1(x) - x} =$$
$$[8x^3 + 4(1+k)x^2 - 2(9 - 2k + k^2)x - (1 + 7k - k^2 + k^3)]$$
$$[8x^3 + 4(1-k)x^2 - 2(9 + 2k + k^2)x - (1 - 7k - k^2 - k^3)]$$

or, for $k = 2\lambda + 1$, $a = -\lambda^2 - \lambda - 2$,

$$\frac{\theta_1^3(x) - x}{\theta_1(x) - x} = [x^3 + (\lambda + 1)x^2 - (\lambda^2 + 2)x - (\lambda^3 + \lambda^2 + 2\lambda + 1)]$$
$$[x^3 - \lambda x^2 - (\lambda^2 + 2\lambda + 3)x + (\lambda^3 + 2\lambda^2 + 3\lambda + 1)].$$

In this way we obtain the general criterion for distinguishing those equations of the third degree the roots of which can be expressed by x, $\theta(x)$, $\theta^2(x)$. In the first place θ must be reducible to the form

$$\theta = x^2 - (\lambda^2 + \lambda + 2)$$

that is, $b^2 - 4c$ must be of the form $4(\lambda^2 + \lambda + 2) = (2\lambda + 1)^2 + 7$; then to every θ there correspond two equations of the required type

$$x^3 + (\lambda + 1)x^2 - (\lambda^2 + 2)x - (\lambda^3 + \lambda^2 + 2\lambda + 1) \doteq 0$$
$$x^3 - \lambda x^2 - (\lambda^2 + 2\lambda + 3)x + (\lambda^3 + 2\lambda^2 + 3\lambda + 1) = 0.$$

It appears at once, however, that θ is unchanged if λ is replaced by $-(\lambda + 1)$, and that the first equation is converted into the second by this same substitution. It is sufficient therefore to retain only one of the two equations.

§ 181. We introduce now the following

DEFINITION. *If all the roots of an equation are rational functions of a single one among them,*

$$x_1, \theta_1(x_1), \theta_2(x_1), \ldots \theta_{n-1}(x_1),$$

then, if these rational relations are such that in every case

$$\theta_\alpha \theta_\beta(x_1) = \theta_\beta \theta_\alpha(x_1),$$

the equation is called an "Abelian Equation." *

We have already seen (§ 173) that, if the roots of an equation are defined by 5), the resolvents

$$\varphi_1 = (1, \theta_1(x_1)), \quad \varphi_2 = (1, \theta_1(x_2)), \ldots \varphi_\nu = (1, \theta_1(x_\nu))$$

satisfy an equation 6) of degree ν the coefficients of which are rationally known. We noted further that this equation is solvable only under special conditions. These conditions are realized in the present case. We proceed to prove

Theorem XI. *Abelian equations are solvable algebraically.* **

In the first place we observe that since φ_1 is symmetric in x_1, $\theta_1(x_1), \ldots \theta_1^{m-1}(x_1)$, every symmetric function of these quantities is rational in φ_1 and the coefficients of 1). If now we consider

$$\varphi_2 = x_2 + \theta_1(x_2) + \theta_1^2(x_2) + \ldots + \theta_1^{m-1}(x_2),$$

and assume that
$$x_2 = \theta_2(x_1),$$
we have
$$\varphi_2 = \theta_2(x_1) + \theta_1 \theta_2(x_1) + \theta_1^2 \theta_2(x_1) + \ldots + \theta_1^{m-1}\theta_2(x_1)$$
$$= \theta_2(x_1) + \theta_2 \theta_1(x_1) + \theta_2 \theta_1^2(x_1) + \ldots + \theta_2 \theta_1^{m-1}(x_1)$$
$$= S[x_1, \theta_1(x_1), \theta_1^2(x_1), \ldots \theta_1^{m-1}(x_1)] = R(\varphi_1).$$

---
* C. Jordan: Traité etc. § 402.
** Abel: Oeuvres complètes, I, No. XI; p. 114-140.

THE ABELIAN EQUATIONS. 211

For from $\theta_1\theta_2(x) = \theta_2\theta_1(x)$ follows also

$$\theta_1^2\theta_2(x) = \theta_1[\theta_1\theta_2(x)] = \theta_1\theta_2[\theta_1(x)] = \theta_2\theta_1^2(x); \ldots$$

The equation $\varphi_2 = R(\varphi_1)$ shows that 6) has in the present case also the property that its roots are all rational functions of a single one.

We write now accordingly

$$\varphi_\alpha = \vartheta_\alpha(\varphi_1) \quad (\alpha = 1, 2, \ldots \nu)$$

Then

$$\vartheta_\alpha(\varphi_1) = \varphi_\alpha = x_\alpha + 0(x_\alpha) + \theta^2(x_\alpha) + \ldots$$
$$= \theta_\alpha(x_1) + \theta(\theta_\alpha(x_1)) + \theta^2(\theta_\alpha(x_1)) + \ldots,$$
$$\vartheta_\alpha\vartheta_\beta(\varphi_1) = \theta_\alpha(\theta_\beta(x_1)) + \theta\theta_\alpha(\theta_\beta(x_1)) + \theta^2\theta_\alpha(\theta_\beta(x_1)) + \ldots$$
$$= \theta_\beta(\theta_\alpha(x_1)) + \theta(\theta_\beta\theta_\alpha(x_1)) + \theta^2(\theta_\beta\theta_\alpha(x_1)) + \ldots$$
$$= \vartheta_\beta(\vartheta_\alpha(\varphi_1)),$$

so that the operations ϑ are again commutative, like the θ's. The equation 6) is therefore itself an Abelian equation, the degree of which is reduced to the mth part of that of 1).

We can then proceed further in the same way until we arrive at equations of the type treated and solved in § 174.

§ 182. The character of a group of an Abelian equation is readily determined as follows:

Suppose that any two substitutions s and t of the group replace an arbitrary element x_1 by

$$x_2 = \theta_2(x_1), \quad x_3 = \theta_3(x_1),$$

respectively. Then st and ts replace x_1 by

$$\theta_2\theta_3(x_1), \quad \theta_3\theta_2(x_1),$$

respectively. But since by assumption $\theta_2\theta_3(x_1) = \theta_3\theta_2(x_1)$, it follows that

$$st = ts.$$

All the substitutions of the group are therefore commutative.

If conversely the group G of an equation $f(x) = 0$ consists of commutative substitutions, we consider first the case where G is intransitive and $f(x)$ is accordingly reducible (§ 156). Suppose that

$$f(x) = f_1(x)f_2(x) \ldots$$

where $f_1(x), f_2(x), \ldots$ are rationally known irreducible functions. If we consider the roots of $f_1(x) = 0$ alone, every rationally known function of these is unchanged by the group G and conversely. Accordingly we obtain the group G_1 belonging to $f_1(x) = 0$ by simply dropping from all the substitutions $s_1, s_2, s_3, \ldots$ of G those elements which are not roots of $f_1(x) = 0$, and retaining among the resulting substitutions $\sigma_1, \sigma_2, \sigma_3, \ldots$ those which are distinct. It is clear that from $s_\alpha s_\beta = s_\beta s_\alpha$ follows also $\sigma_\alpha \sigma_\beta = \sigma_\beta \sigma_\alpha$. The group of every irreducible factor of $f(x)$ is therefore itself composed of commutative substitutions, and is moreover transitive.

We may therefore confine ourselves to the case where G is a transitive group. If now G contains a substitution s_1, which leaves x_1 unchanged and replaces x_2 by x_3, then if we select any substitution s_2 of G which replaces x_1 by x_2, we have

$$s_1 s_2 = (x_1 x_2 \ldots), \quad s_2 s_1 = (x_1 x_3 \ldots).$$

This being inconsistent with the commutative property of the group, it follows that every substitution of G either affects all the elements, or is the identical substitution.

If now any arbitrary root x_1 of $f(x) = 0$ is regarded as known, the group of the equation reduces to those substitutions which leave x_1 unchanged, i. e., to the identical substitution. Consequently every function of the roots is then rationally known; in particular $x_2, x_3, \ldots$ are rational functions of x_1,

$$x_2 = \theta_2(x_1), \quad x_3 = \theta_3(x_1), \ldots$$

Again, if s_α and s_β replace x_1 by $x_\alpha = \theta_\alpha(x_1)$ and $x_\beta = \theta_\beta(x_1)$ respectively, then $s_\alpha s_\beta$ and $s_\beta s_\alpha$ replaces x_1 by $\theta_\alpha \theta_\beta(x_1)$ and $\theta_\beta \theta_\alpha(x_1)$, and, as $s_\alpha s_\beta = s_\beta s_\alpha$, the operations θ are also commutative.

Theorem XII. *The substitutions of the group of an Abelian equation are all commutative. Conversely, if the substitutions of the group of an equation are all commutative, the irreducible factors of the equation are all Abelian equations.*

§ 183. The substitutions of the group of an Abelian equation, as well as the relations between the roots, therefore fulfill the conditions assigned in the investigations of §§ 137–139.

In particular we can arrange the roots in the following system

$$\theta_1^{h_1} \theta_2^{h_2} \theta_3^{h_3} \ldots \theta_k^{h_k}(x_1) \quad (h_i = 0, 1, 2, \ldots n_i - 1),$$
$$n_1 n_2 n_3 \ldots n_k = n,$$

in which every root occurs once and only once. The numbers $n_1, n_2, \ldots n_k$ are such that every one of them is equal to or is contained in the preceding one, and that they are the smallest numbers for which

$$\theta_1^n(x_1) = x_1, \quad \theta_2^{n_2}(x_1) = x_1, \ldots \theta_k^{n_k}(x_1) = x_1,$$

respectively. There is only one substitution of the group of the equation which converts x_1 into $\theta_a(x_1)$. Denoting this by s_a, we can arrange the substitutions of the group also in a system

$$s_1^{h_1} s_2^{h_2} s_3^{h_3} \ldots s_k^{h_k} \quad (h_i = 0, 1, 2, \ldots n_i - 1).$$

$$n_1 n_2 n_3 \ldots n_k = n,$$

where again every substitution occurs once and only once, and corresponding to the properties of the θ's,

$$s_1^{n_1} = 1, \quad s_2^{n_2} = 1, \ldots s_k^{n_k} = 1.$$

The numbers $n_1, n_2, \ldots n_k$ are the same as those for the θ's.

To form a resolvent we take now

$$\psi_1(x_1) = \sum_{h_2, h_3, \ldots h_k} \theta_2^{h_2} \theta_3^{h_3} \ldots \theta^{h_k}(x_1) \quad (h_i = 0, 1, \ldots n_i - 1)$$

and construct the cyclical function

$$\chi_1(x_1) = [\psi_1(x_1) + \omega_1 \theta_1 \psi_1(x_1) + \omega_1^2 \theta_1^2 \psi_1(x_1) + \ldots + \omega_1^{n_1-1} \theta_1^{n_1-1} \psi_1(x_1)]^{n_1},$$

where ω_1 is a primitive n_1^{th} root of unity. Then $\chi_1(x_1)$ is unchanged by the group of the equation. For the substitutions of the subgroup

$$G_2 = \{s_2, s_3, \ldots s_k\}$$

leave $\psi_1(x_1)$ unchanged, and the powers of s_1 convert $\psi_1(x_1)$ into

$$\theta_1 \psi_1(x_1), \quad \theta_1^2 \psi_1(x_1), \ldots \theta_1^{n_1-1} \psi_1(x),$$

respectively, so that these do not affect the value of χ_1. Consequently χ_1 is rational in the coefficients of the given Abelian equations and in ω. From Theorem IV, ψ_1 is therefore a root of a "simplest Abelian equation" of degree n_1. With ψ_1 all the functions which belong to the subgroup G_2 of G are also rationally known.

Again, if we take

$$\psi_{1,2}(x_1) = \sum_{h_3, \ldots h_k} \theta_3^{h_3} \theta_4^{h_4} \ldots \theta_k^{h_k} \quad (h_i = 0, 1, \ldots n_i - 1),$$

and form the cyclical resolvent

$$\chi_{1,2}(x_1) =$$
$$[\psi'_{1,2}(x_1) + \omega_2\theta_2\psi'_{1,2}(x_1) + \omega_2^2\theta_2^2\psi'_{1,2}(x_1) + \ldots + \omega_2^{n_2-1}\theta_2^{n_2-1}\psi'_{1,2}(x_1)]^{n_2},$$

in which ω_2 is a primitive n_2^{th} root of unity, the function $\chi_{1,2}(x_1)$ is unchanged by the group G_2, and is therefore a rational function of ψ'_1. For the substitutions of the group

$$G_3 = \{s_3, s_4, \ldots s_k\}$$

leave $\psi'_{1,2}$ unchanged and the powers of s_2 convert $\psi'_{1,2}$ into

$$\theta_2\psi'_{1,2}, \quad \theta_2^2\psi'_{1,2}, \ldots \theta_2^{n_2-1}\psi'_{1,2},$$

respectively. Applying Theorem IV again, we obtain $\psi'_{1,2}$ from ψ'_1 by the solution of a second simplest Abelian equation of degree n_2.

In general, if we write

$$\psi'_{1,2,\ldots\nu} = \sum_{h_{\nu+1}\ldots h_k} \theta_{\nu+1}^{h_{\nu+1}} \ldots \theta_k^{h_k}(x_1),$$

$$\chi_{1,2,\ldots\nu} =$$
$$[\psi'_{1,2,\ldots\nu} + \omega_\nu\theta_\nu\psi'_{1,2,\ldots\nu} + \omega_\nu^2\theta_\nu^2\psi'_{1,2,\ldots\nu} + \ldots + \omega_\nu^{n_\nu-1}\theta_\nu^{n_\nu-1}\psi_{1,2,\ldots\nu}]^{n_\nu},$$

the value of $\psi'_{1,2,\ldots\nu}$ is determined from that of the similar function

$$\psi'_{1,2,\ldots\nu-1} = \sum_{h_\nu\ldots h_k} \theta_\nu^{h_\nu} \ldots \theta_k^{h_k}(x_1).$$

by the aid of a simplest Abelian equation, as defined by Theorem IV.

By a continued repetition of this process we obtain finally[*]

Theorem XIII. *If the n roots of an Abelian equation are defined by the system*

$$\theta_1^{h_1}\theta_2^{h_2}\ldots\theta_k^{h_k}(x_1) \quad (h_i = 0, 1, 2, \ldots n_i-1)$$
$$n_1 n_2 n_3 \ldots n_k = n,$$

the solution of the equation can be effected by solving successively k "simplest" Abelian equations of degrees

$$n_1, n_2, n_3, \ldots n_k.$$

---

[*] L. Kronecker: Berl. Ber., Nachtrag z. Dezemberheft, 1877; pp. 846-851.

THE ABELIAN EQUATIONS.   215

§ 184. The solution of irreducible Abelian equations can also be accomplished by another method, to which we now turn our attention.

Theorem XIV. *The solution of an irreducible Abelian equation of degree* $n = p_1^{a_1} p_2^{a_2} \ldots$, *where* $p_1, p_2, \ldots$ *are the different prime factors of* n, *can be reduced to that of* k *irreducible Abelian equations of degrees* $p_1^{a_1}, p_2^{a_2}, \ldots$

The proof * is based on the consideration of the properties of the group of the equation. For simplicity we take $n = p_1^{a_1} p_2^{a_2}$.

Since the order of the group is $r = n$, the order of every one of the substitutions is a factor of n, and is therefore of the form $p_1^{a_1} p_2^{a_2}$. Every substitution of the group can accordingly be constructed by a combination of its $(p_2^{a_2})^{\text{th}}$ power, (which is of order $p_1^{b_1}$) and its $(p_1^{a_1})^{\text{th}}$ power (which is of order $p_2^{b_2}$). Consequently we can obtain every substitution of the group G by combining all the

$$t'_1, t'_2, t'_3, \ldots t'_{r_1},$$

the orders of which are a power of p_1, with all the

$$t''_1, t''_2, t''_3, \ldots t''_{r_2},$$

the orders of which are a multiple of p_2. Since the t's are all commutative, the substitutions of G are, then, all of the form

$$s = (t'_a t'_\beta t'_\gamma \ldots)(t''_\delta t''_\epsilon t''_\zeta \ldots).$$

The order of the product in the first parenthesis is a power of p_1, and therefore a factor of $p_1^{a_1}$. For we have

$$t'_a{}^{p_1^{a_1}} t'_\beta{}^{p_1^{a_1}} \ldots = (t'_a t'_\beta \ldots)^{p_1^{a_1}} = 1.$$

Two substitutions

$$(t'_a t'_\beta \ldots)(t''_\delta t''_\epsilon \ldots), \quad (t'_a t'_b \ldots)(t''_d t''_e \ldots)$$

are different unless the corresponding parentheses are equal each to each. For if the two substitutions are equal, we have

$$(t'_a t'_\beta \ldots)(t'_a t'_b \ldots)^{-1} = (t''_\delta t''_\epsilon \ldots)^{-1}(t''_d t''_e \ldots),$$

and since the order of the left hand member is a divisor of $p_1^{a_1}$, and that of the right hand member a divisor of $p_2^{a_2}$, each of these divisors is 1.

---
* C. Jordan; Traité etc. § 405-407.

The number of substitutions s is equal to $n = p_1^{a_1} p_2^{a_2}$. And since the substitutions t' form a group and every substitution of this group is of order $p_1^{m_1}$, the order of the group itself must be $p_1^{m_1}$, (§ 43). Similarly the order of the group formed by the t'''s is equal to $p_2^{m_2}$. It follows then from

$$n = p_1^{a_1} p_2^{a_2} = p_1^{m_1} p_2^{m_2}$$

that $m_1 = a_1$, $m_2 = a_2$.

Suppose now that φ is a function belonging to the group of the t'''s. Then φ has $p_1^{a_1}$ values, and is the root of an equation of degree $p_1^{a_1}$ the group of which is isormorphic with the group of the t'''s. This is therefore an Abelian equation.

Similarly the function ψ belonging to the group of the t''s is a root of an Abelian equation of degree $p_2^{a_2}$.

If now φ and ψ have been determined, the function

$$\chi = a' \varphi + \beta' \psi$$

belongs to the group 1. Every function of the roots, and in particular the roots themselves, are rational functions of χ, and the theorem is proved.

§ 185. **Theorem XV.** *The solution of an irreducible Abelian equation of degree p^a can be reduced to that of a series of Abelian equations the groups of which contain only substitutions of order p and the identical substitution.*

If G is the group of such an equation, the order of every substitution of G is a power of p. Suppose that p^λ is the maximum order of the substitutions of G. Then those substitutions which are of orders not exceeding $p^{\lambda-1}$ form a subgroup H of G. For if t_1 and t_2 are two of these substitutions, then from the commutative property it follows that

$$(t_1 t_2)^{p^{\lambda-1}} = t_1^{p^{\lambda-1}} t_2^{p^{\lambda-1}} = 1,$$

so that $t_1 t_2$ is also of order not exceeding $p^{\lambda-1}$.

If the group H is of order p^a, any function φ belonging to H will take p^{a-a} values and will therefore satisfy an equation of degree p^{a-a}. If we apply to φ the successive powers of any substitution τ of G which does not occur in H, φ will take only p values, since τ^p is contained in H. The substitutions among the values of φ

which are produced by the substitutions of G, and which form a group isomorphic to G, are therefore all of order p. From the isomorphism of the two groups it follows, as in the preceding Section, that the equation for φ is an Abelian equation.

If φ is known the group G of the given Abelian equation reduces to H. We denote the group composed of those substitutions of H which are of order $p^{\lambda-2}$ or less by H_1. If φ_1 is a function belonging to H_1, then φ_1 is determined from φ by an Abelian equation of degree p^{a-a_1}, the group of which again contains only substitutions of order p, and so on.

§ 186. **Theorem XVI.** *The solution of an irreducible Abelian equation of degree p^a, the group of which contains only substitutions of order p and the identical substitution, reduces to that of a irreducible Abelian equations of degree p.*

Although this Theorem is contained as a special case in that obtained in § 183, we will again verify it by the aid of the method last employed.

Let s_1 be any substitution of the group G of the given Abelian equation; then the order of s_1 is p. Again if s_2 is any substitution of G not contained among the powers of s_1, then since $s_1 s_2 = s_2 s_1$, the group $H = \{s_1, s_2\}$ contains at the most p^2 substitutions. It will contain exactly this number, if the equality $s_1^a s_2^b = s_1^a s_2^\beta$ requires that $a = a$, $b = \beta$. But if $s_1^a s_2^b = s_1^a s_2^\beta$, then $s_1^{a-a} = s_2^{\beta-b}$, and for every value of $\beta - b$ different from 0 we can determine a number m such that
$$m(\beta - b) \equiv 1 \quad (mod.\ p).$$
It follows then that
$$s_2 = s_2^{m(\beta-b)} = s_1^{m(a-a)},$$
which is contrary to hypothesis. Accordingly $\beta = b$ and $a = a$.

If $a > 2$, suppose that s_3 is a substitution of G not contained among the p^2 substitution $s_1^a s_2^b$. Since $s_1 s_3 = s_3 s_1$ and $s_2 s_3 = s_3 s_2$, the group $H_3 = \{s_1, s_2, s_3\}$ contains at the most p^3 substitutions. And it contains exactly this number, for if $s_1^a s_2^b s_3^c = s_1^a s_2^\beta s_3^\gamma$, then $s_3^{\gamma-c} = s_1^{a-a} s_2^{b-\beta}$, and so on, as before.

Proceeding in this way, we perceive that all the substitutions of G can be written in the form
$$s_1^{\lambda_1} s_2^{\lambda_2} \ldots s_a^{\lambda_a}, \quad (\lambda_i = 0, 1, \ldots p-1)$$

where every substitution occurs once and only once (cf. § 183). If now we take for the resolvents and the corresponding groups

$$\varphi_a \ (x_1, x_2, \ldots x_n); \quad H'_a = \{s_1, s_2, \ldots s_{a-1}\},$$
$$\varphi_{a-1}(x_1, x_2, \ldots x_n); \quad H'_{a-1} = \{s_1, s_2, \ldots s_{a-2}, s_a\},$$
$$\ldots \ldots \ldots \ldots \ldots \ldots \ldots \ldots \ldots \ldots \ldots \ldots$$
$$\varphi_1 \ (x_1, x_2, \ldots x_n); \quad H'_1 = \{s_2, s_3, \ldots s_a\},$$

then every resolvent depends on an Abelian equation of degree p. The roots of the given equation of degree p^a are rational functions of $\varphi_1, \varphi_2, \ldots \varphi_a$, for the function

$$\zeta' = \beta_1 \varphi_1 + \beta_2 \varphi_2 + \ldots + \beta_a \varphi_a$$

belongs to the group $1 \, (cf. \ \S \ 177)$.

§ 187. The p^a roots of such an equation may be denoted by

$$x_{z_1, z_2, \ldots z_a} \quad (z_\lambda = 0, 1, 2, \ldots p-1)$$

Suppose that $x_{\zeta_1, \zeta_2, \ldots \zeta_a}$ is the root by which $s_1^{\zeta_1} s_2^{\zeta_2} \ldots s_a^{\zeta_a}$ replaces $x_{z_1, z_2, \ldots z_a}$. Then the substitution

$$s_1^{\zeta_1} s_2^{\zeta_2} \ldots s_a^{\zeta_a} \cdot s_1^{\xi_1} s_2^{\xi_2} \ldots s_a^{\xi_a} = s_1^{\zeta_1 + \xi_1} s_2^{\zeta_2 + \xi_2} \ldots s_a^{\zeta_a + \xi_a}$$

by virtue of the left hand form will replace $x_{z_1, z_2, \ldots z_a}$ by the root by which $s_1^{\xi_1} s_2^{\xi_2} \ldots s_a^{\xi_a}$ replaces $x_{\zeta_1, \zeta_2, \ldots \zeta_a}$. But from the right hand form this root is $x_{\zeta_1 + \xi_1, \zeta_2 + \xi_2, \ldots \zeta_a + \xi_a}$. Consequently every substitution $s_1^{\xi_1} s_2^{\xi_2} \ldots s_a^{\xi_a}$ replaces any element $x_{\zeta_1, \zeta_2, \ldots \zeta_a}$ by

$$x_{\zeta_1 + \xi_1, \zeta_2 + \xi_2, \ldots \zeta_a + \xi_a},$$

that is the substitutions of the group are defined analytically by the formula

$$s_1^{k_1} s_2^{k_2} \ldots s_a^{k_a} = |z_1, z_2 \ldots z_a \quad z_1 + k_1, z_2 + k_2, \ldots z_a + k_a| \quad (\mathrm{mod.} \, p).$$

The group of an Abelian equation of degree p^a, the substitutions of which are all of order p, consists of the arithmetic substitutions of degree p^a (mod. p).

§ 188. Finally we effect the transition from the investigations of the present Chapter to the more special questions of the preceding one.

Let n be any arbitrary integer and let the quotient $\dfrac{2\pi}{n}$ be denoted by a. Then, as is well known, the n quantities

$\cos a$, $\cos 2a$, $\cos 3a$, ... $\cos na$

satisfy an equation, the coefficients of which are rational numbers,

C) $\quad x^n - \tfrac{1}{4} n.x^{n-2} + \tfrac{1}{16} \dfrac{n(n-3)}{1 \cdot 2} x^{n-4} - \ldots = 0.$

If now we write $x = \cos a$, then for every integer m

$$\cos ma = \theta(\cos a),$$

where θ is a rational integral function. Similarly if the value $\cos m_1 a$ is denoted by $\theta_1(\cos a)$, we obtain, by replacing a by $m_1 a$ in the last equation, the result

$$\cos(m\, m_1 a) = \theta(\cos m_1 a) = \theta\theta_1 (\cos a)$$

Again if in the equation $\theta_1(\cos a) = \cos(m_1 a)$ the argument a is replaced by ma, the result is

$$\cos(m_1 ma) = \theta_1(\cos ma) = \theta_1 \theta \cos a.$$

Consequently the roots of C) are so connected that every one of them is a rational function of a single one among them, x, and that

$$\theta_1 \theta(x) = \theta \theta_1(x) \quad (x = \cos a).$$

The equation C) is therefore an Abelian equation. Accordingly

$$x = \cos a = \cos \dfrac{2\pi}{n}$$

can be algebraically obtained. We have here an example of § 181.

§ 189. Suppose now that n is an odd prime number, $n = 2\nu + 1$. Then the roots of the equation C) are the following:

$$\cos \dfrac{2\pi}{2\nu+1}, \quad \cos \dfrac{4\pi}{2\nu+1}, \ldots \cos \dfrac{4\nu\pi}{2\nu+1}, \quad \cos 2\pi.$$

Since the last root is equal to 1, the equation C) is divisible by $x - 1$. The other roots coincide in pairs

$$\cos \dfrac{2m\pi}{2\nu+1} = \dfrac{\cos(2\nu+1-m)2\pi}{2\nu+1}.$$

Consequently we can obtain from C) an equation with rational coefficients, the roots of which will be the following

$$\cos \dfrac{2\pi}{2\nu+1}, \quad \cos \dfrac{4\pi}{2\nu+1}, \ldots \cos \dfrac{2\nu\pi}{2\nu+1}.$$

This equation is of the form

$$C_1) \quad x^\nu - \tfrac{1}{3}x^{\nu-1} + \tfrac{1}{4}(\nu-1)x^{\nu-2} - \tfrac{1}{8}(\nu-2)x^{\nu-3} + \tfrac{1}{16}\frac{(\nu-2)(\nu-3)}{1\cdot 2}x^{\nu-4}$$
$$+ \tfrac{1}{32}\frac{(\nu-3)(\nu-4)}{1\cdot 5}x^{\nu-5} - \ldots = 0.$$

With the notation
$$\cos\frac{2\pi}{2\nu+1} = \cos a = x$$
we have then
$$\cos\frac{2m\pi}{2\nu+1} = \theta(x) = \cos m a,$$
so that the equation C_1) has also the roots
$$\theta(x),\ \theta^2(x),\ \theta^3(x),\ \ldots$$
that is
$$\cos a,\ \cos m a,\ \cos m^2 a,\ \cos m^3 a,\ \ldots \cos m^\mu a,\ \ldots$$

If now g is any primitive root (mod. $2\nu+1$) then the ν terms of the series

R_1) $\qquad \cos a,\ \cos g a,\ \cos g^2 a,\ \ldots \cos g^{\nu-1}a,$

are distinct, For from the equation
$$\cos g^\alpha a = \cos g^\beta a \quad (\alpha > \beta;\ \alpha,\ < \nu)$$
it would follow that
$$g^\alpha a = \pm g^\beta a + 2k\pi,$$
or, replacing a by its value $\dfrac{2\pi}{2\nu+1}$,
$$g^\alpha = \pm g^\beta + k(2\nu+1),$$
$$g^\alpha \mp g^\beta = g^\beta(g^{\alpha-\beta} \mp 1) = k(2\nu+1).$$
Dividing both sides of this equation by g^β, and multiplying by $g^{\alpha-\beta} \pm 1$, we obtain the congruence
$$g^{2(\alpha-\beta)} \equiv 1 \pmod{2\nu+1}.$$
But, since since $2(\alpha-\beta) < 2\nu$, this congruence is impossible. Consequently $\cos g^\alpha a$ is different from $\cos g^\beta a$.

Again
$$\cos g^\nu a = \cos a.$$
For since $g^{2\nu-\alpha} - 1 = (g^\nu - 1)(g^\nu + 1)$ is a multiple of $2\nu+1$, one of the two factors is divisible by $2\nu+1$, so that
$$g^\nu = \pm 1 + k(2n+1),$$
and consequently the relation holds

THE ABELIAN EQUATIONS. 221

$$\cos g^\nu a = \cos[\pm 1 + k(2\nu + 1)]a = \cos(\pm a + 2k\pi) = \cos a.$$

It follows then that the ν roots of the equation $C_1)$ are all contained in the series $R_1)$, or again in the series

$$x,\ \theta(x),\ \theta^2(x),\ \ldots \theta^{\nu-1}(x),$$

while $\theta^\nu(x) = 1$. The equation $C_1)$ can therefore be solved algebraically. We have an example of § 174.

If we have $\nu = n_1 n_2 \ldots n_\omega$, it appears that we can divide the circumference of a circle in $2\nu + 1$ equal parts by the solution of ω equations of degrees $n_1, n_2, \ldots n_\omega$. If $n_1, n_2, \ldots n_\omega$ are prime to each other, the coefficients of these equations are rational numbers. (§ 176).

In particular if $\nu = 2^\omega$, we have the theorem on the construction of regular polygons by the aid of the ruler and compass.

CHAPTER XII.

EQUATIONS WITH RATIONAL RELATIONS BETWEEN THREE ROOTS.

§ 190. The method employed in § 183 is also applicable to other cases. We will suppose for example that all the substitutions of a transitive group G are obtained by combination of the two substitutions s_1 and s_2, which satisfy the conditions 1) that the equation $s_1^a = s_2^\beta$ holds only when both sides are equal to identity, and 2) that $s_2 s_1 = s_1^k s_2$ (*Cf.* § 37). If, then, the orders of s_1 and s_2 are n_1 and n_2, all the substitutions of G are represented, each once and only once by

$$s_1^{h_1} s_2^{h_2} \quad (h_i = 0, 1, 2, \ldots n_i - 1).$$

Suppose now that G is the group of an equation $f(x) = 0$. We construct a resolvent $\zeta' = \zeta'_0$ belonging to the group $1, s_1, s_1^2, \ldots s_1^{n_1-1}$, and denote the functions which proceed from ζ'_0 on the application of $s_2, s_2^2, \ldots s_2^{n_2-1}$ by $\zeta'_1, \zeta'_2, \ldots \zeta'_{n_2-1}$. Then all these ζ''s belong to the same group with ζ'_0. For from $s_2 s_1 = s_1^k s_2$ we have

$$s_2 s_1 s_2^{-1} = s_1^k, \quad s_2 s_1^2 s_2^{-1} = s_1^{k^2}, \ldots$$

from which it follows at once that the powers of s_1 form a self-conjugate subgroup of G. The resolvent

$$\chi = \left[\zeta'_0 + \omega_2 \zeta'_1 + \omega_2^2 \zeta'_2 + \ldots \omega_2^{n_2-1} \zeta'_{n_2-1} \right]^{n_2}$$

is therefore unchanged by every s_1^a, and since s_2 permutes the ζ''s cyclically, χ remains unchanged by all the substitutions of the group, and can be rationally expressed in terms of the coefficients of $f(x)$. We can therefore obtain $\zeta'_0, \zeta'_1, \ldots$ by the extraction of an n_2^{th} root, as in the preceding Chapter. The group of the equation then reduces to the powers of s_1, and the equation itself becomes a simplest Abelian equation.

§ 191. Again, if a transitive group consists of combinations of three substitutions s_1, s_2, s_3, for which 1) the equations

$$s_2{}^a = s_1{}^\beta, \quad s_3{}^a = s_1{}^\beta s_2{}^\gamma$$

are satisfied only when both sides of each equation are equal to identity, and 2) the relations hold

$$s_2 s_1 = s_1{}^k s_2, \quad s_3 s_1 = s_1{}^\lambda s_2{}^\mu s_3, \quad s_3 s_2 = s_1{}^\sigma s_2{}^\tau s_3,$$

then all the substitutions of the group are represented, each once and only once by

$$s_1{}^{h_1} s_2{}^{h_2} s_3{}^{h_3} \quad (h_i = 0, 1, 2, \ldots n_i - 1),$$

where n_1, n_2, n_3 are the orders of s_1, s_2, s_3. If now G is the group of an equation, we can show by precisely the same method as before, that the equation can be solved algebraically.

Obviously we can proceed further in the same direction. That groups actually arise in this way which are not contained among those treated in the last Chapter is apparent from the example on p. 39, where $s_2 s_1 = s_1{}^3 s_2$.

§ 192. Returning to the example of § 190, we examine more closely the group there given. If we suppose s_2 to be replaced by its reciprocal, it follows from the second condition that $s_2{}^{-1} s_1 s_2 = s_1{}^k$. From s_1 we can therefore obtain every possible s_2 by the method of § 46. We have only to write under every cycle of s_1 a cycle of $s_1{}^k$ of the same order, and to determine the substitution which replaces every element of the upper line by the element immediately below it. This substitution will be one of the possible s_2's.

We consider separately the two cases 1) where s_1 consists of two or more cycles, and 2) where s_1 has only one cycle.

In the former case the transitivity of the group is secured by s_2. Consequently every cycle of $s_1{}^k$ must contain some elements different from those of the cycle of s_1 under which it is written. It is clear also that all the cycles of s_1 must contain the same number of elements. Otherwise the elements of the cycles of the same order would furnish a system of intransitivity. The order of the cycles can then obviously be so taken that the elements of the second cycle stand under those of the first, those of the third under those of the second, and so on, so that with a proper notation the following order of correspondence is obtained

$$s_1 = (x_1x_2 \ x_3 \ \ldots)(y_1y_2 \ y_3 \ \ldots) \ldots$$
$$s_1^k = (y_1y_{1+k}y_{1+2k}\ldots)(z_1z_{1+k}z_{1+2k}\ldots)\ldots$$

It follows then that

$$s_2 = (x_1y_1z_1\ldots)(x_2y_{1+k}z_{1+k^2}\ldots)\ldots$$

The group is therefore non-primitive, the systems of non-primitivity being $x_1, x_2, \ldots; y_1, y_2, \ldots; z_1, z_2, \ldots$ The substitutions s_1^a leave the several systems unchanged, the substitutions $s_1^a s_2$ permute the systems cyclically one step, $s_1^a s_2^2$ two steps, and so on. Accordingly every substitution of the group except identity affects every element. The group is, in fact, a group Ω (§ 129).

The adjunction of any arbitrary element x_λ reduces the group to the identical substitution. Consequently all the roots are rational functions of any one among them.

The following may serve as an example:

$$s_1 = (x_1x_2x_3)(y_1y_2y_3), \quad s_2 = (x_1y_1)(x_2y_3)(x_3y_2),$$
$$s_2s_1 = (x_1y_2)(x_2y_1)(x_3y_3) = s_1^2s_2.$$

§ 193. In the second case, where s_1 consists of a single cycle, the transitivity is already secured. We may write then, as in Chapter VIII,

$$s_1 = |z \ \ z+1|, \quad s_1^a = |z \ \ z+a|$$

To construct the s_2's we proceed as before and obtain from

$$s_1 = (x_1x_2 \ x_3 \ \ldots),$$
$$s_1^k = (x_ix_{i+k}x_{i+2k}\ldots) \quad (k > 1)$$

the series of substitutions

$$s_2 = |z \ \ kz+i-k|, \quad s_2^\beta = \left|z \ \ k^\beta z + \frac{k^\beta-1}{k-1}(i-k)\right|.$$

Now, in the first place, it is easily shown that the group contains substitutions different from identity, which do not affect all the elements. For among the powers of s_1 there is certainly one s_1^μ which has a sequence of two elements in common with s_2. Then $s_1^\mu s_2^{-1}$ does not affect all the elements.

Again, it can be shown that there is no substitution except identity which leaves the elements unchanged. For we have

$$s_1^a s_2^\beta = \left|z \ \ k^\beta(z+a) + \frac{k^\beta-1}{k-1}(i-k)\right|,$$

and if x_λ and $x_{\lambda+1}$ were not affected by the substitution we should have

$$k^\beta(\lambda + a) + \frac{k^\beta - 1}{k - 1}(i - k) \equiv \lambda,$$

$$k^\beta(\lambda + 1 + a) + \frac{k^\beta - 1}{k - 1}(i - k) \equiv \lambda + 1,$$

and consequently

$$k^\beta \equiv 1.$$

The substitution then becomes

$$s_1{}^a s_2{}^\beta = |z \quad z + a|,$$

and since x_λ and $x_{\lambda+1}$ are unchanged, $a \equiv 0$, and the substitution is identity.

The following is an example of this type:

$$s_1 = (x_1 x_2 x_3 x_4 x_5 x_6), \quad s_2 = (x_1 x_2)(x_3 x_6)(x_4 x_5)$$

$$s_2 s_1 = (x_1 x_3)(x_4 x_6) = s_1{}^5 s_2$$

From the preceding considerations we deduce

Theorem I. *If the group of an equation is of the kind defined in § 190, all the roots of the equation are rational functions of, at the most, two among them, and the equation is solvable algebraically.*

§ 194. We turn now to the converse problem and consider those irreducible equations, the roots of which are rational functions of two among them:

$$x_3 = \varphi_3(x_1, x_2), \quad x_4 = \varphi_4(x_1, x_2), \ldots x_n = \varphi_n(x_1, x_2).$$

If any substitution of the group G of such an equation leaves x_1 and x_2 unchanged, it must leave every element unchanged. Again, if s_a and s'_a are any two substitutions of G which have the same effect on both x_1 and x_2, then $s'_a s_a^{-1}$ leaves x_1 and x_2 unchanged; consequently $s'_a s_a^{-1} = 1$, and $s'_a = s_a$.

Suppose now that the substitutions of G are

$$s_1, s_2, s_3, \ldots s_r.$$

There are $n(n-1)$ different possible ways of replacing x_1 and x_2 from the n elements $x_1, x_2, \ldots x_n$. If any one of these ways is not represented in the line above, let t_2 be any substitution which produces the new arrangement. Then the substitutions of the line

226   THEORY OF SUBSTITUTIONS.

$$t_2s_1, t_2s_2, t_2s_3, \ldots t_2s_r,$$

will replace x_1 and x_2 by pairs of elements which are all different from one another, and none of which correspond to the first line. If $2r$ is still $< n(n-1)$, then there are other pairs of elements which do not correspond to either line. If t_3 is any substitution which replaces x_1 and x_2 by one of these pairs, we can construct a third line

$$t_3s_1, t_3s_2, t_3s_3, \ldots t_3s_r,$$

and so on until all the $n(n-1)$ possibilities are exhausted. We have therefore.

Theorem II. *The order of the group of an irreducible equation of the nth degree, all the roots of which are rational functions of two among them, is a divisor of $n(n-1)$.*

§ 195. The equations of the preceding Section are not yet identified, however, with those previously considered in § 190. This will be clear from an example. The alternating group of four elements contains no substitution except identity which leaves two elements unchanged. For such a substitution could only be a transposition of the two remaining elements. Consequently the roots of the corresponding equation of the fourth degree are all rational functions of any two among them. But the group cannot be written in the form

$$s_1^{h_1}s_2^{h_2},$$

for it contains only substitutions of the two types $(x_1x_2x_3)$ and $(x_1x_2)(x_3x_4)$, so that the orders n_1 and n_2 can only be 2 and 3, while n_1n_2 must be equal to 12.

§ 196. If however, the degree of the equation of § 194 is a prime number p, we have precisely the case treated at the beginning of the Chapter.

To show this we observe that by Theorem II, the transitive group G of the equation is of an order which is a divisor of $p(p-1)$. Since the transitive group is of degree p, its order is also a multiple of p. It contains, therefore, a substitution s_1 of the pth order, and consequently also a subgroup of the same order. If now in § 128, Theorem I, we take $a=1$, and put for r the order $\dfrac{p(p-1)}{\sigma}$, it follows that $k = 0$, that is, G contains no substitutions of order p except the powers of s_1. Consequently we must have $s_2 s_1 s_2{}^{-1} = s_1{}^k$, and this is the assumption made in § 188.

Equations of this kind were first considered by Galois,[*] and have been called Galois equations. We do not however employ this designation, in order to avoid confusion with the Galois resolvent equations, i. e., those resolvent equations of which every root is a rational function of every other one.

If a substitution of the group G of an equation of the present type is to leave any element x_2 unchanged, we must have from § 193

$$(k^\beta - 1)z + ak^\beta + \frac{k^\beta - 1}{k - 1}(i - k) = 0 \quad (\text{mod. } p).$$

Since $k^\beta - 1$ is either $= 0$ (mod. p) or is prime to p, it follows that either every x is unchanged and the substitution is equal to 1, or one element at the most remains unchanged.

Theorem III. *If all the roots of an irreducible equation of prime degree p are rational functions of two among them, the group of the equation contains, besides the identical substitution, $p - 1$ substitutions of order p and substitutions which affect $p - 1$ elements. The solution of the equation reduces to that of two Abelian equations.*

§ 197. The simplest example of the equations of this type is furnished by the binomial equation of prime degree p

$$x^p - A = 0$$

in the case where the real p^{th} root of the absolute value of the real quantity A does not belong to the domain of those quantities which we regard as rational.

The roots of this equation, if x_1 is one of them, are

$$x_1, \omega x_1, \omega^2 x_1, \ldots \omega^{p-1} x_1 \quad (\omega^p = 1).$$

The quotient of any two roots of the equation is therefore a power of the primitive p^{th} root of unity ω. A properly chosen power of this quotient is equal to ω itself. Consequently if any two roots x_β and x_γ are given, every other root x_α is defined by an equation

---

[*] Evariste Galois: Oeuvres mathématiques, edited by Liouville in Vol. 11 of the Journal de mathématiques pures et appliquées, 1846, pp. 381-444.

$$x_\alpha = x_\beta \left(\frac{x_\beta}{x_\gamma}\right)^m = \frac{x_\beta^{m+1}}{x_\gamma^m},$$

that is, x_α is a rational function of x_β and x_γ.

As soon, therefore, as it is shown that the equation

$$x^p - A = 0$$

is irreducible, it is clear that it belongs to the type under discussion.

If the polynomial $x^p - A$ were factorable

$$x^p - A = \varphi_1(x)\,\varphi_2(x)\ldots,$$

then, since p is a prime number, the several factors could only be of the same degree, if they were all of degree 1. The roots would then all be rational. Consequently this possibility is to be rejected.

Suppose then that $\varphi_1(x)$ is of higher degree than $\varphi_2(x)$. Let the roots of $\varphi_1(x)$ and $\varphi_2(x)$ be

$$\varphi_1(x);\quad x'_1, x'_2, x'_3, \ldots x'_{n_1},$$
$$\varphi_2(x);\quad x''_1, x''_2, x''_3, \ldots x''_{n_2}.$$

Then the last coefficient of each of the polynomials φ_1 and φ_2

$$\pm x'_1 x'_2 x'_3 \ldots = \omega^{\sigma_1} x_1^{n_1},$$
$$\pm x''_1 x''_2 x''_3 \ldots = \omega^{\sigma_2} x_1^{n_2},\quad (n_1 > n_2)$$

and consequently their quotient

$$\pm \omega^\tau x_1^m,\quad (m > 0)$$

is rational within the rational domain. Since p is a prime number, it is possible to find an integer μ such that the congruence

$$m\mu \equiv 1$$

or the equation

$$m\mu = \nu p + 1$$

shall be satisfied. Then the quantity

$$(\pm x_1^m \omega^\tau)^\mu = \pm x^{\nu p+1} \omega^{\mu\tau} = \pm A^\nu x_1 \omega^{\mu\tau} = \pm A^\nu x',$$

and consequently x', is rational. From the reducibility of the equation would therefore follow the rationality of a root, which is certainly impossible.

The group of the equation is of order $p(p-1)$. For if we leave one root x_1 unchanged, any other root ωx_1 can still be converted into any one of the $p-1$ roots $\omega x_1, \omega^2 x_1, \omega^3 x_1 \ldots \omega^{p-1} x_1$.

Theorem IV. *The binomial equation*

$$x^p - A = 0,$$

in which A is not the pth power of any quantity belonging to the rational domain, belongs to the type of § 196. *Its group is of order* $p(p-1)$.

§ 198. REMARK. By Theorem III every irreducible equation the roots of which are rational functions of two among them is algebraically solvable. At present we have not the means of proving the converse theorem. It will however be shown in the following Chapter by algebraic considerations, and again at a later period in the treatment of solvable equations by the aid of the theory of groups, that *every equation of prime degree, which is irreducible and algebraically solvable, is either an equation of the type above considered, or an Abelian equation.* Before we pass to such general considerations, we treat first another special case, characterized by rational relations among the roots taken three by three.

§ 199. An equation is said to be of *triad character*, or it is called briefly a *triad equation*,[*] if its roots can be arranged in triads $x_\alpha, x_\beta, x_\gamma$ in such a way that any two elements of a triad determine the third element rationally, *i. e.*, if x_α and x_β determine x_γ, x_β and x_γ determine x_α, and x_γ and x_α determine x_β.

Thus the equations of the third degree are triad equations; for

$$x_1 + x_2 + x_3 = c_1.$$

Of the equations of the higher degree, those of the seventh degree may be of triad character. In this case the following distribution of the roots $x_1, x_2, \ldots x_7$ is possible:

$$x_1, x_2, x_3; \quad x_1, x_4, x_5; \quad x_1, x_6, x_7; \quad x_2, x_4, x_6; \quad x_2, x_5, x_7;$$
$$x_3, x_4, x_7; \quad x_3, x_5, x_6.$$

If the degree of an equation is n, there are $\dfrac{n(n-1)}{2}$ pairs of roots x_α, x_β. With every one of these pairs belongs a third root x_γ. Every such triad occurs three times, according as we take for the original pair of roots x_α, x_β; x_β, x_γ; or x_γ, x_α. There are therefore $\dfrac{n(n-1)}{6}$ triads, and since this number must be an integer, it

---

[*] Noether: Math. Ann. XV, p. 89.

follows that the triad character is only possible when $n = 6m + 1$ or $n = 6m + 3$. The case $n = 6m$ must be excluded, because n must be an odd number, as appears at once if we combine x_1 with all the other elements, which must then group themselves in pairs.

The general question whether every $n = 6m + 1, n = 6m + 3$ furnishes a triad system we do not here consider. It is however easy to establish processes for deducing from a triad system of n elements a second triad system of $2n + 1$ elements, and from two triad systems of n_1 and n_2 elements a third system of $n_1 n_2$ elements. From the existence of the triad character for $n = 3$ follows therefore that for $n = 7, 15, 31, \ldots$; $9, 19, 39, \ldots$; $21, 43, \ldots$ These do not however exhaust all possible cases. There are for example triad systems for $n = 13$, etc.

§ 200. We proceed to develop the two processes above mentioned. In the first place suppose a triad system of n elements $x_1, x_2, \ldots x_n$ given. To these we add $n + 1$ other elements x_0; $x'_1, x'_2, \ldots x'_n$. We retain the $\frac{1}{6}n(n-1)$ triads of the former elements, and also construct from these $\frac{3n(n-1)}{6}$ new triads by accenting in each case every two of the three x's. Finally we form n further triads x_0, x_1, x'_1; x_0, x'_2, x'_2; $\ldots$, and have then in all

$$\frac{4n(n-1)}{6} + n = \frac{(2n+1)2n}{6}$$

triads, which furnish the system belonging to the $2n + 1$ elements.

For example, suppose $n = 3$. We obtain then the following system

x_1, x_2, x_3; x_1, x'_2, x'_3; x'_1, x_2, x'_3; x'_1, x'_2, x_3; x_0, x_1, x'_1;
x_0, x_2, x'_2; x_0, x_3, x'_3,

which agrees, apart from the mere notation, with the triad system for seven elements established in the preceding Section.

§ 201. Again, suppose two triad systems of degrees n_1 and n_2 to be given. The indices of the first system we denote by $a, b, c, \ldots$, those of the second by $\alpha, \beta, \gamma, \ldots$ We may designate a triad by the corresponding indices. Suppose that the triads of the first system are

RATIONAL RELATIONS BETWEEN THREE ROOTS. 231

T_1) $a, b, c;\quad a, d, e;\quad b, d, g;\quad \ldots$

and those of the second

T_2) $a, \beta, \gamma;\quad a, \delta, \varepsilon;\quad a, \zeta, \eta;\quad \ldots$

We denote the elements of the combined system by $x_{aa}, x_{a\beta}, x_{ba}, \ldots$ and form for these a triad system as follows. In the first place, we write after every index of T_1) the index a. In this way there arise $\dfrac{n_1(n_1-1)}{6}$ triads of elements with double indices.

In the same way we write β, then γ, then $\delta, \ldots$ after every index of T_1). We obtain then in every case $\dfrac{n_1(n_1-1)}{6}$ and in all

$$n_2 \frac{n_1(n_1-1)}{6}$$

triads of the elements $x_{aa}, x_{a\beta}, x_{a\gamma}, \ldots x_{ba}, x_{b\beta}, \ldots$ All of these are different from one another. They are

T''_3) $\begin{cases} aa,\ ba,\ ca;\quad aa,\ da,\ ea;\quad ba,\ da,\ ga;\ \ldots \\ a\beta,\ b\beta,\ c\beta;\quad a\beta,\ d\beta,\ e\beta;\quad b\beta,\ d\beta,\ g\beta;\ \ldots \\ a\gamma,\ b\gamma,\ c\gamma;\quad a\gamma,\ d\gamma,\ e\gamma;\quad b\gamma,\ d\gamma,\ g\gamma;\ \ldots \\ \cdot\quad \cdot \qquad\qquad \cdot \end{cases}$

Again, we write every index of the system T_1) before every index of T_2), and obtain

$$n_1 \frac{n_2(n_2-1)}{6}$$

triads among the same $n_1 n_2$ elements with double indices. These are also different from one another and from those of T''_3) They are

T'''_3) $\begin{cases} aa,\ a\beta,\ a\gamma;\quad aa,\ a\delta,\ a\varepsilon;\quad aa,\ a\zeta,\ a\eta;\ \ldots \\ ba,\ b\beta,\ b\gamma;\quad ba,\ b\delta,\ b\varepsilon;\quad ba,\ b\zeta,\ b\eta;\ \ldots \\ ca,\ c\beta,\ c\gamma;\quad ca,\ c\delta,\ c\varepsilon;\quad ca,\ c\zeta,\ c\eta;\ \ldots \\ \cdot\quad \cdot\quad \cdot \qquad\qquad \cdot \end{cases}$

Finally we combine every triad of T_1) with every triad of T_2) by writing after the three indices of a triad of T_1) the three indices of a triad of T_2). With any two given triads this can be done in six ways. For example from b, d, g and a, ζ, η we have

$ba, d\zeta, g\eta;\quad ba, d\eta, g\zeta;\quad b\zeta, da, g\eta;\quad b\zeta, d\eta, ga;\quad b\eta, da, g\zeta;$
$b\eta, d\zeta, ga.$

We obtain therefore from T_1) and T_2)

$$6\frac{n_1(n_1-1)}{6}\frac{n_2(n_2-1)}{6} = n_1n_2\frac{n_1n_2-n_1-n_2+1}{6}$$

such combinations. These are again all different from one another and from those of T'_3) and T''_3). They are the following:

$$T'''_3)\begin{cases} a\,u,\ b\,\beta,\ c\gamma; & a\,u,\ b\gamma,\ c\beta; & a\beta,\ b\,a,\ c\gamma;\ldots a\gamma,\ b\beta,\ c\,u; \\ a\,a,\ b\,\delta,\ c\,\varepsilon; & a\,a,\ b\,\varepsilon,\ c\,\delta; & a\,\delta,\ b\,a,\ c\,\varepsilon;\ldots a\,\varepsilon,\ b\,\delta,\ c\,a; \\ a\,u,\ d\beta,\ e\gamma; & a\,a,\ d\gamma,\ e\beta; & a\beta,\ d\,a,\ e\gamma;\ldots a\gamma,\ d\beta,\ e\,u; \\ \cdot\quad\cdot\quad\cdot\quad\cdot & \cdot\quad\cdot\quad\cdot\quad\cdot & \cdot\quad\cdot\quad\cdot\quad\cdot\quad\cdot\quad\cdot\quad\cdot \end{cases}$$

We have therefore now constructed in all

$$n_2\frac{n_1(n_1-1)}{6} + n_1\frac{n_2(n_2-1)}{6} + n_1n_2\frac{n_1n_2-n_1-n_2+1}{6} = \frac{n_1n_2(n_1n_2-1)}{6}$$

different triads among the elements

$$x_{aa},\,x_{a\beta},\,x_{a\gamma},\,\ldots;\ x_{ba},\,x_{b\beta},\,x_{b\gamma}\ldots;\,\ldots$$

The three tables T_3) therefore form a possible triad system for $n_1 n_2$ elements.

§ 202. The triad group for $n=3$ demands no special notice. It is simply the symmetric group of the three elements.

To determine the group of the triad equation for $n=7$ we proceed as follows, restricting ourselves to irreducible equations of this type.

With this restriction the resulting group of 7 elements is transitive. Its order is therefore divisible by 7, and it consequently contains a circular substitution of the 7th order, which we may assume to be

$$s_1 = (x_1x_2x_3x_4x_5x_6x_7).$$

We determine now conversely the arrangements of the 7 elements in triads, which are not disturbed by the powers of s_1. These must be such that if $x_a,\ x_\beta,\ x_\gamma$ form a triad, the same is true for every x_a+i, $x_\beta+i,\ x_\gamma+i$ ($i=1, 2, \ldots 6$). Again with a proper choice of notation we may take $x_a = x_1,\ x_\beta = x_2$, since a proper power of s_1 will contain the two elements $x_a,\ x_\beta$ in succession. If now we apply the powers of s_1 to the system

$$x_1,\,x_2,\,x_3;\quad x_1,\,x_2,\,x_4;\quad x_1,\,x_2,\,x_5;\quad x_1,\,x_2,\,x_6;\quad x_1,\,x_2,\,x_7;$$

it appears that only the second and the fourth cases give rise to a triad distribution of the required character, viz.

T_1) x_1, x_2, x_4; x_2, x_3, x_5; x_3, x_4, x_6; x_4, x_5, x_7; x_5, x_6, x_1;
       x_6, x_7, x_2; x_7, x_1, x_3;

T_2) x_1, x_2, x_6; x_2, x_3, x_7; x_3, x_4, x_1; x_4, x_5, x_2; x_5, x_6, x_3;
       x_6, x_7, x_4; x_7, x_1, x_5;

The two distributions are not essentially different, each being obtained from the other by interchanging x_2, x_7; x_3, x_6; and x_4, x_5.

We may therefore assume that T_1) is given, and that s_1 belongs to the corresponding group. If there are other substitutions of the 7th order belonging to the group, a proper power of every one of these will contain x_1 and x_2 in succession. We may write the substitution therefore

$$(x_1 x_2 x_{a_3} x_{a_4} x_{a_5} x_{a_6} x_{a_7}) = (1\ 2\ a_3\ a_4\ a_5\ a_6\ a_7)$$

To this substitution correspond, as in the case of s_1, only two triad systems, which proceed respectively from 1, 2, a_4 and 1, 2 a_6. The indices $a_3, \ldots a_6$ must be so taken that the new systems coincide with T_1). In this way we obtain seven new substitutions s. For example, if the seven triads

1, 2, a_6; 2, a_3, a_7; a_3, a_4, 1; a_4, a_5, 2; a_5, a_6, a_3; a_6, a_7, a_4; a_7, 1, a_5

are to coincide respectively with

1, 2, 4; 2. 3, 5; 3, 7, 1; 7, 6, 2; 6, 4, 3; 4, 5, 7; 5, 1, 6,

we must have $a_3 = 3$, $a_4 = 7$, $a_5 = 6$, $a_6 = 4$, $a_7 = 5$, and accordingly $s = (x_1 x_2 x_3 x_7 x_6 x_4 x_5)$. Similarly we obtain for the seven new s's

$s_2 = (x_1 x_2 x_5 x_4 x_3 x_7 x_6)$, $s_3 = (x_1 x_2 x_6 x_4 x_7 x_3 x_5)$, $s_4 = (x_1 x_2 x_7 x_4 x_6 x_5 x_3)$,
$s_5 = (x_1 x_2 x_3 x_7 x_6 x_4 x_5)$, $s_6 = (x_1 x_2 x_5 x_6 x_7 x_4 x_3)$, $s_7 = (x_1 x_2 x_6 x_5 x_3 x_4 x_7)$,
$\qquad s_8 = (x_1 x_2 x_7 x_3 x_5 x_4 x_6)$.

Beside the powers of $s_1, s_2, \ldots s_8$ there can obviously be no other substitutions of the 7th order in the group. We note, without further proof, that it follows from this by the aid of § 76, Theorem XII, that the required group is

$$\{ s_1, s_2, \ldots s_8 \}$$

The same result has been obtained by Kronecker from an entirely different point of view.

Theorem V. *The roots of the most general irreducible triad equation of the 7th degree can be arranged as follows:*

$$x_3 = \vartheta_1(x_0, x_1), \quad x_4 = \vartheta_1(x_1, x_2), \quad x_5 = \vartheta_1(x_2, x_3), \quad x_6 = \vartheta_1(x_3, x_4),$$
$$x_0 = \vartheta_1(x_4, x_5), \quad x_1 = \vartheta_1(x_5, x_6), \quad x_2 = \vartheta_1(x_6, x_0).$$

The group of the equation is the Kronecker group of order 168, defined by

$$\mid z \quad az+b \mid, \quad \mid z \quad a\,\theta(z+b)+c \mid$$
$$(a = 1, 2, 4; \quad b, c = 0, 1, \ldots 6; \quad \theta(z) = -z^2(z^3+1))$$

It is doubly transitive. Those of its substitutions which replace x_0, x_1 by x_1, x_3 are

$$(x_0x_1x_3)(x_2x_5x_4), \quad (x_0x_1x_3)(x_2x_6x_5), \quad (x_0x_1x_3)(x_2x_4x_6), \quad (x_0x_1x_3)(x_4x_5x_6).$$

All these also replace x_3 by x_0. Consequently we have also

$$x_0 = \vartheta_1(x_1, x_3), \quad x_1 = \vartheta_1(x_3, x_0),$$

and similarly

$$x_1 = \vartheta_1(x_2, x_4), \quad x_2 = \vartheta_1(x_4, x_1); \text{ etc.}$$

All the substitutions of the group which interchange x_0 and x_1 are

$$(x_0x_1)(x_2x_5), \quad (x_0x_1)(x_4x_6), \quad (x_0x_1)(x_2x_4x_5x_6), \quad (x_0x_1)(x_2x_6x_5x_4),$$

and since these all leave x_3 unchanged, it follows that

$$x_3 = \vartheta_1(x_0, x_1) = \vartheta_1(x_1, x_0),$$

and the same property holds for all the other triads. Every symmetric function of the roots of a triad is a 7-valued resolvent.

§ 203. We examine also the triad equations for $n = 9$. In the construction of the triads it is easily recognized that there is only one possible system, if we disregard the mere numbering of the elements. We can therefore assume the system to be that constructed in § 201, and designate the elements accordingly by two indices, each

$$00, 10, 20; \quad 01, 11, 21; \quad 02, 12, 22;$$
$$00, 01, 02; \quad 10, 11, 12; \quad 20, 21, 22;$$
$$00, 11, 22; \quad 01, 12, 20; \quad 02, 10, 21;$$
$$00, 12, 21; \quad 01, 10, 22; \quad 02, 11, 20;$$

A characteristic property of every such triad

$$pq, \; p'q', \; p''q''$$

is the condition

B) $\qquad p+p'+p''\equiv q+q'+q''\quad 0\pmod{3}$

From this it follows that every substitution

$$s = |\, p, q \quad ap+bq+a,\, a'p+b'q+a'\,|$$

transforms the triad system into itself. For the indices p, q; p', q'; p'', q'' become respectively

$$ap+bq+a,\quad a'p+b'q+a';$$
$$ap'+bq'+a,\quad a'p'+b'q'+a';$$
$$ap''+bq''+a,\quad a'p''+b'q''+a';$$

and if the condition B) is satisfied by $p, p', p''; q, q', q''$, it is also satisfied by the new indices.

Conversely, every substitution that leaves the triad system unchanged can be written in the form s by a proper choice of the coefficients $a, b, a; a', b', a'$. For if t_1 is any substitution of the triad group which replaces the index $(0, 0)$ by (a, a'), then

$$s_1 = |\, p, q \quad p+a,\, q+a' \,|$$

does the same, and consequently $t_2 = t_1 s_1^{-1}$, which also belongs to the group, leaves $(0, 0)$ unchanged. If now t_2 replaces $(0, 1)$ by (b, b'), then

$$s_2 = |\, p, q \quad ap+bq,\, a'p+b'q\,|,$$

where a and a' are arbitrary, will leave $(0, 0)$ unchanged, and will replace $(0, 1)$ by b, b'. Consequently $t_3 = t_2 s_2^{-1}$ will leave both $(0, 0)$ and $(0, 1)$ unchanged. Again if t_3 replaces $(1, 0)$ by (c, c'), then

$$s_3 = |\, p, q \quad cp,\, c'p+q\,|$$

will leave $(0, 0)$ and $(0, 1)$ unchanged, and will replace $(1, 0)$ by (c, c') consequently $t_4 = t_3 s_3^{-1}$, which belongs to the group of the triad equation, will leave $(0, 0), (0, 1)$ and $(1, 0)$ unchanged. A glance at the triad system shows that we must have $t_4 = 1$, and it follows accordingly that

$$t_1 = s_3 s_2 s_1.$$

Consequently t_1 is actually of the assigned form. Remembering further that we have established in § 145 the necessary and sufficient condition that this form shall actually furnish a substitution, we have the following

Theorem VI. *The group G of the irreducible triad equation of degree 9, consists of all the substitutions*

$$s = |p, q \quad ap + bq + a, a'p + b'q + a'| \quad (\text{mod. } 3)$$
$$ab' - a'b \equiv 0 \quad (\text{mod. } 3)$$

The order of G is, from § 145

$$r = 3^2(3^2 - 1)(3^2 - 3) = 27 \cdot 16$$

The roots of the equation are connected, in accordance with the triad system, as follows:

$$x_{20} = \vartheta(x_{30}, x_{10}), \quad x_{21} = \vartheta(x_{01}, x_{11}), \quad x_{22} = \vartheta(x_{02}, x_{12}); \ldots$$

All the substitutions of G which replace x_{00} and x_{10} by x_{10} and x_{20} are of the form

$$s' = |p, q \quad p + bq + 1, b'q| \quad (\text{mod. } 3),$$

and since these all convert x_{20} into x_{00}, it follows that we have also

$$x_{00} = \vartheta(x_{10}, x_{20}), \quad x_{10} = \vartheta(x_{20}, x_{00}); \ldots$$

All the substitutions of the group which interchange x_{00} and x_{10} are of the form

$$s'' = |p, q \quad 2p + bq + 1, b'q| \quad (\text{mod. } 3),$$

and since these all leave x_{20} unchanged, we have, again,

$$x_{20} = \vartheta(x_{00}, x_{10}) = \vartheta(x_{10}, x_{00}); \ldots$$

§ 204. The arrangement in triads given at the beginning of the preceding Section possesses a peculiarity, which we can turn to account. The triad system is so distributed in four lines that the three triads of every line contain all the 9 elements.

Evidently every substitution of the group permutes the several lines as entities among themselves. We determine now those substitutions which convert every line into itself. If

$$\sigma = |p, q \quad ap + bq + a, a'p + b'q + a'| \quad (\text{mod. } 3)$$

is to convert the first line into itself, the new value of q must depend solely on the old value of q, but not on the value of p. Consequently we must have $a' = 0$. If the substitution is to convert the second line also into itself, we must again for the same reason have $b = 0$. The substitution is therefore of the form

$$\sigma = |p, q \quad ap + a, b'q + a'| \quad \text{mod. } 3).$$

That conversely all these substitutions satisfy these two conditions is obvious. Their number is $3^2 \cdot 2^2$, since $ab' \equiv 0$ (mod. 3).

It is further required that σ shall also leave the third and fourth lines unchanged. The third line has the property that in every triad $(pq, p'q', p''q'')$ the three sums
$$p+q, \ p'+q', \ p''+q''$$
have respectively the values 0, 1, 2 (mod. 3); the fourth that
$$p+q \equiv p'+q' \equiv p''+q'' \quad \text{(mod. 3)}.$$
If now we apply σ to the triad (00, 12, 21) of the fourth line, we obtain
$$(a, a'; \ a+a, \ 2b'+a'; \ 2a+a, \ b'+a') \quad \text{(mod. 3)},$$
and consequently we must have
$$a+a' \equiv a+2b'+a+a' \equiv 2a+b'+a+a' \quad \text{(mod. 3)},$$
that is,
$$a \equiv b' \quad \text{(mod. 3)}.$$
The final form of σ is therefore
$$\sigma = |p, q \ \ ap+a, \ aq+a'|.$$

Conversely all the substitutions of this type convert every one of the four lines into itself. The substitutions form a subgroup H of G of order $2 \cdot 3^2$, since a can only take the values 1 and 2. H is a self-conjugate subgroup of G. For if τ is any substitution of G, then $\tau^{-1} H \tau$ leaves every line unchanged, i. e., $\tau^{-1} H \tau = H$.

The group H of order $2 \cdot 3^2$, being a self-conjugate subgroup of G which is of order $2^4 \cdot 3^4$, we can construct, by § 86, the quotient $T = G:H$ of order $2^3 \cdot 4 = 24$ and of degree 4 (corresponding to the four lines of triads). This group is of course the symmetric group of 4 elements.

If therefore we construct a function φ of the 9 elements x, which belongs to the group H, this four-valued function is the root of a general equation of the fourth degree, the group of which is T.

If this equation of the fourth degree has been solved, the group of the triad equation reduces to H, of order $2 \cdot 3^2$, as is readily apparent. The systematic discussion of this class of questions is however reserved for Chapter XIV.

§ 205. We consider further the subgroup of H which leaves every single triad of the first line of our table unchanged. In order that
$$|\, p,q \quad ap+a,\ aq+a'\,|$$
may have this property, the values $q = 0, 1, 2$, must give again $q = 0, 1, 2$. Consequently $a = 0$, $a = 1$, and we must take
$$\tau = p, q \quad p+a, q\ .$$
The τ's form again a self-conjugate subgroup of I of H of order 3. We construct by § 86 the quotient $U = H:I$. U is of order $3 \cdot 2$ and of degree 3, corresponding to the three triads. U is therefore the symmetric group of three elements. If, then, we construct a function φ of the 9 elements x, which belongs to the group I, this (after adjunction of φ) three-valued function depends on a general equation of the third degree.

If the latter has been solved, the group of the triad equation reduces to I. Accordingly the symmetric functions of
$$x_{00}, x_{10}, x_{20},$$
are known, and therefore these three values depend on an equation of the third degree, the coefficients of which are rationally expressible in terms of φ. This equation is, in fact, an Abelian equation, since its group only permutes the roots cyclically. We have then

Theorem VII. *The irreducible triad equations of degree 9 can be solved algebraically.*

§ 206. In close relation to the above stands the following

Theorem VIII. *If three of the roots of an irreducible equation of the 9th degree are connected by the equations*
$$x_3 = \theta(x_1, x_2) = \theta(x_2, x_1), \quad x_1 = \theta(x_2, x_3) = \theta(x_3, x_2),$$
$$x_2 = \theta(x_3, x_1) = \theta(x_1, x_3),$$
in which θ is a rational function of its two elements, then the equation can be solved algebraically.

We consider the group of the equation. It is transitive, and it replaces the three roots, x_1, x_2, x_3 by three others between which the same relation must exist as between x_1, x_2, x_3 themselves. Suppose the new roots to be x'_1, x'_2, x'_3. If the two systems

RATIONAL RELATIONS BETWEEN THREE ROOTS.

x_1, x_2, x_3, and x'_1, x'_2, x'_3, have two roots in common, then they have also the third root in common. For, if $x_1 = x'_1$, $x_2 = x'_2$, it follows that

$$x'_3 = \theta(x'_1, x'_2) = \theta(x_1, x_2) = x_3,$$

and if x'_3 and x_3 are not the same root, the given equation, having equal roots, would be reducible.

If x_4 is a root different from x_1, x_2, x_3, there is a substitution in the group which replaces x_1 by x_4. If this substitution leaves no element unchanged, we obtain an entirely new system x_4, x_5, x_6. But if one element, for example x_2, remains unchanged, we have for a new system x_2, x_4, x_7. Proceeding in this way, and examining the possible effects of the substitutions, it is seen that all the roots arrange themselves in the triad system of 9 elements. Comparing this result with Theorem VI, it appears that the equation is exactly one of the triad equations just treated.

It is known * that the nine points of inflection of a plane curve of the third order lie by threes on straight lines. These lines are twelve in number, and four of them pass through every point of inflection. Any two of the nine points determine a third one, so that the points form a triad system, as considered above. The abscissas or the ordinates of the nine points therefore satisfy a triad equation of the 9th degree, and this equation, belonging to the type above discussed, is algebraically solvable.

It can, in fact, be shown that if x_1, x_2, x_3 are the abscissas or the ordinates of three points of inflection lying on the same straight line, then

$$x_3 = \theta(x_1, x_2), \quad x_1 = \theta(x_2, x_3), \quad x_2 = \theta(x_3, x_1),$$

where θ is a rational and symmetric function of its two elements. The discussion of this matter belongs however to other mathematical theories and must be omitted here.

----
* O. Hesse: Crelle XXVIII, p. 68; XXXIV, p. 191. Salmon: Crelle XXXIX, p. 365.

CHAPTER XIII.

THE ALGEBRAIC SOLUTION OF EQUATIONS.

§ 207. In the last three Chapters various equations have been treated for which certain relations among the roots were *à priori* specified, and which in consequence admitted the application of the theory of substitutions.

In general questions of this character, however, a doubt presents itself which, as we have already pointed out, must be disposed of first of all, if the application of the theory of substitutions to general algebraic questions is to be admissible. The theory of substitutions deals exclusively with *rational* functions of the roots of equations. If therefore in the algebraic solution of algebraic equations *irrational* functions of the roots occur, we enter upon a region in which even the idea of a substitution fails. The fundamental question thus raised can of course only be settled by algebraic means; the application to it of the theory of substitutions would beg the question. To cite a single special example, proof of the impossibility of an algebraic solution of general equations above the fourth degree can never be obtained from the theory of substitutions alone.

§ 208. In the discussion of algebraic questions it is essential first of all to define the territory the quantities lying within which are to be regarded as *rational*.

We adopt the definition* that all rational functions with integral coefficients of certain quantities $\Re'$, $\Re''$, $\Re'''$, ... constitute the *rational domain* $(\Re, \Re'' \Re''', ...)$. If among any functions of this domain the operations of addition, subtraction, multiplication, division, and involution to an integral power are performed, the resulting quantities still belong to the same rational domain.

The extraction of roots on the other hand will in general lead

---
* L. Kronecker: Berl. Ber. 1879, p. 205 ff.; cf. also: arithm. Theorie d. algeb. Grössen.

to quantities which lie outside the rational domain. We may limit ourselves to the extraction of roots of prime order, since an $(mn)^{\text{th}}$ root can be replaced by an mth root of an nth root.

All those functions of $\Re'$, $\Re''$, $\Re'''$, ... which can be obtained from the rational functions of the domain by the extraction of a single root or of any *finite* number of roots are designated, collectively, as the *algebraic functions of the domain* $(\Re', \Re'', \Re''', \ldots)$. In proceeding from the rational to the algebraic functions of the domain, the first step therefore consists in extracting a root of prime order p_ν of a rational, integral or fractional function $F_\nu(\Re', \Re'', \Re''', \ldots)$ which in the domain $(\Re', \Re'', \Re''', \ldots)$ is not a perfect p_ν^{th} power. Suppose the quantity thus obtained to be V_ν so that

$$V_\nu^{p_\nu} = F_\nu(\Re', \Re'', \Re''', \ldots).$$

We will now extend the rational domain by adding or *adjoining* to it the quantity V_ν, so that we have from now on for the rational domain $(V_\nu; \Re', \Re'', \Re''', \ldots)$, *i. e.*, all rational, integral or fractional functions of $V_\nu, \Re', \Re'', \Re''', \ldots$ are regarded as rational. The present domain includes the previous one. With this extension goes a like extension of the property of reducibility. Thus the function $x^p - F_\nu(\Re', \Re'', \ldots)$ was originally irreducible: it has now become reducible and has, in the extended domain $(V_\nu; \Re', \Re''_1, \ldots)$, the rational factor $x - V_\nu$.

The new domain can be extended again by the extraction of a second root of prime order. We construct any rational function which is not a perfect $(p_{\nu-1})^{\text{th}}$ power within $(V_\nu; \Re', \Re'', \ldots)$, and denote its $(p_{\nu-1})^{\text{th}}$ root by $V_{\nu-1}$, so that

$$V_{\nu-1}^{p_{\nu-1}} = F_{\nu-1}(V_\nu; \Re', \Re'', \ldots).$$

It is not essential here that V_ν should occur in $F_{\nu-1}$. If now we adjoin $V_{\nu-1}$, we obtain the further extended rational domain $(V_{\nu-1}, V_\nu; \Re', \Re'', \ldots)$. Similarly we construct

$$V_{\nu-2}^{p_{\nu-2}} = F_{\nu-2}(V_{\nu-1}, V_\nu; \Re', \Re'', \ldots),$$
$$V_{\nu-3}^{p_{\nu-3}} = F_{\nu-3}(V_{\nu-2}, V_{\nu-1}, V_\nu; \Re', \Re'', \ldots),$$
$$\cdots \cdots \cdots \cdots \cdots \cdots \cdots \cdots$$
$$V_1^{p_1} = F_1(V_2, V_3, \ldots V_\nu; \Re', \Re'', \ldots),$$

where the F's denote rational functions of the quantities in parentheses, and $p_1, p_2, \ldots p_{\nu-2}$ are prime numbers.

Any given algebraic expression can therefore be represented in conformity with the preceding scheme, by treating it in the same way in which the calculation of such an expression involving only numerical quantities is accomplished.

§ 209. The F_a's are readily reduced to a form in which they are integral in the corresponding V's, that is $V_{a+1}, V_{a+2} \ldots V_\nu$, and are fractional only in the $\Re', \Re'', \ldots$

Thus, suppose that

$$F_a = \frac{G_0 + G_1 V_{a+1} + G_2 V_{a+1}^2 + \cdots}{H_0 + H_1 V_{a+1} + H_2 V_{a+1}^2 + \cdots},$$

where $G_0, G_1, G_2, \ldots; H_0, H_1, H_2, \ldots$ are rational in $V_{a+2}, V_{a+3}, \ldots V_\nu; \Re', \Re'', \ldots$ If now ω is a primitive $(p_{a+1})^{\text{th}}$ root of unity, the product

$$P) \qquad \prod_{\lambda=0}^{p_{a+1}-1} [H_0 + H_1 \omega^\lambda V_{a+1} + H_2 \omega^{2\lambda} V_{a+1}^2 + \cdots]$$

is a rational function of $V_{a+2}, V_{a+3}, \ldots V_\nu; \Re', \Re'', \ldots$ For on the one hand the product is rational in the H's, and on the other it is integral and symmetric in the roots of

$$V_{a+1}^{p_{a+1}} = F_{a+1}(V_{a+2}, \ldots V_\nu; \Re', \Re'', \ldots)$$

and is therefore rational in the coefficient F_{a+1} of this equation.

Again, if we omit from the product P) the factor $H_0 + H_1 V_{a+1} + H_2 V_{a+1}^2 + \cdots$, the resulting product

$$P_1) \qquad \prod_{\lambda=1}^{p_{a+1}-1} [H_0 + H_1 \omega^\lambda V_{a+1} + H_2 \omega^{2\lambda} V_{a+2}^2 + \cdots]$$

is integral in V_{a+1} and rational in $V_{a+2}, \ldots V_\nu; \Re', \Re'', \ldots$ Moreover, since ω does not occur in P) or in the omitted factor, it does not occur in P_1).

If now we multiply numerator and denominator of F_a by P_1), the resulting denominator is a rational function of $V_{a+2}, \ldots V_\nu$; $\Re', \Re'', \ldots$ alone, while the numerator is rational in these quantities and in V_{a+1}. Dividing the several terms of the numerator by the denominator, we have for the reduced form of F_a

$$F_a = J_0 + J_1 V_{a+1} + J_2 V_{a+1}^2 + \ldots,$$

where the coefficients $J_0, J_1, J_2, \ldots$ are all rational functions of $V_{a+2}, \ldots V_\nu; \Re', \Re'', \ldots$ On account of the equations

$$V_{a+1}^{p_{a+1}} = F_{a+1}, \quad V_{a+1}^{p_{a+1}+1} = F_{a+1} V_{a+1}, \ldots$$

we may assume that the reduced form of F_a contains no higher power of V_{a+1} than the $(p_{a+1}-1)^{\text{th}}$.

The several coefficients J can now be reduced in the same way as F_a above. By multiplying numerator and denominator of their fractional forms by proper factors, all the J's can be converted into integral functions of V_{a+2} of a degree not exceeding $p_{a+2}-1$, and with coefficients which are rational in $V_{a+3}, \ldots V_\nu; \Re', \Re'', \ldots$ In this way we can continue to the end.

§ 210. We have now at the outset to establish a preliminary theorem which will be of repeated application in the investigation of the algebraic form peculiar to the roots of solvable equations.*

Theorem I. *If $f_0, f_1, \ldots f_{p-1}$; F are functions within a definite rational domain, the simultaneous existence of the two equations*

.A) $\qquad f_0 + f_1 w + f_2 w^2 + \ldots + f_{p-1} w^{p-1} = 0,$
B) $\qquad w^p - F \qquad\qquad\qquad = 0,$

requires either that one of the roots of B) belongs to the same rational domain with $f_0, f_1, \ldots f_{p-1}$; F, or that

$$f_0 = 0, \; f_1 = 0, \ldots f_{p-1} = 0.$$

If all the $f_0, f_1, \ldots f_{p-1}$ are not equal to 0, the equations A) and B) have at least one root w in common. In the greatest common divisor of the polynomials A) and B) the coefficient of the highest power of w is unity, from the form of B). Suppose the greatest common divisor to be

C) $\qquad \varphi_0 + \varphi_1 w + \varphi_2 w^2 + \ldots + w^\nu.$

Equated to 0, this furnishes ν roots of B). If one of these is deno-

---
*This theorem was originally given by Abel: Oeuvres complètes II, 196. Kronecker was the first to establish it in the full importance: Berl. Ber. 1879, p. 206.

244   THEORY OF SUBSTITUTIONS.

ted by w_1, and a primitive p^{th} root of unity by ω, then all the roots of C) can be expressed by

$$w_1, \omega^\alpha w_1, \omega^\beta w_1, \omega^\gamma w_1, \ldots$$

Apart from its algebraic sign, φ_0 is the product of these roots

$$\varphi_0 = \pm \omega^\delta w_1^\nu.$$

Now since p is a prime number, it is possible to find two numbers u and v, for which
$$pu + \nu v = 1,$$
and consequently
$$(\pm \varphi_0)^v = \omega^{v\delta} w_1^{1-pu},$$
$$\omega^{v\delta} w_1 = F^u(\pm \varphi_0)^v.$$

One root, $\omega^{v\delta} w_1$, of the equation B) therefore belongs to the given rational domain.

§ 211. We apply Theorem I first to the further reduction of

$$F_a = J_0 + J_1 V_{a+1} + J_2 V_{a+1}^2 + \ldots + J_{p_a-1-1} V_{a+1}^{p_a+1-1}.$$

If J_κ is any one of the coefficients $J_1, J_2, \ldots$, which does not vanish, we determine a new quantity W_{a+1} by the equation

A_1) $\qquad W_{a+1} - J_\kappa V_{a+1}^k = 0,$

annex to this the equation of definition for V_{a+1}

B_1) $\qquad V_{a+1}^{p_a+1} - F_{a+1} = 0,$

and fix for the rational domain

R) $\qquad (W_{a+1}; V_{a+2}, V_{a+3}, \ldots; \mathfrak{R}', \mathfrak{R}'', \ldots),$

It follows then, if A) and B) of Theorem I are replaced by A_1) and B_1), that, since the possibility $W_{a+1} = 0$, $J_\kappa = 0$ is excluded, we must have

C_1) $\qquad \omega V_{a+1} = R(W_{a+1}; V_{a+2}, \ldots V_\nu; \mathfrak{R}', \mathfrak{R}'', \ldots)$

where ω is a $(p_{a+1})^{\text{th}}$ root of unity.

We can therefore introduce into the expression for F_a in the place of V_{a+1} the function W_{a+1}, provided we adjoin the $(p_{a+1})^{\text{th}}$ root of unity, ω, to the rational domain. From A_1) and C_1) it is clear that $(W_{a+1}; V_{a+2}, \ldots; \mathfrak{R}'.\mathfrak{R}'', \ldots)$ and $(V_{a+1}, V_{a+2}, \ldots; \mathfrak{R}'. \mathfrak{R}'', \ldots)$, define the same rational domain, and the equation

B') $\qquad W_{a+1}^{p_a+1} = J_\kappa^{p_a} + V_{a+1}^{k p_a+1} = \varPhi_{a+1}(V_{a+2}, V_{a+3}, \ldots V_\nu; \mathfrak{R}', \mathfrak{R}'', \ldots)$

THE ALGEBRAIC SOLUTION OF EQUATIONS.

can be taken in the scheme of § 208 in place of B_1). The equations of definition for V_a, V_{a-1}, ... V_1 are not essentially affected by this change. We have only to substitute in the functions F_a, F_{a-1}, ... F_1 in the place of V_{a+1} the value taken from C_1). The expression for F_a then becomes simplified

$$F_a = J_0 + W_{a+1} + L_2 W_{a+1}^2 + L_3 W_{a+1}^3 + \ldots + L_{p_{a+1}-1} W_{a+1}^{p_{a+1}-1}.$$

We may suppose this reduction to have been effected in the case of every F_a.

§ 212. We pass now to the investigation of the form of the roots of algebraically solvable equations. Given an algebraic equation

1) $$f(x) = 0$$

of degree n, the requirement that this shall be algebraically solvable can be stated in the following terms:—from the rational domain $(\mathfrak{R}', \mathfrak{R}'', \ldots)$, which includes at least the coefficients of 1), we are to arrive at the roots of 1) by a finite number of algebraic operations, viz. addition, subtraction, multiplication, division, raising to powers, and extraction of roots of prime orders. One of the roots of 1) can therefore be exhibited by the following scheme:

2) $$\begin{cases} V_\nu^{p_\nu} = (\mathfrak{R}', \mathfrak{R}'', \ldots), \\ V_{\nu-1}^{p_{\nu-1}} = F_{\nu-1}(V_\nu; \mathfrak{R}', \mathfrak{R}'', \ldots) \\ V_{\nu-2}^{p_{\nu-2}} = F_{\nu-2}(V_{\nu-1}, V_\nu; \mathfrak{R}', \mathfrak{R}'' \ldots) \\ \cdot \cdot \cdot \cdot \cdot \cdot \cdot \cdot \cdot \cdot \\ V_1^{p_1} = F_1(V_2, V_3, \ldots V_\nu; \mathfrak{R}', \mathfrak{R}'', \ldots); \end{cases}$$

3) $$x_0 = G_0 + G_1 V_1 + G_2 V_1^2 + \ldots + G_{p_1-1} V^{p_1-1},$$

where $G_0, G_1, G_2, \ldots$ are integral functions of $V_2, V_3, \ldots V_\nu$ and rational functions of $\mathfrak{R}', \mathfrak{R}'', \ldots$, and G_1 may be assumed to be 1(§ 211).

Taking the powers of x_0 and reducing in every case those powers of V_1 above the $(p_1 - 1)^{th}$, we obtain for every ν

$$x_0^\nu = G_0^{(\nu)} + G_1^{(\nu)} V_1 + G_2^{(\nu)} V_1^2 + \ldots + G_{p_1-1}^{(\nu)} V_1^{p_1-1}.$$

If these powers of x_0 are substituted in 1), we have

4) $$f(x_0) = H_0 + H_1 V_1 + H_2 V_1^2 + \ldots + H_{p_1-1} V^{p_1-1},$$

where the H's are formed additivily from the $G^{(\nu)}$'s and the coefficients of 1). Joining with $A)$ the equation of definition of V_1

B)
$$V_1^{p_1} - F_1(V_2, V_3 \ldots) = 0,$$

and applying Theorem I to A) and B), we have only two possibilities: either a root of B) is rational in the domain $(V_2, V_3, \ldots V_\nu;$ $\Re', \Re'', \ldots)$, or

$$H_0 = 0, \; H_1 = 0, \; H_2 = 0, \ldots H_{p_1-1} = 0.$$

Both cases actually occur. In the former the scheme 2), by which we passed from the original rational domain to the root x_0, can be simplified by merely suppressing the equation

$$V_1^{p_1} = F_1(V_2, V_3, \ldots),$$

and adding the p_1^{th} root of unity to the rational domain.

§ 213. As an example of this case we may take the equation of the third degree

$$f(x) \equiv x^3 - 3ax - 2b = 0,$$

the rational domain being formed from the coefficients a and b. By Cardan's formula

$$x_0 = \sqrt[3]{b + \sqrt{b^2 - a^3}} + \sqrt[3]{b - \sqrt{b^2 - a^3}}.$$

This algebraic expression can be arranged schematically as follows:

$$V_3^2 = b^2 - a^3,$$
$$V_2^3 = b + V_3,$$
$$V_1^3 = b - V_3,$$
$$x_0 = V_2 + V_1.$$

The expression for $\tfrac{1}{2} f(x_0)$, formed as in the preceding Section, then becomes

$$\tfrac{1}{2} f(x_0) \equiv -aV_2 + (V_2^2 - a)V_1 + V_2 V_1^2 = 0.$$

Comparing this with

$$V_1^3 - (b - V_3) = 0,$$

and determining V_1 from the last two equations, we obtain

$$V_1 = \frac{a(b + V_3) - a^2 V_2 + (b - V_3) V_2^2}{a^2 + (b + V_3) V_2 - a V_2^2},$$

so that V_1 is already contained in the rational domain $(V_2, V_3; a, b)$. If we now transform V_1 into an integral function of V_2 by the process of § 200, we obtain from the relations

$[a^2+(b+V_3)\omega V_2-a\omega^2 V_2^2][a^2+(b+V_3)\omega^2 V_2-a\omega V_2^2]$
$$=2b(b+V_3)(a+V_2^2),$$
$[a^2+(b+V_3)V_2-aV_2^2][2b(b+V_3)(a+V_2^2)]=[2b(b+V_3)]^2,$
$[a(b+V_3)-a^2 V_2+(b-V_3)V_2^2][2b(b+V_3)(a+V_2^2)]$
$$=4ab^2(b+V_3)V_2^2,$$

where ω is a primitive cube root of unity, the simpler form

$$V_1 = \frac{4ab^2(b+V_3)V_2^2}{4b^2(b+V_3)^2} = \frac{aV_2^2}{b+V_3}.$$

Removing V_3 from the denominator by multiplying both terms of the fraction by $b-V_3$, we have finally

$$V_1 = \frac{b-V_3}{a^2}V_2^2,$$

and herewith the reduced form of x_0

$$x_0 = V_2 + \frac{b-V_3}{a^2}V_2^2.$$

V_1 can therefore be suppressed in the scheme above.

§ 214. We return now to the results of § 212 and examine the second possible case. In

$$f(x_0) = H_0 + H_1 V_1 + H_2 V_1^2 + \ldots + H_{p_1-1} V_1^{p_1-1}$$

suppose that V_1 is not rational in the domain $(V_2, V_3, \ldots; \mathfrak{R}', \mathfrak{R}'', \ldots)$. Then from Theorem I

$$H_0 = 0, \ H_1 = 0, \ H_2 = 0, \ \ldots H_{p_1-1} = 0.$$

If now, in analogy to 3), we form the expressions

3') $x_k = G_0 + G_1 \omega_1^k V_1 + G_2 \omega_1^{2k} V_1^2 + \ldots G_{p_1-1} \omega_1^{(p_1-1)k} V_1^{p_1-1},$
$$(k=0, 1, \ldots p_1 - 1)$$

in which ω_1 is a primitive p_1^{th} root of unity, it follows, with the same notation and process as in § 212, that

$$x_k^\nu = G_0^{(\nu)} + G_1^{(\nu)} \omega_1^k V_1 + G_2^{(\nu)} \omega_1^{2k} V_1^2 + \ldots,$$
$$f(x_k) = H_0 + H_1 \omega_1 V_1 + H_2 \omega_1^2 V_1^2 + \ldots$$

Since the H's vanish identically, the latter expression is also equal to 0, i. e., x_k is a root of $f(x) = 0$ for $k = 1, 2, \ldots p_1 - 1$.

For example, in the case of the equations of the third degree

$$x^3 - 3ax - 2b = 0,$$
where the first of the two possibilities above has been excluded by reducing x_0 to the form
$$x_0 = V_2 + \frac{b-V_3}{a^2} V_2^2,$$
the other two roots are
$$x_1 = \omega V_2 + \frac{b-V_3}{a^2} \omega^2 V_2^2,$$
$$x_2 = \omega^2 V_2 + \frac{b-V_3}{a^2} \omega V_2^2. \qquad \left(\omega = \frac{-1+\sqrt{-3}}{2}\right)$$

§ 215. If now we make the allowable assumption (§ 211) that $G_1 = 1$, (whereupon V_1 may possibly take a new form different from its original one), we obtain by linear combination of the p_1 equations for $x_0, x_1, \ldots x_{p_1-1}$
$$x_0 = G_0 + V_1 + G_2 V_1^2 + \ldots + G_{p_1-1} V_1^{p_1-1},$$
$$x_1 = G_0 + \omega V_1 + G_2 \omega_1^2 V_1^2 + \ldots + G_{p_1-1} \omega_1^{p_1-1} V_1^{p_1-1},$$
$$\cdots$$
$$x_{p_1-1} = G_0 + \omega^{p_1-1} V_1 + G_2 \omega^{2(p_1-1)} V_1^2 + \ldots + G_{p_1-1} \omega^{(p_1-1)^2} V_1^{p_1-1},$$
the value of V_1:
$$V_1 = \frac{1}{p_1} \sum_{k=0}^{p_1-1} \omega_1^{-k} x_k.$$

The irrational function V_1 of the coefficients is therefore a rational and, in fact, a linear function of the roots $x_0, x_1, \ldots x_{p_1-1}$ as soon as the primitive p_1^{th} root of unity ω_1 is adjoined to the rational domain.

§ 216. In the construction of the scheme 2) it is not intended to assert that F_a necessarily contains $V_{a-1}, V_{a-2}, \ldots$ If V_{a-1} is missing in F_a, another arrangement of 2) is possible; we can replace the order
$$V_{a+1}^{p_{a+1}} = F_{a+1}(V_{a+2}, \ldots), \quad V_a^{p_a} = F_a(V_{a+2}, \ldots), \quad V_{a-1}^{p_{a-1}} = F_a(V_a, \ldots)$$
by the order
$$V_a^{p_a} = F_{a+1}(V_{a+2}, \ldots), \quad V_{a+1}^{p_{a+1}}(V_{a+2}, \ldots), \quad V_{a-1}^{p_{a-1}} = F_{a-1}(V_a, \ldots).$$

It is therefore possible, for example, that different V's occur at the end of the series 2). In this case different constructions 3) for the root x_0 are possible, and the theorem proved in the preceding Section holds for the last V of 2) in every case.

To prove the same theorem for all V's which occur, not in the last, but in the next to the last place in 2), we will simply assume that F_1 actually contains V_2. The proof (§ 215) of the theorem for V_1 was based on the fact that an expression

$$G_0 + V_1 + G_2 V_1^2 + \ldots$$

satisfied an equation with rational coefficients. We demonstrate the same property for an expression

$$L_0 + V_2 + L_2 V_2^2 + \ldots$$

If we suppose all the permutations of the roots of the equation 1) to be performed on

$$y - \Big[\frac{1}{p_1}\sum_{k=0}^{p_1-1} \omega_1^{-k} x^k\Big]^{p_1},$$

the product of the resulting expressions is an integral function of y, with coefficients which are symmetric in the x's and are therefore rational functions of $\Re', \Re'', \ldots$

If we denote this function by $\varphi(y)$, the coefficients of the equation

$$\varphi(y) = 0$$

belong to the domain $(\Re', \Re'', \ldots)$, and one of the roots, with possibly an unessential modification of the meaning of V_2 (cf. § 211) is

$$y_0 = \Big[\frac{1}{p_1}\sum \omega_1^{-k} x_k\Big]^{p_1} = V_1^{p_1} = F_1(V_2, \ldots) = L_0 + V_2 + L_2 V_2^2 + \ldots$$
$$+ L_{p_2-1} V_2^{p_2-1}.$$

It is therefore essential that V_2 should actually occur in F_1. We can now apply to $\varphi(y) = 0$ with the root y_0 the same process which we applied above to $f(x) = 0$ with the root x_0. If we assume, as is allowable, that the series $V_\nu, V_{\nu-1}, \ldots V_3, V_2$ is so chosen that V_2 is not rational in the preceding V's, it follows that

$$V_2 = \frac{1}{p_2}\sum_{k=0}^{p_2-1} \omega_2^{-k} y_k,$$

where ω_2 is a primitive p_2^{th} root of unity. Every y_k is produced

16a

from y_0 by certain substitutions among the $x_0, x_1, \ldots x_{n-1}$; consequently V_2 is a rational integral function of the p_1^{th} degree of the roots of $f(x) = 0$, provided the quantities ω_1 and ω_2 are adjoined to the rational domain.

In the same way every V can be treated which occurs in the next to the last but not in the last place in 2). Proceeding upward in the series we have finally.

Theorem II. *The explicit algebraic function x_0, which satisfies a solvable equation $f(x) = 0$, can be expressed as a rational integral function of quantities*

$$V_1, V_2, V_3, \ldots V_\nu,$$

with coefficients which are rational functions of the quantities $\Re', \Re''$. The quantities V_λ are on the one hand rational integral functions of the roots of the equation $f(x) = 0$ and of primitive roots of unity, and on the other hand they are determined by a series of equations

$$V_a{}^{p_a} = F(V_{a-1}, V_{a-2}, \ldots V_\nu; \Re', \Re'', \ldots).$$

In these equations the $p_1, p_2, p_3, \ldots p_\nu$ are prime numbers, and $F_1, F_2, \ldots F_\nu$ are rational integral functions of their elements V and rational functions of the quantities $\Re', \Re'', \ldots$, which determine the rational domain.

§ 217. This theorem ensures the possibility of the application of the theory of substitutions to investigation of the solution of equations. It furnishes further the proof of the fundamental proposition:

Theorem III. *The general equations of degree higher than the fourth are not algebraically solvable.*

For if the n quantities $x_1, x_2, \ldots x_n$, which in the case of the general equation are independent of one another, could be algebraically expressed in terms of $\Re', \Re'', \ldots$, then the first introduced irrational function of the coefficients, V_ν, would be the p_ν^{th} root of a rational function of $\Re', \Re'', \ldots$ Since, from Theorem II, V_ν is a rational function of the roots, it appears that V_ν, as a p_ν-valued function of $x_1, x_2, \ldots x_n$, the p_ν^{th} power of which is symmetric, is either the square root of the discriminant, or differs from the latter only by

THE ALGEBRAIC SOLUTION OF EQUATIONS. 251

a symmetric factor. Consequently we must have $p_\nu = 2$ (§ 56). If we adjoin the function $V_\nu = S_1 \sqrt{} \sqrt{J}$ to the rational domain, the latter then includes all the one-valued and two-valued functions of the roots. If we are to proceed further with the solution, as is necessary if $n > 2$, there must be a rational function $V_{\nu-1}$ of the roots, which is $(2p_{\nu-1})$-valued, and of which the $(p_{\nu-1})^{\text{th}}$ power is two-valued. But such a function does not exist if $n > 4$ (§ 58). Consequently the process, which should have led to the roots, cannot be continued further. The general equation of a degree above the fourth therefore cannot be algebraically solved.

§ 218. We return now to the form of the roots of solvable algebraic equations

3) $\qquad x_0 = G_0 + V_1 + G_2 V_1^2 + \ldots + G_{p_1-1} V_1^{p_1-1}.$

We adjoin to the rational domain the primitive $p_1^{\text{th}}, p_2^{\text{th}}, \ldots$ roots of unity, and assume that the scheme which leads to x_0 is reduced as far as possible, so that for instance V_a is not already contained in the rational domain $(V_{a-1} \ldots V_\nu; \mathfrak{R}', \mathfrak{R}'', \ldots; \omega_1, \omega_2, \ldots)$. We have seen that the substitution of

$$\omega_1^k V_1 \quad (k = 1, 2, \ldots p_1 - 1)$$

for V in 3) produces again a root of $f(x) = 0$. We proceed to prove the generalized theorem:

Theorem IV. *If in the scheme* 2), *which leads to the expression* 3) *for* x_0, *any* V_a *is multiplied by any root of unity, the values* $V_{a-1}, V_{a-2}, \ldots V_2, V_1$ *will in general be converted into new quantities* $v_a, v_{a-1}, \ldots v_2, v_1$. *If the latter are substituted in the place of the former in the expression for* x_0, *the result is again a root of* $f(x) = 0$.

We may, without loss of generality, assume that $f(x)$ is irreducible in the domain $(\mathfrak{R}', \mathfrak{R}'', \ldots)$.

Starting now from 3), and denoting by ω_τ a primitive τ^{th} root of unity, we construct

$$\prod_{\lambda=0}^{p_1-1}(x - x_\lambda) = \prod_{\lambda=0}^{p_1-1}[x - (G_0 + \omega_1^\lambda V_1 + G_2 \omega_1^{2\lambda} V_1^2 + \ldots)].$$

In this product V_1 certainly vanishes. Possibly other V's vanish also. Suppose that V_a ($a \geq 2$) is the lowest V that actually occurs. Then

4) $\displaystyle\prod_{\lambda=0}^{p_1-1}(x-x_\lambda) = f_a(x; V_a, V_{a+1}, \ldots) = a_0 + a_1 V_a + a_2 V_a^2 + \ldots$

The a's which occur here belong to the domain $(V_{a+1}, \ldots)$. We construct further

5) $\displaystyle\prod_{\lambda=0}^{p_a-1} f_a(x; \omega_a^\lambda V_a, V_{a+1}\ldots) = f_b(x; V_b, V_{b+1}, \ldots) = \beta_0 + \beta_1 V_b + \beta_2 V_b^2 + \ldots$

where V_b ($b \geq a+1$) is again the lowest V that actually occurs. Similarly let

6) $\displaystyle\prod_{\lambda=0}^{p_b-1} f_b(x; \omega_b^\lambda V_b, V_{b+1}, \ldots) = f_c(x; V_c, V_{c+1}\ldots) = \gamma_0 + \gamma_1 V_c + \gamma_2 V_c^2 + \ldots,$

and finally, supposing the series to end at this point.

7) $\displaystyle\prod_{\lambda=0}^{p_c-1} f_c(x; V_c, V_{c+1}, \ldots) = f_d(x; \mathfrak{R}', \mathfrak{R}'', \ldots),$

where f_d is rational in $\mathfrak{R}', \mathfrak{R}'', \ldots$, all the $V_{c+1}, \ldots V_\nu$ disappearing with V_c.

We assume now, reserving the proof for the moment, that the functions

$$f_a(x; V_a, \ldots), \ f_b(x; V_b, \ldots), \ f_c(x; V_c, \ldots), \ f_d(x; \mathfrak{R}', \ldots)$$

are irreducible in the domains

$$(V_a, V_{a+1} \ldots), \ (V_b, V_{b+1} \ldots), \ (V_c, V_{c+1}, \ldots), \ (\mathfrak{R}', \mathfrak{R}'' \ldots),$$

respectively. Then $f_d(x; \mathfrak{R}', \ldots) = 0$ and $f(x) = 0$ have in the domain $(\mathfrak{R}', \mathfrak{R}'', \ldots)$ one root $x = x_0$ in common, since $x - x_0$ occurs as a common factor of $f_d(x)$ and $f(x)$. Both these functions being by assumption irreducible, it follows that

$$f_d(x; \mathfrak{R}', \mathfrak{R}'', \ldots) = f(x).$$

If now we assign to V_ν any arbitrary value v_ν consistent with $V_\nu^{p_\nu} = F_\nu(\mathfrak{R}', \ldots)$ and to $V_{\nu-1}$ any value $v_{\nu-1}$ consistent with $V_{\nu-1}^{p_{\nu-1}} = F_{\nu-1}(v_\nu; \mathfrak{R}', \ldots)$, and continue in this way, we have the series

THE ALGEBRAIC SOLUTION OF EQUATIONS. 253

2')
$$\begin{cases} v_\nu^{p_\nu} = F_\nu(\Re', \Re'', \ldots), \\ v_{\nu-1}^{p_{\nu-1}} = F_{\nu-1}(v_\nu; \Re', \Re'', \ldots), \\ v_{\nu-2}^{p_{\nu-2}} = F_{\nu-2}(v_{\nu-1}, v_\nu; \Re', \Re'' \ldots), \\ \cdot \cdot \cdot \cdot \cdot \cdot \cdot \cdot \cdot \cdot \cdot \cdot \\ v_1^{p_1} = F_1(v_2, v_3, \ldots v_\nu; \Re', \Re'', \ldots); \end{cases}$$

3') $\quad \xi_0 = g_0 + g_1 v_1 + g_2 v_1^2 + \ldots + g_{p_1-1} v_1^{p_1-1}$.

3') being obtained from 3) by putting the v's in place of the V's. The product

$$\prod_{\lambda=0}^{p_1-1}[x-(g_0+\omega_1^\lambda v_1+g_2\omega_1^{2\lambda}v_1^2+\ldots)]$$

will only differ from those obtained above by the introduction of the g's and v's in place of the G's and V's, since in all the reductions 2') replaces 2). Consequently this product is equal to $f_a(x; v_a, v_{a+1}, \ldots)$ and similarly

$$\prod_{\lambda=0}^{p_a-1} f_a(x; \omega_a^\lambda v_a, v_{a+1}, \ldots) = f_b(x; v_b, v_{b+1}, \ldots),$$

$$\prod_{\lambda=0}^{p_b-1} f_b(x; \omega_b^\lambda v_b, v_{b+1}, \ldots) = f_c(x; v_c, v_{c+1}, \ldots),$$

$$\prod_{\lambda=0}^{p_c-1} f_c(x; \omega_c^\lambda v_c, v_{c+1}, \ldots) = f_d(x; \Re', \Re'', \ldots) = f(x).$$

This furnishes the proof that ξ_0 is a root of $f(x) = 0$.

We have still to prove the irreducibility of $f_a(x), f_b(x), \ldots$ in the rational domains $(V_a, V_{a+1}, \ldots), (V_b, V_{b+1}, \ldots) \ldots$, respectively.

Assuming the irreducibility of $f_a(x)$ in the domain $(V_a, V_{a+1}, \ldots)$, we proceed to demonstrate that of $f_b(x)$ in the domain $(V_b, V_{b+1}, \ldots)$. The method employed applies in general.

If $\varphi(x; V_b, \ldots)$ is one of the irreducible factors of $f_b(x)$, so chosen that it contains $f_a(x; V_a, \ldots)$ as a factor, then we have in the domain $V_a, V_{a+1}, \ldots$ the equation

8) $\quad \varphi(x; V_b, \ldots) = f(x; V_a, \ldots) \cdot \varphi'(x; V_a, \ldots),$

which can be rewritten in the form

$A)$ $\quad \chi_0 + \chi_1 V_a + \chi_2 V_a^2 + \ldots + \chi_{p_a-1} V_a^{p_a-1}.$

where the χ's are rational functions of $V_{a+1}, \ldots$ Since moreover

$B)$ $\quad V_a^{p_a} - F_a(V_{a+1}, \ldots) = 0,$

it follows from Theorem I that either V_a is rational in $V_{a+1}, \ldots$, which is to be excluded, or all the χ's vanish, so that $A)$ and with it the equation 8) above still hold, if V_a is replaced by $\omega_a V_a, \omega_a^2 V_a, \ldots$ Again, $f_a(x; \omega_a^\alpha V_a, \ldots)$ is different from $f_a(x; \omega_a^\beta V_a, \ldots)$. For if we write

$$f_a(x; V_a, \ldots) = \varepsilon_0 + \varepsilon_1 V_a + \varepsilon_2 V_a^2 + \ldots,$$

it would follow from the equality of the two functions f_a that

$A_1)$ $\quad \varepsilon_1(\omega_a^\alpha - \omega_b^\beta) V_a + \varepsilon_2(\omega_a^{2\alpha} - \omega_a^{2\beta}) V_a^2 + \ldots = 0,$

and consequently, from the equation of definition

$B_1)$ $\quad V_a^p - F_a(V_{a+1}, \ldots) = 0,$

that V_a must be rational in the domain $(V_{a+1}, \ldots \Re', \ldots \omega_1, \ldots)$, since $\alpha \neq \beta$.

Accordingly $f_a(x; V_a, \ldots), f_a(x; \omega_a V_a, \ldots), \ldots$ are all divisors of φ. All these functions are different from one another, and they are all irreducible in the domain $(V_a, V_{a+1}, \ldots)$. Consequently φ contains their product, which, on account of the degrees of f_a and f_b in x, is possible only if φ and f_b coincide.

Since the foregoing proof holds for every irreducible factor of 1), it still holds if we drop the assumption of irreducibility.

§ 219. At the beginning of the preceding Section we remarked that in the product construction with V_1 other V's might vanish. This possibility is however excluded in the case of certain V's, as we shall now show.

We designate any V_τ of 2) as an *external radical* when the following $V_{\tau+1}, V_{\tau+2}, \ldots$ i. e., $F_{\tau+1}, F_{\tau+2}, \ldots$ do not contain V_τ. Every such external radical can be brought to the last position of 2), and the expression of x_0, as given in 3), can be arranged in terms of every external radical present. We shall see that in the product construction with V_1 no other external radical can be missing. Thus, if V_τ is missing in

$$f_a(x; V_a, \ldots) = \prod_{\lambda=0}^{p_1-1} [x - (G_0 + \omega_1^\lambda V_1 + G_2 \omega_1^{2\lambda} V_1^2 + \ldots)],$$

THE ALGEBRAIC SOLUTION OF EQUATIONS. 255

then f_a cannot be changed if we replace V_τ in the fundamental radical expression by $\omega_\tau^\kappa V_\tau$, without thereby changing V_1. If, as a result, the G's are converted into the g's, we should then have also

$$f_a(x; V_a, \ldots) = \prod_{\lambda=0}^{\mu_1-1} [x - (g_0 + \omega_1^\lambda V_1 + g_2 \omega_1^{2\lambda} V_1^2 \ldots)].$$

Every linear factor in x of this last expression must therefore be equal to some factor of the preceding expression

A) $g_0 + V_1 + g_2 V_1^2 + \ldots = G_0 + \omega_1^\lambda V_1 + G_2 \omega_1^{2\lambda} V_1^2 + \ldots$

Taking into account the equation of definition

B) $V_1^{\rho_1} - F_1(V_2, \ldots) = 0,$

it follows from Theorem I that either V_1 is rational in $V_2, V_3, \ldots \omega_\tau$, which may be excluded, since otherwise 2) could be reduced further on the adjunction of ω_τ, or that

$$g_0 = G_0, \quad g_2 = G_2, \ldots$$

In some one of these equations V_τ must actually occur. Developing this equation according to powers of V_τ, we have

A_1) $K_0 + K_1 V_\tau + K_2 V_\tau^2 + \ldots = K_0 + K_1 \omega_\tau V_\tau + K_2 \omega_\tau^2 V_\tau^2 + \ldots,$

and combining with this

B_1) $V_\tau^{\rho_\tau} - F_\tau(V_{\tau+1}, \ldots) = 0,$

the impossibility of both alternatives of Theorem I appears at once. Consequently V_τ could not have been missing in the product construction.

If we consider only f_a (§ 218, 4)), the series 2) ending with V_a can also contain external radicals, in fact possibly such as are not external in respect to the entire series. These also cannot vanish in the further product construction. The irreducibility of f_a being borne in mind, the proof is exactly the same as the preceding.

Theorem V. *In the product construction of the preceding Section no external radicals can disappear from f_a except V_1. The same is true for f_b in respect to the external radicals occurring among $V_\nu, V_{\nu+1}, \ldots V_a$, and so on.*

If several external radicals occur in x_0 or in one of the expressions $f_a, f_b, f_c, \ldots$, the product of all the corresponding exponents is a factor of n.

Theorem VI. *If an irreducible equation of prime degree p is algebraically solvable, the solution will contain only one external radical. The index of the latter is equal to p, and if ω is a primitive p^{th} root of unity, the polynomial of the equation is*

$$f(x) = \prod_{\lambda=0}^{p-1} [x - (G_0 + \omega^\lambda V_1 + G_2 \omega^{2\lambda} V_1^2 + \ldots + G_{p-1} \omega^{(p-1)\lambda} V_1^{p-1})].$$

Theorem VII. *If the algebraic expression*

3) $\qquad x_0 = G_0 + V_1 + G_2 V_1^2 + \ldots G_{p_1-1} V^{p_1-1}$

is a root of an equation $f(x) = 0$, which is irreducible in the domain $(\Re', \Re'', \ldots)$, and if we construct the product of the p_1 factors, in which V_1 is replaced by $\omega_1 V_1, \omega_1^2 V_1 \ldots$

$$f_a(x; V_a, \ldots) = \prod (x - x_k),$$

where V_a is the lowest V present, and again the product $f_b(x; V_b, \ldots)$ of the p_a factors $f_a(x; \omega_a^\lambda V_a, \ldots)$, and so on, we come finally to the equation $f(x) = 0$, the degree of which is

$$n = p_1 p_a p_b \ldots$$

The functions $f_a, f_b, \ldots$ are irreducible in the domains $(V_a, V_{a+1}, \ldots)$, $(V_b, V_{b+1}, \ldots), \ldots$

§ 220. We examine now further those radicals which vanish in the first product construction. The remaining $V_a, V_{a+1}, \ldots$ are not altered in the product construction. We may therefore add these to the rational domain, or, in other words, we may consider an irreducible equation $f(x) = f_a(x; V_a, \ldots)$ in the rational domain $(V_a, V_{a+1}, \ldots; \Re', \Re'', \ldots)$.

Here all the $V_1, V_2, \ldots V_{a-1}$ already vanish in the first product construction.

We examine now what is the result of assigning to V_{a-1} any arbitrary value consistent with its equation of definition, then with this basis assigning any arbitrary value of V_{a-2} consistent with its equation of definition, and so on. Suppose that the functions

$$V_{a-1}, V_{a-2}, \ldots V_2, V_1; G_0, G_1, \ldots G_{p-1}$$

are thereupon converted into

THE ALGEBRAIC SOLUTION OF EQUATIONS. 257

$$v_{a-1}, v_{a-2}, \ldots v_2, v_1; \ g_0, g_2, \ldots g_{p-1}.$$

The new value assumed by x_0 is then

$$\xi_0 = g_0 + g_1 v_1 + g_2 v_1^2 + g_3 v_1^3 + \ldots + g_{p-1} v_1^{p-1}.$$

From § 218 ξ_0 is again a root, and this together with the system $\xi_1, \xi_2, \ldots \xi_{p-1}$, which arises from ξ_0 when v_1 is replaced by ωv_1, $\omega^2 v_1, \ldots \omega^{p-1} v_1$, gives again the complex of all the roots. We can therefore take

$$g_0 + v_1 + g_2 v_1^2 + \ldots = G_0 + \omega' V_1 + G_2 \omega'^2 V_1^2 + \ldots,$$
$$g_0 + \omega v_1 + g_2 \omega^2 v_1^2 + \ldots = G_0 + \omega'' V_1 + G_2 \omega''^2 V_1^2 + \ldots,$$
$$g_0 + \omega^2 v_1 + g_2 \omega^4 v_1^2 + \ldots = G_0 + \omega''' V_1 + G_2 \omega'''^2 V_1^2 + \ldots,$$
$$\cdots \cdots \cdots \cdots \cdots \cdots \cdots \cdots$$

where $\omega', \omega'', \omega''', \ldots$ are the pth roots of unity $\omega, \omega^2, \omega^3, \ldots$, apart from their order. By addition of these equations we obtain

$$g_0 = G_0,$$

so that G_0 is unaffected by the modifications of $V_{a-1}, V_{a-2}, \ldots V_2$. Also $p G_0$ is the sum of all the roots, and is therefore a rational function in the domain $(\mathfrak{R}', \mathfrak{R}'', \ldots)$.

Again we obtain from the system above the equation

$$p v_1 = G_0(1 + \omega^{-1} + \omega^{-2} + \ldots) + V_1(\omega' + \omega'' \omega^{-1} + \omega''' \omega^{-2} + \ldots)$$
$$+ G_2 V_1^2(\omega'^2 + \omega''^2 \omega^{-1} + \omega'''^2 \omega^{-2} + \ldots)$$
$$+ \ldots$$

Here the first term on the right vanishes. We denote the parentheses in the following terms briefly by $p\Omega_1, p\Omega_2, p\Omega_3, \ldots$, and write

9) $\qquad v_1 = \Omega_1 V_1 + \Omega_2 G_2 V_1^2 + \Omega_3 G_3 V_1^3 + \ldots$

On raising this to the pth power

A) $\qquad v_1^p = F_1(v_2, v_3, \ldots v_{a-1}; \mathfrak{R}', \ldots) = [\Omega_1 V_1 + \Omega_2 G_2 V_1^2 + \ldots]^p$
$$= A_0 + A_1 V_1 + A_2 V_2^2 + \ldots,$$

and annexing the equation of definition

B) $\qquad V_1^p = F_1(V_2, V_3, \ldots V_{a-1}; \mathfrak{R}', \ldots),$

it follows from Theorem I that either V_1 is rational in

$$V_2, V_3, \ldots V_{a-1}; \ v_2, v_3, \ldots v_{a-1}; \ \mathfrak{R}', \mathfrak{R}'', \ldots,$$

or that

$$V_1^p = A_0, \quad A_1 = 0, \quad A_2 = 0, \ldots A_{p-1} = 0.$$

17

We consider now the first of these alternatives. In the rational expression of V_1 in terms of $V_2, V_3, \ldots; v_2, v_3, \ldots; \Re' \, \Re'', \ldots$ all the $v_2, v_3, \ldots v_{a-1}$ cannot vanish; otherwise V_2 should have been suppressed in 2). If then we define V_1, as in §§ 208 and 212 by a system of successive radicals, some V_κ will occur last among the v's and some V_λ last among the v's. If we substitute the expression for V_1 in x_0, we have

$$x_0 = R(V_1, \ldots V_{a-1}; \Re', \ldots) = R_1(V_2, \ldots V_{a-1}; v_2, \ldots v_{a-1}; \Re', \ldots)$$

Here all the v's cannot vanish, as we have just seen. For the same reason all the V's cannot vanish, since we might have started out from ξ_0. But V_κ and v_λ are two external radicals, and the product of their exponents must therefore be a factor of p (Theorem V). This being impossible, the first alternative is excluded.

Accordingly we must have in A)

$$V_1^p = A_0, \quad A_1 = 0, \quad A_2 = 0, \ldots A_{p-1} = 0.$$

The question now arises what the form of 9) must be in order that its pth power may take the form $V_1^p = A_0$. The equation A) is

$$v_1^p = A_0 + A_1 V_1 + A_2 V_1^2 + \ldots = [\Omega_1 V_1 + \Omega_2 G_2 V_1^2 + \ldots]^p.$$

The result just obtained shows that the left member is unchanged if V_1 is replaced by $\omega V_1, \omega^2 V_1^2, \ldots$ Consequently

$$[\Omega_1 \omega V_1 + \Omega_2 G_2 \omega^2 V_1^2 + \ldots]^p = v_1^p,$$
$$[\Omega_1 \omega^2 V_1 + \Omega_2 G_2 \omega^4 V_1^2 + \ldots]^p = v_1^p,$$
$$\cdots \cdots \cdots \cdots \cdots \cdots$$

and on the extraction of the pth root we have

$$\Omega_1 \omega V_1 + \Omega_2 G_2 \omega^2 V_1^2 + \ldots = \omega^\kappa v_1.$$

But from 9) follows also

$$\Omega_1 \omega^\kappa V_1 + \Omega_2 G_2 \omega^\kappa V_1^2 + \ldots = \omega^\kappa v_1,$$

and equating the two left members and applying Theorem I as usual, we have

$$\Omega_1 = 0, \quad \Omega_2 G_2 = 0, \ldots \Omega_{\kappa-1} G_{\kappa-1} = 0, \quad \Omega_{\kappa+1} G_{\kappa+1} = 0, \ldots,$$

that is, 9) reduces to the single term

9') $$V_1 = \Omega_\kappa G_\kappa V_1^\kappa.$$

THE ALGEBRAIC SOLUTION OF EQUATIONS. 259

Substituting this result, together with $g_0 = G_0$ in the expression for ξ_0, we have
$$\xi_0 = G_0 + \Omega_\kappa G_\kappa V_1^\kappa + g_2(\Omega_\kappa G_\kappa V_1^\kappa)^2 + g_3(\Omega_\kappa G_\kappa V_1^\kappa)^3 + \ldots$$

On the other hand the root ξ_0, which is contained among $x_0, x_1, \ldots$, can also be expressed in the form
$$\xi_0 = G_0 + \omega' V_1 + G_2 \omega'^2 V_1^2 + G_3 \omega'^3 V_1^3 + \ldots,$$

and, comparing the two right members, it follows from Theorem I that terms with equal or congruent exponents (mod. p) are identical. In particular we have
$$\Omega_\kappa G_\kappa V_1^\kappa = G_\kappa \omega'^\kappa V_1^\kappa,$$
and therefore
$$\Omega_\kappa = \omega'^\kappa,$$
10) $$v_1 = \omega'^\kappa G_\kappa V_1^\kappa.$$

Theorem VIII. *If, in the explicit expression* 3) *of the root x_0 of an irreducible equation of prime degree p, the irrationalities V are modified in any way consistent with the equations of definition, then V_1^p is converted at the same time into $(G_\lambda V_1^\lambda)^p$.*

§ 221. The relations arising from such a transformation are most readily discussed by the introduction of a *primitive congruence root* e in the place of the prime number p. We write
$$V_1 = \sqrt[p]{R_0}, \quad G_e V_1^e = \sqrt[p]{R_1}, \quad G_{e^2} V_1^{e^2} = \sqrt[p]{R_2}, \ldots,$$
where however every e^a is to be reduced to its least not negative remainder (mod. $p-1$). Then the quantities
$$V_1, \quad G_2 V_1^2, \quad G_3 V_1^3, \ldots G_{p-1} V_1^{p-1}$$
coincide, apart from their order, with the quantities
$$\sqrt[p]{R_0}, \quad \sqrt[p]{R_1}, \quad \sqrt[p]{R_2}, \ldots \sqrt[p]{R_{p-2}},$$
and we have
11) $$x_0 = G_0 + \sqrt[p]{R_0} + \sqrt[p]{R_1} + \ldots + \sqrt[p]{R_{p-2}}.$$

The changes in the values of the radicals, considered above, which replace V_1^p by $(G_\lambda V_1^\lambda)^p$ and consequently $[G_a V_1^a]$ by $[G_{a_\lambda} V_1^{a_\lambda}]^p$, where $a_\lambda < p$ and
$$a_\lambda \equiv \lambda a \pmod{p},$$
have therefore the effect of replacing every R_a by

$$R_{a+\kappa} \quad (a = 0, 1, \ldots p-2),$$

where κ is $< p-1$ and is defined by the congruence

$$c^\kappa \equiv \lambda \quad (\text{mod. } p).$$

Consequently the quantities

I) $\qquad R_0, \ R_1, \ R_2, \ldots R_{p-2}$

are converted in order into

$$R_\kappa, \ R_{\kappa+1}, \ R_{\kappa+2}, \ldots R_{\kappa+p-2},$$

and, if the same operation is performed a times, I) is replaced by

$$R_{a\kappa}, R_{a\kappa+1}, R_{a\kappa+2}, \ldots + R_{a\kappa+p-2},$$

where the indices are of course to be reduced (mod. $p-1$).

If there is another modification of the radicals, which converts R_0 into R_μ, this on being repeated β times converts the series I) into

$$R_{\beta\mu}, \ R_{\beta\mu+1}, \ R_{\beta\mu+2}, \ldots R_{\beta\mu+p-2}.$$

Finally if we apply the first operation a times and the second β times, I) becomes

$$R_{a\kappa+\beta\mu}, \ R_{a\kappa+\beta\mu+1}, \ R_{a\kappa+\beta\mu+k-2}.$$

Here a and β can be so chosen that $a\kappa + \beta\mu$ gives the greatest common divisor of κ and μ. Consequently if R_k is the R of lowest index which is obtainable from R_0 by alteration of the radicals, every other R obtainable from R_0 in this way will have for its index a multiple of k, so that the permutations of the R's take place only within the systems

$$R_0, \ R_k, \ R_{2k} \ldots R_{\left(\frac{p-1}{k}-1\right)k},$$

$$R_1, \ R_{k+1}, \ R_{2k+1}, \ldots R_{\left(\frac{p-1}{k}-1\right)k+1},$$

$$\cdot \ \cdot \ \cdot \ \cdot \ \cdot \ \cdot \ \cdot \ \cdot \ \cdot \ \cdot$$

Here k is a divisor of $p-1$.

There are then alterations in the meaning of the radicals which produce the substitution

$$(R_0 \, R_k \, R_{2k} \ldots) \, (R_1 \, R_{k+1} \, R_{2k+1} \ldots) \ldots$$

§ 222. The preceding developments enable us to determine the group of the irreducible solvable equations 1) of prime degree p.

Every permutation of the x's can only be produced by the alter-

THE ALGEBRAIC SOLUTION OF EQUATIONS. 261

ations in the radicals $V_1, V_2, \ldots V_{a-1}$, and consequently only such permutations of the x's can occur in the group as are produced by alterations of the V's. From the result of the preceding Section V_1 can be converted into $\omega^\tau G_{e^k} V_1^{e^k}$, and the possible alterations in V_1 do not change this form. Substituting this in the table of § 215, we have

$$\xi_0 = G_0 + \omega^\tau G_{e^k} V_1^{e^k} + \ldots,$$
$$\xi_1 = G_0 + \omega^{\tau+1} G_{e^k} V_1^{e^k} + \ldots,$$
$$\cdot \quad \cdot \quad \cdot$$

We examine now whether any root x_μ can remain unchanged in this transformation. In that case we must have

$$G_0 + \ldots + \omega^{\mu e^k} G_{e^k} V_1^{e^k} + \ldots = G_0 + \omega^{\tau+\mu} G_{e^k} V_1^{e^k} + \ldots,$$

and from the method which we have repeatedly employed it follows, as a necessary and sufficient condition, that

$$\omega^{\mu e^k} = \omega^{\tau+\mu}$$
$$\mu e^k \equiv \mu + \tau \quad (\text{mod. } p).$$

If $e^k \equiv 1$ (mod. p), then for $\tau \not\equiv 0$ there is no solution μ, and therefore no root x_μ which remains unchanged. But for $\tau = 0$, every μ satisfies the condition, and the substitution reduces to identity.

If $e^k \not\equiv 1$, then for every τ there is a single solution μ, and the corresponding substitution leaves only one element unchanged.

Theorem IX. *The group of a solvable irreducible equation of prime degree is the metacyclical group (§ 134) or one of its subgroups.*

§ 223. Since now, as we saw in § 221, all the substitutions of the group permute the values $R_0, R_k, R_{2k}, \ldots$ only among themselves, the symmetric functions of these values are known, and the values themselves are the roots of an equation of degree $\dfrac{p-1}{k}$. The latter is an Abelian equation since the group permutes the values $R_0, R_k, R_{2k}, \ldots$ only cyclically. Consequently every $R_k, R_{2k}, R_{3k}, \ldots$ is a rational function of R_0. But the same is true of every R_a. For the form of the substitution at the end of § 221 shows that after the adjunction of R_0 all the other R_a's are known, since the group reduces to 1.

Finally it appears that if
$$R_a = F_a(R_0)$$
then
$$R_{a+km} = F_a(R_{km}).$$
For the application of properly chosen substitutions of the group converts the first equation into the second.

We consider now all the substitutions of the group of $f(x) = 0$ which leave $R_0 = V_1^p$ unchanged and accordingly can only convert V_1 into some $\omega^\nu V_1$. Then x_0 is replaced by x_ν. But since R_1 is a rational function of R_0, it appears that $R_1 = G_e^p V^\varphi$ is also unchanged, so that $G_e V_1^e$ is converted into some $G_e \omega^\mu V_1^e$. The power ω^μ can be determined from x_ν; for the expression for x_ν contains the term $G_e V_1^e \omega^{\nu e}$, and this must be identical with $G_e \omega^\mu V_1^e$. Consequently $\mu = \nu e$, and $G_e V_1^e$ becomes
$$G_e \omega^{\nu e} V_1^e,$$
while at the same time V_1^e becomes
$$\omega^{\nu e} V_1^e,$$
so that the factor G_e remains unchanged. That is, every substitution of the group, which leaves R_0 unchanged, leaves G_e unchanged also. Accordingly G_e is a rational function of R_0. The same is true of all the other G's. We can therefore write

12) $\quad x_0 = G_0 + V_1 + \varphi_2(V_1^p) \cdot V_1^2 + \varphi_3(V_1^p) \cdot V_1^3 + \ldots + \varphi_{p-1}(V_1^p) \cdot V_1^{p-1},$

where $\varphi_2, \varphi_3, \ldots$ are rational functions of V_1^p in the domain $(\Re', \Re'', \ldots)$. From this it appears that in 11) the radicals $\sqrt[p]{R_0}, \sqrt[p]{R_1}, \ldots$ do not admit of multiplying every term by an arbitrary root of unity, as indeed is already evident á priori since otherwise x_0 would have not p, but p^p values.

A still further transformation of 12) is possible. We have
$$\sqrt[p]{R_1} = G_e \sqrt[p]{R_0^e} = \varphi_1(R_0) \cdot \sqrt[p]{R_0^e}.$$
From § 221 there are alterations in the $V_1, V_2, \ldots V_{a-1}$ which convert R_0 into R_k and consequently $\sqrt[p]{R_0}$ into $\omega^{\sigma^k} \sqrt[p]{R_k}$. The form of the exponent of ω evidently involves no limitation. At the same time the x_0 becomes

THE ALGEBRAIC SOLUTION OF EQUATIONS.   263

$$x_\sigma = G_0 + \omega^\sigma \sqrt[p]{R_0} + \ldots + \omega^{\sigma^k} \sqrt[p]{R_k} + \omega^{\sigma^{k+1}} \sqrt[p]{R_{k+1}} + \ldots,$$

and since R_1 becomes R_{k+1}, it follows that $\sqrt[p]{R_1}$ becomes

$$\omega^{\sigma^{k}+1} \sqrt[p]{R_{k+1}}.$$

If now we apply these transformations to the equations above, we obtain

$$\omega^{\sigma^k+1} \sqrt[p]{R_{k+1}} = \psi_1(R_k) \cdot (\omega^{\sigma^k} \sqrt[p]{R_k})^\epsilon,$$

$$\sqrt[p]{R_{k+1}} = \psi_1(R_k) \cdot \sqrt[p]{R_k^\epsilon}.$$

We can therefore also write

$$x_0 = G_0 + \sqrt[p]{R_0} \qquad + \sqrt[p]{R_k} \qquad + \sqrt[p]{R_{2k}} + \ldots$$
$$+ \psi_1(R_0) \cdot \sqrt[p]{R_0^\epsilon} + \psi_1(R_k) \cdot \sqrt[p]{R_k^\epsilon} + \psi_1(R_{2k}) \cdot \sqrt[p]{R_{2k}^\epsilon} + \ldots$$
$$+ \psi_2(R_0) \cdot \sqrt[p]{R_0^{\epsilon^2}} + \psi_2(R_k) \cdot \sqrt[p]{R_k^{\epsilon^2}} + \psi_2(R_{2k}) \cdot \sqrt[p]{R_{2k}^{\epsilon^2}} + \ldots$$

. . . .

Theorem X. *The roots of a solvable equation of prime degree p can be written in either of the two forms* 12) *or* 13). *In* 13) $\sqrt[p]{R_k}, \sqrt[p]{R_{2k}}, \ldots$ *are rational functions of* $\sqrt[p]{R_0}$. *The values*

$$R_0, R_k, R_{2k}, \ldots R_{(\frac{p-1}{k}-1)k}$$

are roots of a simplest Abelian equation, the group of which is composed of the powers of

$$\sigma = (R_0 \; R_k \; R_{2k} \ldots)$$

Its roots are connected by the relations

14) $\sqrt[p]{R_k} = f(R_0) \cdot \sqrt[p]{R_0^\epsilon}, \quad \sqrt[p]{R_{2k}} = f(R_k) \cdot \sqrt[p]{R_k^\epsilon},$
        $\sqrt[p]{R_{3k}} = f(R_{2k}) \cdot \sqrt[p]{R_{2k}^\epsilon}, \ldots$

where $G_{\sigma^k} = f(R_0)$.

§ 224. The form 13), together with the relations 14) between $R_0, R_k, \ldots$, is not only necessary but also sufficient for a root x_0 of an irreducible solvable equation of prime degree p. For 14) shows that all the possible permutations among the R's are simply powers of σ. If now σ^a converts R_0 into R_{ak}, then $\sqrt[p]{R_0}$ will become some $\omega^\mu \sqrt[p]{R_0}$, and from 14)

$$\sqrt[p]{R_k} = f(R_0) \cdot \sqrt[p]{R_0^\epsilon} \text{ becomes } \omega^{\mu\epsilon} f(R_0) \cdot \sqrt[p]{R_0^\epsilon}, \text{ etc.}$$

Consequently 13) becomes

$$x_\mu = G_0 + \quad \omega^\mu \sqrt[p]{R_0} \quad + \omega^{\mu e^k} \sqrt[p]{R_k}$$
$$+ \psi_1(R_0) \cdot \omega^{\mu e} \sqrt[p]{R_0^e} + \psi_1(R_k) \cdot \omega^{\mu e^{k+1}} \sqrt[p]{R_k^e} + \ldots$$
$$+ \ldots$$

That is, x_0 has only the p values $x_0, x_1, \ldots x_{p-1}$.

§ 225. We will examine further the Abelian equation of degree $\dfrac{p-1}{k} = m$

$$\varphi(R) = 0,$$

which is satisfied by $R_0, R_k, \ldots R_{(m-1)k}$. From 14) we have

$$R_k = f^p(R_0) \cdot R_0^{e^k},$$
$$R_{2k} = f^p(R_k) \cdot f^{pe^k}(R_0) \cdot R_0^{e^{2k}},$$
$$R_{3k} = f^p(R_{2k}) \cdot f^{pe^k}(R_k) \cdot f^{pe^{2k}}(R_0) \cdot R_0^{e^{3k}},$$
$$\cdot \quad \cdot \quad \cdot \quad \cdot \quad \cdot \quad \cdot \quad \cdot \quad \cdot \quad \cdot$$

and since $R_{mk} = R_0$, it follows from these equations that

$$1 = R_0^{e^{p-1}-1} [f^{e^{p-1-k}}(R_0) \cdot f^{e^{p-1-2k}}(R_k) \ldots f(R_{p-1-k})]^p,$$
$$1 = R_k^{e^{p-1}-1} [f^{e^{p-1-k}}(R_k) \cdot f^{e^{p-1-2k}}(R_{2k}) \ldots f(R_0)].$$
$$\cdot \quad \cdot \quad \cdot \quad \cdot \quad \cdot \quad \cdot \quad \cdot \quad \cdot \quad \cdot$$

Now the primitive congruence root e for p can be so chosen that $e^{p-1}-1$ is divisible by no power of p higher than the first. For if

$$e^{p-1} \equiv 1 \quad (\text{mod. } p^2)$$

then

$$(p-e)^{p-1} \equiv e^{p-1} - (p-1)pe^{p-2} \equiv e^{p-1} + pe^{p-2} \equiv 1 + pe^{p-2}$$
$$(\text{mod. } p^2),$$

so that $(p-e)^{p-1}$ is divisible only by p, and we can therefore take $p-e$ in place of e. In $e^{p-1}-1 = p \cdot q$, then, q is prime to p. Consequently we can determine t so that

$$tq + 1 \equiv 0 \quad (\text{mod. } p).$$

Suppose that

$$tq = sp - 1.$$

Substituting this in equation 15), and taking the pth root of the tth power, we have

$$1 = R_0^{sp-1} [f^{e^{p-1-k}}(R_0) \cdot f^{e^{p-1-2k}}(R_k) \ldots]^t,$$
$$1 = R_k^{sp-1} [f^{e^{p-1-k}}(R_k) \cdot f^{e^{p-1-2k}}(R_{2k}) \ldots]^t,$$
$$\cdot \quad \cdot \quad \cdot \quad \cdot \quad \cdot \quad \cdot \quad \cdot \quad \cdot \quad \cdot$$

THE ALGEBRAIC SOLUTION OF EQUATIONS. 205

Again, if we write
16) $$f^\iota(R_{ak}) = a_a$$
and take again the pth root we have

$$\sqrt[p]{\overline{R_0}} = R_0{}^s [a_0{}^{e^{p-1-k}} \cdot a_1{}^{e^{p-1-2k}} \ldots a_{m-1}]^{\frac{1}{p}},$$

$$\sqrt[p]{\overline{R_k}} = R_k{}^s [a_1{}^{e^{p-1-k}} \cdot a_2{}^{e^{p-1-2k}} \ldots a_0]^{\frac{1}{p}},$$

. . . .

Since by equations 16) the a_a's are rational functions of the roots of an Abelian equation, the a_a's are themselves roots of an Abelian equation. The substitution

$$\sigma = (R_0 R_k R_{2k} \ldots)$$

of the former corresponds to the substitution

$$\tau = (a_0\, a_1\, a_2 \ldots)$$

of the latter. If the roots $a_0, a_1, a_2, \ldots$ are different from one another, then R_{ak} is a function of a_a

$$R_{ak} = \Phi(a_a),$$

and this function is, in fact, the same for all values of a (§ 189).

Theorem XI. *The quantities $\sqrt[p]{\overline{R}}$ can be reduced to the form*

$$\sqrt[p]{\overline{R_0}} = \Phi(a_0) \cdot [a_0{}^{e^{p-1-k}} \cdot a_1{}^{e^{p-1-2k}} \ldots a_{m-1}]^{\frac{1}{p}},$$

17) $$\sqrt[p]{\overline{R_k}} = \Phi(a_1) \cdot [a_1{}^{e^{p-1-k}} \cdot a_2{}^{e^{p-1-2k}} \ldots a_0]^{\frac{1}{p}},$$

. . . .

This form contains the roots $a_0, a_1, \ldots a_{m-1}$ of a simplest Abelian equation. Φ is an arbitrary function. The form 17) is not only necessary but also sufficient.

The last statement remains to be proved.

In the first place the R's, as rational functions of the (distinct) roots of an Abelian equation, are themselves roots of an Abelian equation with the group $1, \sigma, \sigma^2, \ldots \sigma^{m-1}$, which corresponds to the group $1, \tau, \tau^2, \ldots \tau^{m-1}$. Again the first two equations of 17) give

$$R_k = \left[\frac{\Phi(a_1) \cdot a^{-q}}{\Phi(a_0) \cdot e^k}\right]^p \cdot R_0{}^{e^k}$$

$$= f(R_0) \cdot R_0{}^{e^k},$$

so that we are brought back to the characteristic equations 14).*

---
*Cf Abel: Oeuvres, II pp. 217 ff. (Edition of Sylow and Lie); and Kronecker: Monatsber. d. Berl. Akad., 1853, June 20.

CHAPTER XIV.

THE GROUP OF AN ALGEBRAIC EQUATION.

§ 226. We have already seen in Chapter IX, § 153 that every special, or affect, equation $f(x) = 0$ is completely characterized by a *single* relation between its coefficients or between its roots. Suppose that in any particular case the relation is

$$\varphi(x_1, x_2, \ldots x_n) = 0.$$

More accurately speaking, it is not the function φ itself, but the family of φ and the corresponding group G, which characterize the equation. Only those substitutions among the roots are permissible, which belong to G. For this group we have the fundamental theorem:

Theorem I. *Given an equation $f(x) = 0$ and a corresponding rational domain $\Re$, all rational integral functions of the roots of the equation which are rational within $\Re$ are unchanged by the group G of the equation, i. e., they belong to the family of G or to an included family. Conversely, all integral functions of the roots which are unchanged by G are rational within $\Re$.*

The algebraic character of a given equation, for example one with numerical coefficients, is therefore by no means determined by the knowledge of the cofficients alone; but, as was first indicated by Abel, and then systematically elaborated by Kronecker, the boundaries of the rational domain must also be designated. The solution of the equation $x^2 - 2 = 0$, for example, requires very different means, according as $\sqrt{2}$ is or is not included in the rational domain. The rational domain can be defined on the one hand by assigning the elements $\Re'$, $\Re''$, ..., from which it is constructed. Or we may construct the Galois resolvent equation and determine one of its irreducible factors in the rational domain. The latter does not, to be sure, entirely replace the assignment of $\Re'$, $\Re''$, ..., but it furnishes everthing which is of importance from the algebraic standpoint for the equation considered.

The determination of the irreducible factor gives at once the group of the equation; in the $n!$ factors

$$\xi - (u_{i_1} x_1 + u_{i_2} x_2 + \ldots + u_{i_n} x_n)$$

of which the Galois resolvent is composed, we have only to regard the u's as undetermined quantities, and to form the group of the u's which permute the factors of the irreducible factor among themselves.

It must be always borne in mind that from the algebraic standpoint only those equations have a special character, according to Kronecker an affect, for which the Galois resolvent of the $(n!)^{\text{th}}$ degree is factorable.

§ 227. On account of the intimate connection between an equation and its group, we may carry over the expressions "transitive," "primitive" and "non-primitive," "simple" and "compound" from the group to the equation. Accordingly we shall designate equations as transitive, primitive or non-primitive, simple or compound, when their groups possess these several properties. Conversely, we apply the term "solvable," which is taken from the theory of equations, also to groups, and speak of solvable groups as those whose equations are solvable. Since, however, an infinite number of equations belong to a single group, this usage must be justified by a proof that the solution of all the equations belonging to a given group is furnished by that of a single one among them. This proof will be given presently (Theorem V).

In the first place we attempt to reproduce the properties of the groups in the form of equivalent algebraical properties of their equations. We have already (§ 156)

Theorem II. *If an equation is irreducible, its group is transitive; conversely, if the group of an equation is transitive, the equation is irreducible.*

§ 228. To determine under what form the non-primitivity of the group reappears as a property of the equation, we recur to the treatment of those irreducible equations one root of which was a rational function of another. The equation of degree $m\nu$ reduced to ν equations of degree m, the coefficients of which were rationally expressible in terms of the roots of an equation of degree ν (§ 174). We arrive in the present case at a similar result.

Suppose that the group G of the equation $f(x)=0$ is non-primitive; then the roots of the equation can be distributed into ν systems of m roots each

$$x_{11}, x_{12}, \ldots x_{1m}; \; x_{21}, x_{22}, \ldots x_{2m}; \ldots x_{\nu 1}, x_{\nu 2}, \ldots x_{\nu m}; \quad (\nu m = n),$$

such that every substitution of the group which converts one root of any system into a root of another system converts the entire former system into the latter. We take now for a resolvent any arbitrary symmetric function of all the roots of the first system

1) $$y_1 = S(x_{11}, x_{12}, \ldots x_{1m})$$

and apply to S all the substitutions of G. Since G is non-primitive, the entire system $x_{11}, x_{12} \ldots x_{1m}$ is converted either into itself or into one of the other systems. There are therefore only ν values of y

2) $$\begin{cases} y_1 = S(x_{11}, x_{12}, \ldots x_{1m}), \\ y_2 = S(x_{21}, x_{22}, \ldots x_{2m}), \\ y_3 = S(x_{31}, x_{32}, \ldots x_{3m}), \\ \cdot \quad \cdot \quad \cdot \quad \cdot \quad \cdot \quad \cdot \\ y_\nu = S(x_{\nu 1}, x_{\nu 2}, \ldots x_{\nu m}). \end{cases}$$

Consequently y is a root of an equation of degree ν

3) $$\varphi(y) = 0,$$

the coefficients of which are unchanged by all the substitutions of G, and which are therefore, from Theorem I, rationally known. If $\varphi(y)=0$ has been solved, *i. e.*, if all its roots $y_1, y_2, \ldots y_\nu$ are known, then all the symmetric functions of every individual system are also known. For each of these functions belongs to the same group as the corresponding y, and can therefore be rationally expressed in terms of the latter and of the coefficients of $f(x)$. If we denote, in particular, the elementary symmetric functions of $x_{a1}, x_{a2}, \ldots x_{am}$ by

$$S_1(y_a), \; S_2(y_a), \ldots S_m(y_a),$$

then the quantities $x_{a1}, x_{a2}, \ldots x_{am}$ are the roots of the equation

4) $$x^m - S_1(y_a) x^{m-1} + S_2(y_a) x^{m-2} - \ldots \pm S_m(y_a) = 0$$

Consequently $f(x)$ can be obtained by eliminating y from 3) and 4), and we have

$$f(x) \equiv \prod_{a=1}^{\nu} [x^m - S_1(y_a) x^{m-1} + S_2(y_a) x^{m-2} - \ldots \pm S_m(y_a)] = 0.$$

Conversely, if we start from the last expression, as the result of eliminating y from 3) and 4), then the group belonging to $f(x) = 0$ is non primitive, if we assume that 3) and 4) are irreducible. For we form first a symmetric function of the roots of 4). This is rational in y_a. We denote it by $F(y_a)$. Again we form the product

5) $\qquad [u - F(y_1)] [u - F(y_2)] \ldots [u - F(y_\nu)]$

for all the roots of 4). This product is rationally known; for its coefficients are symmetric in $y_1, y_2, \ldots y_\nu$, and are therefore rationally expressible in the coefficients of 3). Accordingly 5) remains unchanged by all the substitutions of the group, i. e., every substitution of the group interchanges the linear factors of this product only among themselves. If therefore $F(y_a)$ can be expressed in terms of the x's in only one way, it follows that the group converts the symmetric functions of $x_{a1}, x_{a2}, \ldots x_{am}$ into those of another system. The group is therefore non-primitive. But if the roots of $f(x)$ are different from one another, the assumption in regard to $F(y_a)$ can be realized by § 111.

Theorem III. *The group of an equation of degree $m\nu$, which is obtained by the elimination of y from the two irreducible equations*

3) $\qquad \varphi(y) \equiv y^\nu - A_1 y^{\nu-1} + \ldots \pm A_\nu = 0,$
4) $\qquad x^m - S_1(y) x^{m-1} + S_2(y) x^{m-2} - \ldots \pm S_m(y) = 0$

is non primitive; and conversely every equation, the group of which is non-primitive is the result of such an elimination.

§ 229. The properties of an equation the group of which is compound do not present themselves in so apparent a form as in the case of the transitivity or non-primitivity of the group. We can however replace the problem of the solution of the equation by another equivalent problem in which the compound or the simple character of the group has an easily observed effect on the equation itself.

For this purpose we have only to take in the place of the general equation

6) $$f(x) = 0$$

its Galois resolvent equation

7) $$F(\xi) = 0.$$

$F(\xi)$ is irreducible. We have first to examine more closely the latter equation and its properties.

Given a general equation 6), there is a linear function of the roots of 6), formed with n undetermined parameters

8) $$\xi_1 = a_1 x_1 + a_2 x_2 + \ldots + a_n x_n,$$

which has $n!$ values; so that all the substitutions of the symmetric group G belonging to 6) convert ξ_1 into $n!$ different values

$$\xi_1, \xi_2, \xi_3, \ldots \xi_{n!}$$

The permutations among the $\xi_1, \xi_2, \ldots \xi_{n!}$ produced by G form a new group among the $n!$ elements ξ, which we denote by Γ. Γ is simply isomorphic to G, and is the group of the equation 7). Γ has the property that its order is equal to its degree, as appears either from the method of its construction, or from the fact that every ξ is a rational function of every other one. The equation 7), which is identical with

7') $$(\xi - \xi_1)(\xi - \xi_2) \ldots (\xi - \xi_{n!}) = 0.$$

therefore requires for its complete solution only the determination of a single root. The solution of 7) is equivalent to that of 6).

The question arises, how these relations are modified, if we pass from the general equation 6) to a special equation. Every special equation is characterized by a single relation between the roots

9) $$\varphi(x_1, x_2, \ldots x_n) = 0.$$

If φ belongs to a group G of the order r, then only the substitutions belonging to G can be applied to the roots. For if a substitution were admitted which converted φ into

$$\varphi_1(x_1, x_2, \ldots x_n) = 0,$$

where φ_1 is different from φ, then all rational functions of

$$\alpha \varphi(x_1, x_2, \ldots x_n) + \beta \varphi_1(x_1, x_2, \ldots x_n)$$

would be rationally known. The rational domain thus determined would be more extensive than that derived from φ. Consequently 9) would not represent all the relations which exist between the roots.

We can now obtain a resolvent of our special equation in either of two ways. Either we proceed from

8) $$\xi_1 = a_1 x_1 + a_2 x_2 + \ldots + a_n x_n,$$

apply to ξ_1 all the r substitutions of G, obtain

$$\xi_1, \xi_2, \xi_3 \ldots \xi_r,$$

and form the resolvent of the rth degree

10) $$F_1(\xi) \equiv (\xi - \xi_1)(\xi - \xi_2) \ldots (\xi - \xi_r) = 0;$$

or we proceed from the expression $F(\xi)$, already given in 7) and 7'), and observe that $F(\xi)$ becomes reducible on the adjunction of 9) and that $F_1(\xi)$ is one of the irreducible factors. The other factors are, like F_1, of the rth degree. They differ from each other only in the constants a. Every one of them is obtained by multiplying together all the factors $\xi - \xi_a$, which arise from the application of the group G to a single one among them. The group of $F_1(\xi) = 0$, regarded as a group among the ξ's, is of degree and order r. It is simply isomorphic to the group G of degree n and order r belonging to φ.

The groups of all the factors $F_1(\xi), \ldots$ of $F(\xi)$ therefore differ from one another only in the particular designation of their elements.

Theorem IV. *If a special equation $f(x) = 0$ is characterized by the family of*

9) $$\varphi(x_1, x_2, \ldots x_n) = 0,$$

with a group G of order r, then the general Galois resolvent decomposes into $\rho = \dfrac{n!}{r}$ factors

10) $$F_1(\xi) = 0,$$

every one of which can serve as the Galois resolvent of the special equation. All the roots of 10) *are rational functions of every one among them, and in terms of these all the roots of $f(x) = 0$ can be rationally expressed. The transition from $f(x) = 0$ to $F_1(\xi) = 0$ has its counterpart in the transition from G to the simply isomorphic group Ω (§ 129) of $F_1(\xi)$.*

Since the construction of 10) depends only on the group G, and not on the particular nature of 9), this same resolvent belongs to all equations which are characterized by functions of the same family with 9). If one of these equations has been solved, then $x_1, x_2, \ldots x_n$ and consequently ξ_1 are known. The equation 10) is therefore solved, and with it every other equation of this sort. We have then the proof of the theorem stated in § 226:

Theorem V. *Given an equation $f(x) = 0$, the coefficients of which belong to any arbitrary rational domain, the adjunction of either $\varphi_1 = 0$ or $\varphi_2 = 0$, where φ_1 and φ_2 belong to the same family of the roots $x_1, x_2, \ldots x_n$, leads, as regards solvability, to the same special equation.*

§ 230. We have treated in earlier Chapters cases where such relations between the roots either were directly given or were easily recognized as involved in the data. Frequently, however, the conditions are such that, instead of a known function, $\varphi(x_1, x_2 \ldots x_n)$ being directly designated as adjoined, φ presents itself implicitly as a root of an equation which is *regarded* as solvable. For example, in the problem of the algebraic solution of equations the auxiliary equation is of the simple form

$$y^p - A(x_1, x_2, \ldots x_n) = 0.$$

Here y is regarded as known, *i. e.*, we extend the rational domain of $f(x) = 0$ by adjoining to it every rational function of the roots of which any power belongs to the domain. The actual solution of the auxiliary equations does not enter into consideration.

It is a natural step, when an irreducible auxiliary equation is regarded as solvable, to adjoin not one of roots φ, but all of its roots to the domain of $f(x) = 0$. These roots are the different values which $\varphi(x_1, x_2, \ldots x_n)$ assumes within the rational domain. For to find the auxiliary equation which is satisfied by φ_1 we apply to φ_1 all the r substitutions of the group G and obtain, for example, m distinct values

11) $\qquad \varphi_1, \varphi_2, \varphi_3, \ldots \varphi_m.$

The symmetric functions of these values, and therefore the coefficients of the equation

THE GROUP OF AN ALGEBRAIC EQUATION.

12) $$g(\psi) \equiv (\psi - \psi_1)(\psi - \psi_2) \ldots (\psi - \psi_m) = 0$$

are known within the rational domain of $f(x) = 0$, and 12) is the required auxiliary equation, the solution of which is regarded as known.

Now given the equation $f(x) = 0$, characterized by the group G, or by any function $\varphi(x_1, x_2, \ldots x_n)$ belonging to G, we adjoin to it *all* the roots of 12), or, what amounts to the same thing, a linear combination of these m roots

$$\chi = a_1 \psi_1 + a_2 \psi_2 + \ldots + a_m \psi_m,$$

where the a's are undetermined constants. The question then arises, what the group of $f(x) = 0$ becomes under the new conditions.

The adjoined family of functions was originally that of φ. Now it is that of

$$\varphi + \chi = \varphi + a_1 \psi_1 + a_2 \psi_2 + \ldots + a_m \psi_m.$$

The group was originally G. Now it is that subgroup of G, which is also contained in all the groups

$$H_1, H_2, \ldots H_m,$$

of $\psi_1, \psi_2, \ldots \psi_m$. Suppose that K is the greatest common subgroup of these m groups. Then K belongs to the function χ.

If now we apply all the substitutions of G to the series $\psi_1, \psi_2, \ldots \psi_m$, the result is in every case the same series in a new order; for $\psi_1, \psi_2, \ldots \psi_m$ are all the values which G produces from ψ_1. Consequently the series $H_1, H_2, \ldots H_m$ is also reproduced by transformation with respect to G; and K is therefore unchanged by transformation with respect to G. We have then

$$G^{-1} K G = K.$$

Again we denote by I' the greatest subgroup of G which is contained in K. I' therefore belongs to $\varphi + \chi$, and accordingly characterizes the family which belongs to $f(x) = 0$ after the adjunction of all the roots of 12). I', like K, *is also commutative with* G; for on transforming I' with respect to G, the result must belong to both G and K, and is therefore I' itself. I' is, then, a self-conjugate subgroup of G, and in fact is the most comprehensive of those which are also common to $H_1, H_2, \ldots H_m$.

If I' does not reduce to the identical operation, G is a compound

18

group. If G is simple, $\varGamma$ is necessarily identity, and the group of $f(x)=0$ is reduced by the solution of 12) to 1, i. e., after the solution of 12) all the roots of $f(x)=0$ are known; or, in other words, the solution of 12) furnishes that of $f(x)=0$ also. We have then the following

Theorem VI. *Given any arbitrary equation $f(x)=0$ with the group G, if we adjoin to it all the roots*

11) $\qquad \psi_1, \psi_2, \psi_3, \ldots \psi_m$

of an irreducible equation of the mth degree

12) $\qquad g(\psi) \equiv \psi^m - A\psi^{m-1} + \ldots = 0,$

the coefficients of which are rational in the rational domain of $f(x)$, and the roots rational functions of $x_1, x_2, \ldots x_n$, then G reduces to the largest self-conjugate subgroup $\varGamma$ of G which leaves $\psi_1, \psi_2, \ldots \psi_m$ all unchanged. If G is a simple group, $\varGamma = 1$. Only in case G is compound is it possible by the solution of an auxiliary equation to reduce the group to a subgroup different from identity, and consequently to divide the Galois resolvent equation into non-linear factors.

§ 231. We consider these results for a moment. If the general equation of the nth degree $f(x)=0$ is given, the corresponding group G is of order $r = n!$. This group is compound, the only actual self-conjugate subgroup being the alternating group (§ 92). If we take for a resolvent

$$\psi_1 = \sqrt{\varDelta},$$

where $\varDelta$ denotes, as usual, the discriminant of $f(x)$, then the resolvent equation becomes

12') $\qquad \psi^2 - \varDelta = 0,$

and $\varGamma$ is the alternating group. After adjunction of the two roots of 12') the previously irreducible Galois resolvent equation divides into two conjugate factors of degree $\frac{1}{2}n!$, and only such substitutions can be applied to the resolvent

$$\xi = a_1 x_1 + a_2 x_2 + \ldots + a_n x_n$$

as leave $\sqrt{\varDelta}$ unchanged and therefore belong to the alternating group.

For $n > 4$ the alternating group is simple. If there is an m-

valued resolvent ψ, its values $\psi'_1, \psi'_2, \ldots \psi'_m$ are obtained by the solution of an equation of the mth degree. On the adjunction of these values, or of

$$\chi = \beta_1 \psi'_1 + \beta_2 \psi'_2 + \ldots + \beta_m \psi'_m$$

the group of the given equation reduces, by Theorem VI, to the identical substitution. The equation $f(x) = 0$ is therefore solved; for all functions are known which belong to the group 1 or to any other group. The investigations of Chapter VI show, however, that no reduction of the degree of the equation to be solved can be effected in this way, since if $n > 4$, the number m of the values of ψ, if it exceeds 2, is greater than n or equal to n. In the latter case, if $n \doteq 6$, the function ψ is always symmetric in $n-1$ elements, so that we can take directly $\psi'_1 = x_1$, and the resolvent equation is identical with the original $f(x) = 0$.

Theorem VII. *The general equation of the nth degree ($n > 4$) is solved, as soon as any arbitrary resolvent equation of a degree higher than the second is solved. There are, however, no resolvent equations the degree of which is greater than 2 and less than n. Moreover, if $n \doteq 6$, there is no resolvent equation of the nth degree essentially different from $f(x) = 0$. For $n = 6$ there is a distinct resolvent equation of degree 6.*

One other result of our earlier investigations, as reinterpreted from the present point of view, may be added here:

Theorem VIII. *The general equation of the fifth degree has a resolvent equation of the sixth degree.*

§ 232. We return now, from the incidental results of the preceding Section, to Theorem VI, and examine the group of the equation

12) $\qquad g(\psi) \equiv (\psi - \psi'_1)(\psi - \psi'_2) \ldots (\psi - \psi'_m) = 0$,

the roots $\psi'_1, \psi'_2, \ldots \psi'_m$ of which were all adjoined to the equation $f(x) = 0$.

The order of the group of 12) is most easily found from the fact that it is equal to the degree of the irreducible equation of which

$$\omega = \gamma_1 \psi'_1 + \gamma_2 \psi'_2 + \ldots + \gamma_m \psi'_m$$

is a root. We must therefore apply to ω all the r substitutions of

G. The values thus obtained may partly coincide. The number of distinct values gives the order of the group of 12). Now if we retain the designations of Theorem VI, $\varGamma$ includes all substitutions of $x_1, x_2, \ldots x_u$ which leaves all the ψ's unchanged. Suppose that the order of $\varGamma$ is r'. Then the required number is $\nu = r : r'$.

From this we perceive that, if the group G is simple and $\varGamma$ accordingly of order $r' = 1$, the order ν of the group of every resolvent equation is the same as that of $f(x) = 0$, so that no simplification can be effected in this way.

We actually obtain the group of 12) by the consideration that it contains all and only those substitutions among the x's which do not alter the nature of $f(x) = 0$. If therefore we apply to

$$\psi'_1, \psi'_2, \psi'_3, \ldots \psi'_m$$

all the substitutions of G, the resulting permutations of the ψ's form the group required. All the r substitutions thus obtained are not however necessarily different; for all the substitutions of $\varGamma$ leave all the elements ψ unchanged. From this, again, it follows that the order of the group K of 12) is $\nu = r : r'$. In the same way we recognize that K is $(1-r)$ fold-isomorphic to G. With the notation of § 86, K is the quotient of G and $\varGamma$; $K = G : \varGamma$.

Theorem IX. *If the group G of $f(x) = 0$ is of order r, and contains a self-conjugate subgroup $\varGamma$ of order r', and if G reduces to $\varGamma$ on the adjunction of all the roots* 11) *of*

12) $\qquad\qquad g(\psi) = 0,$

then the group K of the latter equation is of order $\nu = r : r'$. K is the quotient of G and $\varGamma$ and is $(1-'r)$-fold isomorphic to G.

By a proper choice of the resolvent ψ we can give the equation 12) a very special character.

We choose as a resolvent a function χ belonging to the self-conjugate subgroup $\varGamma$. Then χ is a root of an equation of degree $\nu = r : r'$, all the roots of which are rationally expressible in terms of any one among them; for $\chi_1, \chi_2, \ldots \chi_\nu$ all belong to the same group $\varGamma$, (§ 100, Theorem VIII). The group of 12) is therefore a group $\varOmega$; for it is transitive, since $g(\chi)$ is irreducible. We have therefore

Theorem X. *If the group G of the equation $f(x) = 0$ is*

of order r, and contains a self-conjugate subgroup Γ of order r', and if χ_1 is a function of the roots $x_1, x_2, \ldots x_n$, belonging to Γ, then an irreducible resolvent equation of degree $\nu = r : r'$

$$h(\chi) \equiv \chi^\nu - A_1 \chi^{\nu-1} + A_2 \chi^{\nu-2} - \ldots \pm A_\nu = 0$$

can be constructed, the roots of which are all rational functions of a single one among them, and which possesses the property that the adjunction of one of its roots to $f(x) = 0$ reduces the group G to Γ.

§ 233. **Theorem XI.** *If Γ is a maximal self-conjugate subgroup of G, then the group of $h(\chi) = 0$ is a transitive, simple group. Conversely, if Γ is not a most extensive self-conjugate subgroup of G, then the group of $h(\chi) = 0$ is compound.*

We denote the group of $h(\chi) = 0$ by G'. Its order is $\nu = r : r'$. We assume that G' contains a self-conjugate subgroup Γ'', of order r_1. From Theorem IX G' is r'-fold isomorphic to G. From the results of § 73 it follows that the subgroup J of G, which corresponds to the group Γ'', is a self-conjugate subgroup of G and is of order $\nu' r'$. J is, then, like Γ, a self-conjugate subgroup of G, and their orders are respectively $\nu' r'$ and r'. We show that Γ is contained in J. This follows directly from the construction of G' (§ 232), in accordance with which the substitution 1 of G' corresponds to all the substitutions of G which leave the series 11) unaltered. Γ in G therefore corresponds to the one substitution 1 of G'. Accordingly if G' is compound, then Γ is not a maximal self-conjugate subgroup of G.

The converse theorem is similarly proved from the properties of isomorphic groups.

In these last investigations we have dealt throughout with the group of the equation, but never with the particular values of the coefficients. If therefore two equations of degree n have the same group, the reductions of the Theorem X are *entirely independent of the coefficients of the equations*. The coefficients of $h(\chi)$ will of course be different in the two cases, but the different equations $h(\chi) = 0$ all have the same group, and every root of any one of these equations is a rational function of every one of its roots. *This common property relative to reduction, which holds also for the further investigations of the present Chapter, is the chief reason for the collection of all equations belonging to the same group into a family.*

§ 234. We observe further that with every reduction of the group there goes a decomposition of the Galois resolvent equation, while the equation $f(x) = 0$ need not resolve into factors.

Collecting the preceding results we have the following

Theorem XII. *If the group G of an equation $f(x) = 0$ is compound, and if*

$$G, G_1, G_2, \ldots G_\nu, 1$$

is a series of composition belonging to G, so that every one of the groups $G_1, G_2, \ldots G_\nu, 1$ is a maximal self-conjugate subgroup of the preceding one, further if the order of the several groups of the series are

$$r, r_1, r_2, \ldots r_\nu, 1,$$

then the problem of the solution of $f(x) = 0$ can be reduced as follows. We have to solve in order one equation of each of the degrees

$$\frac{r}{r_1}, \frac{r_1}{r_2}, \frac{r_2}{r_3}, \ldots \frac{r_{\nu-1}}{r_\nu}, r_\nu,$$

the coefficients of which are rational in the rational domain determined by the solution of the preceding equation. These equations are irreducible and simple, and of such a character that all the roots of any one of them are expressible rationally in terms of any root of the same equation. The orders of the groups of the equations are respectively

$$\frac{r}{r_1}, \frac{r_1}{r_2}, \frac{r_2}{r_3}, \ldots \frac{r_{\nu-1}}{r_\nu}, r_\nu.$$

The groups are the quotients

$$G : G_1, \quad G_1 : G_2, \quad G_2 : G_3, \ldots G_{\nu-1} : G_\nu, \quad G_\nu : 1.$$

The equations being solved, the Galois resolvent equation, which was originally irreducible and of degree r, breaks up successively into

$$\frac{r}{r_1}, \frac{r}{r_2}, \frac{r}{r_3}, \ldots \frac{r}{r_\nu}, r$$

factors. After the last operation $f(x) = 0$ is therefore completely solved

§ 235. The composition of the group G of an equation $f(x) = 0$ is therefore reflected in the resolution of the Galois resolvent equation into factors. We turn our attention for a moment to the question, when a resolution of the equation $f(x) = 0$ itself occurs. It is readily seen that, in passing from G_a to G_{a+1} in the series of composition of G, a separation of $f(x)$ into factors can only occur when G_{a+1} does not connect all the elements transitively which are connected transitively by G_a. The resulting relations are determined by § 71. G_a is non-primitive in respect to the transitively connected elements which G_{a+1} separates into intransitive systems.

Starting from G, with an irreducible $f(x) = 0$, suppose now that $G_1, G_2, \ldots G_a$ are transitive, but that G_{a+1} is intransitive, so that by § 71 G_a is non-primitive. Then at this point $f(x)$ separates into as many factors as there are systems of intransitivity in G_{a+1}. But (again from § 71), *all the elements occur in* G_{a+1}. We arrange, then, the substitutions of G_a in a table based on the systems of intransitivity of G_{a+1}. Suppose that there are μ such systems, so that $f(x)$ divides into μ factors. Then we take for the first line of the table all and only those substitutions of G_a, which do not convert the elements of the first system of intransitivity into those of another system. The substitutions of this line form a group, which is contained in G_a as a subgroup. Its order is therefore $k r_{a+1}$. The second line of the table consists of all the substitutions of G_a which convert the first system of intransitivity into the second. The number of these is also $k r_{a+1}$. There are μ such lines, and they include all the substitutions of G_a. Consequently

$$\mu k r_{a+1} = r; \quad \mu = \frac{1}{k} \frac{r}{r_{a+1}},$$

i. e., the number μ of the factors into which $f(x)$ divides is a divisor of the number $\dfrac{r}{r_{a+1}}$ *of the factors into which the Galois resolvent equation divides at the same time. A similar result obviously occurs in every later decomposition.*

The decomposition can therefore only take place according to the scheme of Theorem III. The several irreducible factors are all of the same order.

§ 236. Thus far we have adjoined to the given equation $f(x) = 0$

the root ψ of a second irreducible equation only when the ψ's were *rational* functions of $x_1, x_2, \ldots x_n$. This seems a strong limitation. We will therefore now adjoin to the equation $f(x) = 0$ all the roots of an irreducible equation

13) $$g(z) = 0$$

without making this special assumption. The only case of interest is of course that in which the adjunction produces a reduction in the group G of $f(x) = 0$.

In the first instance we adjoin only a single root z_1 of $g(z) = 0$. Suppose that G then reduces to its subgroup H_1. If the rational function $\varphi_1(x_1, x_2, \ldots x_n)$ belongs to H_1, then the same reduction of G can be produced by adjoining φ_1 instead of z_1. Suppose that under the operation of G, the function φ_1 takes the conjugate values $\varphi_1, \varphi_2, \ldots \varphi_m$, with the groups $H_1, H_2, \ldots H_m$ respectively. These values satisfy an irreducible equation

$$k(\varphi) \equiv (\varphi - \varphi_1)(\varphi - \varphi_2) \ldots (\varphi - \varphi_m) = 0.$$

The adjunction of z_1 to the rational domain has, by the mediation of H_1, made φ_1 rational, so that φ_1 is a rational function of z_1

$$\varphi_1(x_1, x_2, \ldots x_n) = \psi(z_1).$$

It appears therefore that, in order that the adjunction of z_1 may produce a reduction of G, it is necessary and sufficient that there should be a rational non-symmetric function of the roots $\varphi_1(x_1, x_2, \ldots x_n)$ which is rationally expressible in terms of z_1.

Suppose that the roots of the irreducible equation $g(z) = 0$ are $z_1, z_2, \ldots z_\mu$, and that its group is Γ. Since $k[\psi(z)] = 0$ is satisfied by z_1, all the

$$\psi(z_i) \quad (i = 1, 2, 3, \ldots \mu)$$

are roots of $k(\varphi) = 0$; that is $\psi(z_1), \psi(z_2), \ldots \psi(z_\mu)$ are the conjugate values of φ_1. On the other hand the coefficients of the product

$$k_1(\varphi) = [\varphi - \psi(z_1)][\varphi - \psi(z_2)] \ldots [\varphi - \psi(z_\mu)]$$

are symmetric functions of the roots of $g(z) = 0$ and are therefore rationally known; and the equation $k_1 = 0$ has all its roots in common with $k = 0$. Consequently $k_1(\varphi)$ is a power of $k(\varphi)$

$$k_1(\varphi) = k^q(\varphi),$$

and the μ values $\psi(z_1), \psi(z_2), \ldots \psi(z_\mu)$ coincide in sets of q each.

With a slight change in the notation for the z's we can therefore write

$$\mathfrak{T}) \quad \begin{aligned} \varphi_1 &= \varphi(z_1') = \varphi(z_2') = \ldots = \varphi(z_q'), \\ \varphi_2 &= \varphi(z_1'') = \varphi(z_2'') = \ldots = \varphi(z_q''), \end{aligned} \quad (qm = \mu)$$

. . . .

Since $\varphi(z_\alpha^{(\gamma)}) - \varphi(z_\beta^{(\gamma)}) = 0$, this quantity is rationally known. It is therefore unchanged by all the substitutions of $\varGamma$, i. e., $\varGamma$ interchanges the lines of $\mathfrak{T}$), and therefore gives rise to a group T of the elements φ which is isomorphic to $\varGamma$. To the substitution 1 in T correspond in $\varGamma$ the substitutions of the subgroup $\varDelta$ of order d_1 which only interchange $z'_1, z'_2, \ldots z'_q$ among themselves, $z''_1, z''_2, \ldots z''_q$ among themselves, and so on. $\varGamma$ and T are $(1-d_1)$-fold isomorphic.

If we coordinate all the substitutions of G and T which leave φ_1 unchanged, and again one substitution each from G and T which converts φ_1 into φ_2, one which converts φ_1 into φ_3, and so on, an isomorphism is also established between G and T. To the substitution 1 in T correspond in G the substitutions of the subgroup D of order d which is the maximal common subgroup of $H_1, H_2, \ldots H_m$.

Accordingly G and $\varGamma$ are also isomorphic, and in fact their isomorphism is $(d-d_1)$-fold, as shown by the preceding considerations, and again $(r-r_1)$-fold, as appears from the orders of G and $\varGamma$. Consequently

$$r : r_1 = d : d_1; \quad \frac{r}{d} = \frac{r_1}{d_1}.$$

If now we adjoin to the equation $f(x) = 0$ all the roots of $g(z) = 0$, then $\varphi_1, \varphi_2, \ldots \varphi_m$ are rationally known. G reduces to the subgroup D of order d common to the groups $H_1, H_2, \ldots H_m$. To D belongs the function

14) $\quad \rho(x_1, x_2, \ldots x_n) = a_1 \varphi_1 + a_2 \varphi_2 + \ldots + a_m \varphi_m = \omega(z_1, z_2, \ldots z_\mu),$

and every function of $x_1, x_2, \ldots x_n$ which can be rationally expressed in terms of $z_1, z_2, \ldots z_\mu$ belongs to the family of ρ or to an included family. For every such function is rationally known, as soon as the $z_1, z_2, \ldots z_\mu$ are adjoined to the equation $f(x) = 0$.

Conversely, if we adjoin to the equation $g(z) = 0$ all the roots of $f(x) = 0$, it follows by the same reasoning that there is a function

15) $\quad \omega_0(z_1, z_2, \ldots z_\mu) = \rho_0(x_1, x_2, \ldots x_n),$

such that *every function of* $z_1, z_2, \ldots z_\mu$ *which can be rationally expressed in terms of* $x_1, x_2, \ldots x_n$ *belongs to the family of* ω_0 *or to an included family.*

Since now ρ_0 is rational in the z's, it follows from the above property that

16) $\qquad \rho_0 = R(\rho),$

where R is a rational function; and since ω is rational in the x's, it follows that

17) $\qquad \omega = R_0(\omega_0)$

or, which is the same thing,

17') $\qquad \rho = R_0(\rho_0).$

From 16) and 17') it follows *that ρ and ρ_0 belong to the same family.* The adjunction of all the roots of $f = 0$ to $g = 0$ therefore gives rise to the same rational domain as the adjunction of all the roots of $g = 0$ to $f = 0$.

It is obvious at once that the first adjunction, since it made $\varphi_1, \varphi_2, \ldots \varphi_m$ rational, also furnished ω, so that Γ reduces to Δ. But the proof just given was necessary to exclude the possibility of any further reduction.

If we write $\dfrac{r}{d} = \dfrac{r_1}{d_1} = \nu$, it follows that if the second adjunction reduces the order r of G to its νth part, then the first adjunction also reduces the order r_1 of Γ to its νth part.

Theorem XIII. *The effect of the adjunction of all the roots of any arbitrary equation* 13) *on the reduction of the group G of $f(x) = 0$ can be equally well produced by the adjunction of all the roots of an equation* 12) *which is satisfied by rational functions of $x_1, x_2, \ldots x_n$.*

In spite of removal of apparent limitations, we have therefore not departed from the earlier conditions, where only the adjunction of rational functions of the roots was admitted.

§ 237. **Theorem XIV.** *If*

$$f(x) = 0, \quad g(z) = 0$$

are two equations, the coefficients of which belong to the same rational domain, and which are of such a nature that the solution of the second and the adjunction of all its roots to the first reduces the group of $f(x) = 0$ to a self-conjugate subgroup contained in it of an order ν times as small, then conversely the solution of the first equation reduces the group of the second to its νth part. The group of $f(x) = 0$, like that of $g(z) = 0$, is compound, and ν is a factor of composition. Those rational functions of the roots of one of the two equations, by which the same reduction of its group is accomplished as by the solution of the other equation are rational in the roots of the latter.

As we see, the group of $f(x) = 0$ can be reduced by the solution of an equation $g(z) = 0$, although the roots of the latter are not rational functions of $x_1, x_2, \ldots x_n$. It is only necessary that there should be rational functions of $z_1, z_2, \ldots z_\mu$ which are also rational functions of $x_1, x_2, \ldots x_n$.

From the preceding Theorem follow at once the Corollaries

Corollary I. *If the group G of the equation $f(x) = 0$ is simple, the equation can only be solved by the aid of equations with groups the orders of which are multiples of the order of G.*

For since G reduces to 1, the ν of Theorem XIV must be taken equal to the order of G.

Corollary II. *If the group G of $f(x) = 0$ can be reduced by the solution of a simple equation $g(z) = 0$, then $z_1, z_2, \ldots z_\mu$ are rational functions of the roots of $f(x) = 0$.*

For in this case ν is equal to the order of the group of $g(z) = 0$. After the reduction this is equal 1. Consequently

$$\Psi_1 = a_1 z_1 + a_2 z_2 + \ldots + a_\mu z_\mu = \psi_1(x_1, x_2, \ldots x_n),$$

where Ψ_1 is the Galois resolvent of $g(z) = 0$.

Corollary III. *If the adjunction of the roots of $g(z) = 0$ is to produce a reduction of the group of $f(x) = 0$, then the orders r and r_1 cannot be prime to each other.*

The preceding Sections contain a proof and an extension, resting entirely on considerations belonging to the theory of substitutions, of Theorem II, § 216, where the subject was treated purely arithmetically. For if we retain the notation of § 216, it follows that since $V_\nu^{\rho\nu} = V'_\nu$ is a simple equation, V_ν is a rational function of the roots of $f(x) = 0$, and so on.

The proof of the impossibility of the algebraic solution of the equations of higher degree might therefore be based on the present considerations.

§ 238. As the adjunction of the roots of a new equation $g(z) = 0$ to $f(x) = 0$ leads to nothing more than the adjunction of rational functions of the roots of $f(x) = 0$, so no new result is obtained, if the roots of both equations are connected by a rational relation. We prove

Theorem XV. *If*
$$f(x) = 0, \quad g(z) = 0$$
are two irreducible equations, the roots of which are connected with each other by rational relations
$$\varphi_1(x_1, x_2, \ldots x_n; z_1, z_2, \ldots z_\mu) = 0,$$
the latter can all be obtained from a single relation of the form
$$\psi_1(x_1, x_2, \ldots x_n) = \chi(z_1, z_2, \ldots z_\mu),$$
in which the roots of the two equations are separated.

For, if we denote the corresponding Galois resolvents by ξ and ζ and the irreducible resolvent equations of $f(x) = 0$ and $g(z) = 0$ by
$$F(\xi) = 0, \quad G(\zeta) = 0,$$
the degrees r and r^k of F and G are equal to the orders of the respective groups.

Now $x_1, x_2, \ldots x_n$ can be rationally expressed in terms of ξ, and $z_1, z_2, \ldots z_\mu$ in terms of ζ, so that
$$\varphi(x_1, x_2, \ldots x_n; z_1, z_2, \ldots z_\mu) = \Phi(\xi, \zeta) = 0.$$
The expression Φ can be so reduced by the aid of $F = 0$ and $G = 0$ that its degree becomes less than r in ξ and less than r' in ζ. Then the two equations
$$\Phi(\xi, \zeta) = 0, \quad F(\xi) = 0$$
have a common root. Consequently, if we add ζ to the rational domain, the resolvent $F(\xi)$ becomes reducible, since otherwise the irreducible equation of the rth degree would have a root in common

with an equation of a degree less than r. The only exception occurs when $\Phi(\xi, \zeta)$ is identically 0.

If this does not happen, the adjunction of all the roots of $g(z) = 0$ or that of ζ breaks up the resolvent of $f(x) = 0$ into factors, and we have therefore the case of the last Section. We can effect the same reduction by the adjunction of a rational function χ of $x_1, x_2, \ldots x_n$, and we have

$$\chi(x_1, x_2, \ldots x_n) = \psi(z_1, z_2, \ldots z_\mu).$$

If several such relations exist, they can all be deduced from one and the same equation. The latter can be easily found, if we select a function χ such that all the others belong to an included family.

On the other hand if $\Phi(\xi, \zeta)$ is identically 0, it follows that the coefficients in the polynomial $\Phi(\xi, \zeta)$ arranged according to powers of ξ vanish, so that we have equations of the form

$$\chi_1(\zeta) - \chi_2(z_1, z_2, \ldots z_\mu) = 0,$$

and similarly, if $\Phi(\xi, \zeta)$ is arranged in powers of ζ,

$$\psi_2(x_1, x_2, \ldots x_n) = 0.$$

These equations can actually make $\Phi = 0$. But this amounts to only an apparent, not an actual dependence of the roots of $f(x) = 0$ and $g(z) = 0$. The function $\chi_2 = 0$ belongs to the group of $g(z) = 0$, and $\psi_2(x_1, x_2, \ldots x_n)$ belongs to the group of $f(x) = 0$.

CHAPTER XV.

ALGEBRAICALLY SOLVABLE EQUATIONS.

§ 239. In § 234 we have established the following theorem. If the group G of the equation $f(x) = 0$ has the series of composition

1) $\qquad G, G_1, G_2, \ldots G_\nu, 1,$

and if the orders of these several groups are

$$r, r_1, r_2, \ldots r_\nu, 1,$$

then the solution of $f(x) = 0$ can be effected by solving a series of simple, irreducible equations of degrees

$$\frac{r}{r_1}, \frac{r_1}{r_2}, \frac{r_2}{r_3}, \ldots \frac{r_{\nu-1}}{r_\nu}, r_\nu,$$

the first of which has for its coefficients functions belonging to G and for its roots functions belonging to G_1, the second coefficients belonging to G_1 and roots belonging to G_2, and so on. All these equations

$$\chi_1 = 0, \quad \chi_2 = 0, \ldots \chi_\nu = 0, \quad \chi_{\nu+1} = 0$$

have the property that the roots of any one of them are all rational functions of one another, so that the order of the corresponding group is equal to its degree $i.~e.$, the group is of the type Ω (§ 129).

We have now to examine under what circumstances all these equations $\chi_\nu = 0$ become binomial equations of order p_λ

$$\chi_\lambda \equiv z^{p_\lambda} - H_\lambda = 0,$$

where H_λ is rational in the quantities belonging to the family of $G_{\lambda-1}$. In other words, we have to determine the necessary and sufficient condition that $f(x) = 0$ shall be algebraically solvable.

For this result it is necessary that the factors of composition $\frac{r}{r_1}, \frac{r_1}{r_2}, \frac{r_2}{r_3}, \ldots$ should all be prime numbers, $p_1, p_2, p_3, \ldots$ For these quotients give the degrees of the equations $\chi_1 = 0, \chi_2 = 0, \chi_3 = 0, \ldots$

This condition is also sufficient, as has already been shown in §§ 110, 111, Theorems X and XII. Not that every function belonging to G_λ on being raised to the $(p_\lambda)^{\text{th}}$ power gives a function belonging to $G_{\lambda-1}$; but some function can always be found which has this property, as soon as the condition above is fulfilled.

We have then

Theorem I. *In order that the algebraic equation $f(x) = 0$ may be algebraically solvable, it is necessary and sufficient that the factors of composition of its group should all be prime numbers.*

§ 240. By the aid of Theorem XII, § 110 we can give this theorem another form

Theorem II. *In order that the algebraic equation $f(x) = 0$ may be algebraically solvable, it is necessary and sufficient that its group should consist of a series of substitutions*

$$1, t_1, t_2, t_3, \ldots t_\nu, t_{\nu+1}$$

which possess the two following properties: 1) *the substitutions of the group* $G_\lambda = \{1, t_1, t_2, \ldots t_{\lambda-1}, t_\lambda\}$ *are commutative, except those which belong to the group* $G_{\lambda-1} = \{1, t_1, t_2, \ldots t_{\lambda-2}, t_{\lambda-1}\}$, *and* 2) *the lowest power of* t_λ, *which occurs in* $G_{\lambda-1}$ *has for its exponent a prime number* (*cf. also* § 91, *Theorem XXIV*).

§ 241. Again the investigations of § 94 enable us to state Theorem I in still a third form. It was there shown that if the principal series of G

2) $\qquad\qquad G, H, J, K, \ldots 1$

does not coincide with the series of composition, then 1) can be obtained from 2) by inserting new groups in the latter, for example between H and J the groups

$$H', H'', \ldots H^{(\lambda)}.$$

Then the factors of composition which correspond to the transitions from H to H', from H' to H'', ... from $H^{(\lambda)}$ to J are all equal. Accordingly, if all the factors of composition belonging to 1) are not equal, then G has a principal series of composition 2). We saw further (§ 95) that, if, in passing successively from $H, H', H'', \ldots$ to the following group, the corresponding factors of composition were all prime numbers, (which then, as we have just seen, are all equal to each other), and only in this case, the substitutions of H are commutative, except those which belong to J. From this follows

Theorem III. *In order that the algebraic equation $f(x) = 0$ may be algebraically solvable, it is necessary and sufficient that its principal series of composition*

$$G,\ H,\ J,\ K,\ \ldots\ 1$$

should possess the property that the substitutions of every group are commutative, except those which belong to the next following group.

The substitutions of the last group of the series, that which precedes the identical group, are therefore all commutative.

§ 242. Before proceeding further with the theory, we give a few applications of the results thus far obtained.

Theorem IV. *If a group Γ is simply isomorphic with a solvable group G, then Γ is also a solvable group.*

From § 96 the factors of composition of G coincide with those of Γ. Consequently Theorem IV follows at once from Theorem I.

Theorem V. *If the group Γ is multiply isomorphic with the solvable group G, and if to the substitution 1 of G corresponds the subgroup Σ of Γ; finally if Σ is a solvable group, then Γ is also solvable.*

The factors of composition of Γ consist, from § 96, of those of G and those of Σ. Reference to Theorem I shows at once the validity of the present theorem.

Theorem VI. *If a group G is solvable, all its subgroups are also solvable.*

We write as usual

$$\xi_1 = a_1 x_1 + a_2 x_2 + \ldots + a_n x_n,$$

apply to ξ_1 all the substitutions of G, obtain $\xi_1, \xi_2, \ldots \xi_r$, and form

$$g(\xi) \equiv (\xi - \xi_1)(\xi - \xi_2) \ldots (\xi - \xi_r).$$

It is characteristic for the solvability of G that $g(\xi)$ can be resolved into linear factors by the extraction of roots.

If now H of order r is a subgroup of G, and if the application of H to $\bar{z}_1$ gives rise to the values $\bar{z}_1, \bar{z}_2, \ldots \bar{z}_{r_1}$, then these are all contained among $\bar{z}_1, \bar{z}_2, \ldots \bar{z}_r$. Consequently

$$h(\bar{z}) \equiv (\bar{z} - \bar{z}_1)(\bar{z} - \bar{z}_2) \ldots (\bar{z} - \bar{z}_{r_1})$$

is a divisor of $g(\bar{z})$. Then $h(\bar{z})$ is also resolvable algebraically into linear factors, i. e., H is a solvable group.

We might also have proved this by showing that all the factors of composition of H occur among those of G.

Theorem VII. *If the order of a group G is a power of a prime number p, the group is solvable.*

The group G is of the same type as a subgroup of the group which has the same degree n as G and for its order the highest power p^f which is contained in $n!$ (cf. §§ 39 and 49). That the latter group is solvable follows from its construction (§ 39), all of its factors of composition being equal to the prime number p. It follows then from Theorem VI that G is also solvable.

Theorem VIII. *If the group G is of order*

$$r = p_1{}^\alpha p_2{}^\beta p_3{}^\gamma p_4{}^\delta \ldots$$

where $p_1, p_2, p_3, p_4, \ldots$ are different prime numbers such that

$$p_1 > p_2{}^\beta p_3{}^\gamma p_4{}^\delta \ldots, \quad p_2 > p_3{}^\gamma p_4{}^\delta \ldots, \quad p_3 > p_4{}^\delta \ldots,$$

then G is solvable.*

We make use of the theorem of § 128, and write $r = p_1{}^\alpha q$, where then $p_1 > q$. G contains at least one subgroup H of the order $p_1{}^\alpha$. If we denote by $kp_1 + 1$ the total number of subgroups of order $p_1{}^\alpha$ contained in G, and by $p_1{}^\alpha i$ the order of the maximal subgroup of G which is commutative with H, then $r = p_1{}^\alpha i(kp_1 + 1)$. Since $r = p_1{}^\alpha q$ and $q < p_1$, we must take $k = 0$ and $r = p_1{}^\alpha i$. That is, G is itself commutative with H. By the solution of an auxiliary equation of degree q, with a group of order q, we arrive therefore at a function belonging to the family of H, and the group G reduces to H (§ 232, Theorem X). From Theorem VII the latter group is solvable. Accordingly, if the auxiliary equation is solvable, the group G is solvable also.

The group of the auxiliary equation with the order $q = p_2{}^\beta p_3{}^\gamma \ldots$ admits of the same treatment as G. Its solvability therefore follows

---
* L. Sylow: Math. Ann. V, p. 585.

from that of a new auxiliary equation with a group of order $p_3{}^\gamma p_4{}^\delta \ldots$, and so on.

§ 243. We return to the general investigations of § 241. -
The transition from G to G_1 decomposes the Galois resolvent equation into $\dfrac{r}{r_1} = p_1$ factors. The transition from G_1 to G_2 decomposes each of these previously irreducible factors into $\dfrac{r_1}{r_2} = p_2$ new factors, and so on.

Since $f(x) = 0$ was originally irreducible, but is finally resolved into linear factors, it follows from § 235 that once or oftener a resolution of $f(x)$ or of its already rationally known factors will occur simultaneously with the resolution of the Galois resolvent equation or of its already known rational factors. The number of factors into which $f(x) = 0$ resolves, which is of course greater than 1, must from § 235, be a divisor of the number of factors into which the Galois resolvent equation divides. In the case of solvable equations the latter is always a prime number $p_1, p_2, p_3, \ldots$ Consequently the same is true of $f(x) = 0$. *All prime factors of the degree n of the solvable equation $f(x) = 0$ are factors of composition of the group G, and in fact each factor occurs in the series of composition as often as it occurs in n.*

To avoid a natural error, it must be noted that if in passing from G to G_λ the polynomial $f(x)$ resolves into rational factors one of which is $f'_\lambda(x)$, this factor does not necessarily belong to the group G_λ. It may belong to a family included in that of G_λ. The number of values of $f'_\lambda(x)$ is therefore not necessarily equal to $r:r_\lambda$. It may be a multiple of this quotient. And the product $f'_\lambda(x) \cdot f''_\lambda(x) \ldots$ of all the values of $f'_\lambda(x)$ is not necessarily equal to $f(x)$, but may be a power of this polynomial.

We will now assume that n is not a power of a prime number p, so that n includes among its factors different prime numbers. Then different prime numbers also occur among the factors of composition of the series for G, and consequently (§ 94, Corollary I) G has a principal series
$$G, H, J, K, \ldots M, 1.$$
Suppose that in one of the series of composition belonging to G other groups

3) $$H', H'', \ldots H^{(\lambda)}$$
occur between H and J. Since n includes among its factors at least two different prime numbers, $f(x)$ must resolve into factors at least twice in the passage from a group of the series of composition to the following one. Since the number of the factors of $f(x)$ is the same as the factor of composition, and since the latter is the same for all the intermediate groups 3), the two reductions of $f(x)$ cannot both take place in the *same* transition from a group H of the principal series to the next following group J. It is to be particularly noticed, that all the resolutions of $f(x)$ cannot occur in the transition from the last group M to 1, that is, within the groups

$$M', M'', \ldots M^{(\kappa-1)}, 1,$$

following M in the series of composition. At least one of the resolutions must have happened before M. Suppose, for example, that the first resolution occurs between H' and H''. Then it follows from § 235 that H' is non-primitive in those elements which it connects transitively, and that H'' is intransitive, the systems of intransitivity coinciding with the system of non-transitivity of H'. The same intransitivity then occurs in all the following groups $H''', \ldots H^{(\lambda)}$, and likewise in the next group J of the principal series, which by assumption is different from 1.

Suppose that J distributes the roots in the intransitive systems

$$x'_1, x'_2, \ldots x'_i; \quad x''_1, x''_2 \ldots x''_i; \quad \ldots \quad x_1^{(m)}, x_2^{(m)}, \ldots x_i^{(m)},$$

these systems being taken as small as possible. Then the expression

$$f'_\lambda(x) = (x-x'_1)(x-x'_2)\ldots x-x'_i)$$

becomes a rationally known factor of $f(x)$, which does not contain any smaller rationally known factor. Since from the properties of the groups of the principal series

$$G^{-1} J G = J,$$

all the values of $f'_\lambda(x)$ belong to the same group J. They are therefore all rationally known with $f'_\lambda(x)$. Of the values of $f'_\lambda(x)$ we know already

$$f'_\lambda(x) = (x-x'_1)\ (x-x'_2)\ \ldots (x-x'_i),$$
$$f''_\lambda(x) = (x-x''_1)\ (x-x''_2)\ \ldots (x-x''_i),$$
$$\cdots\cdots\cdots\cdots\cdots\cdots\cdots\cdots\cdots$$
$$f_\lambda^{(m)}(x) = (x-x_1^{(m)})\ (x-x_2^{(m)})\ldots (x-x_i^{(m)}).$$

If there were other values, these must have roots in common with some $f_\lambda^{(a)}(x)$. Then $f_\lambda^{(a)}(x)$ and consequently $f'_\lambda(x)$ would resolve into rational factors. This being contrary to assumption, $f'_\lambda(x)$ has only $m = \dfrac{n}{i}$ values, and is therefore a root of an equation of degree m. If this equation is

4) $\quad \varphi(y) \equiv (y-f'_\lambda)(y-f''_\lambda) \ldots (y-f_\lambda^{(m)}) = 0$,

then $f(x)$ is the result of elimination between 4) and

5) $\quad f'_\lambda(x) = x^i - \psi_1(y')x^{i-1} + \psi_2(y')x^{i-2} - \ldots = 0$,

where

$$\psi_1(y') = x'_1 + x'_2 + x'_3 + \ldots + x'_i,$$
$$\psi_2(y') = x'_1 x'_2 + x'_1 x'_3 + \ldots + x'_{i-1} x'_i,$$
$$\ldots \ldots \ldots \ldots$$

so that $\psi_1, \psi_2, \ldots$ are rationally expressible in terms of f'_λ. Since $f(x)$ is the eliminant of 4) and 5), it follows from § 228 that the group of $f(x) = 0$ is non-primitive.

These conclusions rest wholly on the circumstance that J belongs to the principal series of G, and that accordingly $G^{-1}JG = J$. It is only under this condition that *all* the values of $f'_\lambda(x)$ which occur in the rational domain of $f(x) = 0$ are rationally known. This shows itself very strikingly in an example to be presently considered.

Theorem IX. *If the degree n of an irreducible algebraic equation is divisible by two different prime numbers, then n can always be divided into two factors $n = im$, such that the given equation $f(x) = 0$ resolves into m new ones*

$$f'_\lambda(x) = 0, \ f''_\lambda(x) = 0, \ \ldots \ f_\lambda^{(m)}(x) = 0,$$

which are all of degree i, and the coefficients of which are obtainable from known quantities by the solution of an equation of degree m.[*] *The group of the equation $f(x) = 0$ is non-primitive.*

For the purpose of comparison we consider the solution of the general equation of the fourth degree, to which, since $4 = 2^2$, the preceding results are not applicable. It appears at once that both of the resolutions of the polynomial into linear factors take place in the domain belonging to the last group of the principal series $M, M', M'', \ldots 1$. The series of the equation consists of the following groups:

---

[*] Abel: Oeuvres complètes II, p. 191.

ALGEBRAICALLY SOLVABLE EQUATIONS. 293

1) the symmetric group;
2) the alternating group;
3) $[1, (x_1x_2)(x_3x_4), (x_1x_3)(x_2x_4)(x_1x_4)(x_2x_3)]$;
4) $[1, (x_1x_2)(x_3x_4)]$, 4') $[1, (x_1x_3)(x_2x_4)]$, or 4'') $[1, (x_1x_4)(x_2x_3)]$;
5) the group 1.

The principal series consists of the groups 1), 2), 3), 5). The passage from 3) to 4) and that from 4) to 5) both give the prime factor 2. The group 4) is the first intransitive one. For this $f(x)$ resolves into the two factors $(x-x_1)(x-x_2)$ and $(x-x_3)(x-x_4)$. But since the group 4) does not belong to the principal series, all the six values of $(x-x_1)(x-x_2)$ are not known. If we had chosen the group 4') instead of 4), we should have had the two factors $(x-x_1)(x-x_3)$ and $(x-x_2)(x-x_4)$, and so on. The product of these six values give the third power of $f(x) \equiv (x-x_1)(x-x_2)(x-x_3)(x-x_4)$.

We can therefore, to be sure, resolve $f(x)$ into a product of two factors of the second degree. But the coefficients of every such factor are the roots not of an equation of degree $\frac{4}{2} = 2$, but of an equation of degree 6.

If we consider further the irreducible solvable equations of the sixth degree, it appears that these are of one of two types, according as we eliminate y from

$$x^2 - f_1(y)x + f_2(y) = 0, \quad y^3 - c_1 y^2 + c_2 y - c_3 = 0,$$

or from

$$x^3 - f_1(y)x^2 + f_2(y)x - f_3(y) = 0, \quad y^2 - c_1 y + c_2 = 0.$$

§ 244. The preceding results enable us to limit our consideration to those equations $f(x) = 0$ the degree of which is a power of a prime number p. For otherwise the problem can be simplified by regarding the equation as the result of an elimination. Furthermore we may assume that such a resolution into factors as was considered in the preceding Section does not occur in the case of our present equations of degree p^λ, since otherwise the same simplification would be possible. We assume therefore that the group of the equation is primitive, thus excluding both the above possibilities.

With this assumption we proceed to the investigation of the group. Suppose that the degree of the equation is p^λ and that its principal series of composition is

2) $\qquad G, H, J, K, \ldots M, 1,$

In passing from G through $H, J, \ldots$ to M, no resolution of $f(x)$ into factors can occur. Otherwise we should have the case of the last Section, and G would be non-primitive. The passage from G to M "prepares" the equation $f(x)$ for resolution, but does not as yet resolve $f(x)$ into factors. The λ resolutions of the equation of degree p^λ therefore occurs in passing from the last group of the principal series to identity, that is, in

$$M, M', M'', \ldots M^{\kappa-1}, 1.$$

Accordingly we must have $\kappa \geq \lambda$. The application of § 94, Corollary IV shows that all the substitutions of M are commutative. The equation characterized by the family of M is therefore an Abelian equation of degree p^λ (§ 182). From § 94 there belongs to every transition from one group to the next in the last series the factor of composition p, so that the order of M is equal to p^κ. Again M can be obtained by combining κ groups which have only the identical operation in common, which are similar to each other, and are of order p. Suppose that these are

$$M^{(\kappa-1)}, M_1^{(\kappa-1)}, M_2^{(\kappa-1)}, \ldots M_{\kappa-1}^{(\kappa-1)}.$$

From the above properties it appears that every one of these groups is composed of the powers of a substitution of order p

$$s, s_1, s_2, \ldots s_{\kappa-1},$$

and that on account of the commutativity of the groups (cf. § 95) we must also have

$$s_\alpha^\mu s_\beta^\nu = s_\beta^\nu s_\alpha^\mu \quad (\alpha, \beta = 0, 1, \ldots \kappa-1).$$

Consequently every substitution of M can be expressed by

$$s^\lambda s_1^\mu s_2^\nu \ldots s_{\kappa-1}^\tau,$$

and from the same commutative property

$$(s^\lambda s_1^\mu s_2^\nu \ldots s_{\kappa-1}^\tau)^p = s_\bullet^{\lambda p} s_1^{\mu p} s_2^{\nu p} \ldots s_{\kappa-1}^{\tau p} = 1.$$

Every substitution of the group M is of order p. Our Abelian equation therefore belongs to the category treated in § 186, and its substitutions are there given in the analytic form

$$t \equiv |z_1, z_2, \ldots z_\kappa \quad z_1 + a_1, z_2 + a_2, \ldots z_\kappa + a_\kappa| \quad (\text{mod. } p).$$

The symmetric occurrence of all the indices $z_1, z_2, \ldots z_\kappa$ already shows that in the reduction of M to 1 exactly $\varkappa$ resolutions of the polynomial $f(x)$ will occur, as is also recognized if we write for example

$M' \equiv |z_1, z_2, z_3, \ldots z_\kappa \quad z_1, z_2 + a_2, z_3 + a_3, \ldots z_\kappa + a_\kappa|$ (mod. p),
$M'' \equiv |z_1, z_2, z_3, \ldots z_\kappa \quad z_1, z_2, z_3 + a_3, \ldots z_\kappa + a_\kappa|$ (mod. p),
. . . .

Accordingly $\varkappa = \lambda$, and we have as a first result

Theorem X. *The last group of the principal series of a primitive, solvable equation of degree p^κ consists of the p^κ arithmetic substitutions*

$t \equiv |z_1, z_2, \ldots z_\kappa \quad z_1 + a_1, z_2 + a_2, \ldots z_\kappa + a_\kappa|$ (mod. p),

the roots of the equation being denoted by

$$x_{z_1, z_2, \ldots z_\kappa} \quad (z_\lambda = 0, 1, 2, \ldots p-1).$$

Since G, the group of the equation, is commutative with M, it follows from § 144 that G is a combination of arithmetic and geometric substitutions. We have therefore as a further result

Theorem XI. *The group G of every solvable primitive equation of degree p^κ consists of the group of the arithmetic substitutions of the degree p^κ, combined with geometric substitutions of the same degree*

$u \equiv |z_1, z_2, \ldots z_\kappa \quad a_1 z_1 + b_1 z_2 + \ldots + c_1 z_\kappa, a_2 z_1 + b_2 z_2 + \ldots + c_2 z_\kappa, \ldots|$
(mod. p).

§ 245. Before proceeding further with the general investigation, we consider particularly the cases $\varkappa = 1$ and $\varkappa = 2$, the former of which we have already treated above.

We consider first the solvable, primitive equations of prime degree p. We may omit the term "primitive," since non primitivity is impossible with a prime number of elements.

The group of the most general solvable equation of degree p must then coincide with or be contained in

$G \equiv |z \quad az + \alpha|$ $(a = 1, 2, \ldots p-1; \alpha = 0, 1, \ldots p-1)$ (mod. p).

We prove that the former is the case, by constructing the groups of composition from G to M and showing that all the factors of composition which occur are prime numbers. We divide $p-1$ into its prime factors: $p-1 = q_1 q_2 \ldots$, and construct the subgroup

$$H \equiv |z \ a_1 q_1 z + a_1| \ (a_1 = 1, 2, \ldots \frac{p-1}{q_1}; \ a_1 = 0, 1, \ldots p-1),$$

then the subgroup

$$J \equiv |z \ a_2 q_1 q_2 z + a_2| \ (a_2 = 1, 2, \ldots \frac{p-1}{q_1 q_2}; \ a_2 = 0, 1, \ldots p-1),$$

and so on. Then $H, J, \ldots$ all belong to the principal series of G. Thus we have, for example $G^{-1} J G = J$. For, if we take

$$t \equiv |z \ az + a|,$$

then

$$t^{-1} \equiv \left|z \ \frac{1}{a}(z-a)\right|,$$

$$\left|z \ \frac{1}{a}(z-a)\right| \cdot \left|z \ a_2 q_1 q_2 z + a_2\right| \cdot \left|z \ az + a\right|$$

$$\equiv \left|z \ a_2 q_1 q_2 z + \frac{a_2 q_1 q_2 a - a + a_2}{a}\right|,$$

so that the transformation of a substitution of J with respect to any substitution of G leads to another substitution of J. Evidently the principal series coincides here with the series of composition. The factors of composition $q_1, q_2, \ldots$ are all prime numbers. The proof is then complete.

If a substitution of G leaves two roots x_λ and x_μ unchanged, then it leaves all the roots unchanged. For from $\lambda \equiv a\lambda + a$, $\mu \equiv a\mu + a$ follows necessarily $a \equiv 1$, $a \equiv 0$, and the substitution becomes identical: $1 = |z \ z|$.

If a substitution of G leaves one root x_λ unchanged and if it converts $x_{\lambda+1}$ into x_μ, then every x_ν becomes $x_{(\mu-\lambda)(\nu-\lambda)+\lambda}$. For from $\lambda \equiv a\lambda + a$, $\mu \equiv a(\lambda+1) + a$, follows $a \equiv \mu - \lambda$, $a \equiv \lambda(\lambda - \mu - 1)$, and the substitution is of the form $|z \ (\mu - \lambda)z + \lambda(\lambda - \mu + 1)|$.

If a substitution of G leaves no root unchanged, and if it converts x_λ into x_μ, then every x_ν is converted into $x_{\nu+\mu-\lambda}$. For only in this case is there no solution λ of the congruence $\lambda \equiv a\lambda + a$, when $a \equiv 1$. If $\lambda + 1$ is to become μ, then we must have $\mu = \lambda + a$. This gives $a = \mu - \lambda$, and the substitution is $|z \ z + \mu - \lambda|$.

These are precisely the same results which the earlier algebraic method furnished us.

Theorem XII. *The general solvable equations of prime degree p are those of § 196. Their group is of order $p(p-1)$ and consists of the substitutions of the form*

$$s \equiv |z \quad az+\alpha| \quad (a = 1, 2, \ldots p-1;\ \alpha = 0, 1, \ldots p-1) \quad (\text{mod. } p).$$

Its factors of composition are all prime divisors of $p-1$, each factor ocurring as many times as it occurs in $p-1$, and beside these p itself.

§ 246. We pass to the general solvable primitive equations of degree p^2. As a starting point we have the arithmetic substitutions

$$t \equiv |z_1, z_2 \quad z_1+a_1, z_2+a_2| \quad (\text{mod. } p),$$

which form the last group M of the corresponding principal series. To arrive at the next preceding group, we must determine a substitution s which has the following properties. Its form is

$$s \equiv |z_1, z_2 \quad a_1 z_1 + b_1 z_2, a_2 z_1 + b_2 z_2| \quad (\text{mod. } p),$$

and the lowest power of s which occurs in M, and is therefore of the form t, must have a prime number as exponent. Since now all the powers of s are of the same form as s itself, the required power must be $|z_1, z_2 \quad z_1, z_2| = 1$. That is, the order of the substitution s must be a prime number.

From these and other similar considerations we arrive at the following results,[*] the further demonstration of which we do not enter upon.

Theorem XIII. *The general solvable, primitive equations of degree p^2 are of three different types.*

The first type is characterized by a group of order $2p^2(p-1)^2$, the substitutions of which are generated by the following:

$$\begin{aligned}&|z_1, z_2 \quad z_1+a_1, z_2+a_2| \quad (a_1, a_2 \equiv 0, 1, 2, \ldots p-1),\\ &|z_1, z_2 \quad a_1 z_1, a_2 z_2| \quad\quad\quad (a_1, a_2, \equiv 1, 2, 3, \ldots p-1),\\ &|z_1, z_2 \quad z_2, z_1|.\end{aligned} \quad (\text{mod. } p),$$

The groups belonging to the second type are of order $2\,p^2(p^2-1)$, and their substitutions are generated by the following:

---
[*] C. Jordan: Liouville, Jour. de Math. (2) XIII, pp. 111-135.

$|z_1, z_2 \quad z_1+a_1, z_2+a_2|$ $(a_1, a_2 \equiv 0, 1, 2, \ldots p-1)$,
$|z_1, z_2 \quad az_1+bez_2, bz_1+az_2|$ $(a, b \equiv 0, 1, \ldots p-1;$ but not $a, b \equiv 0)$,
$|z_1, z_2 \quad z_1, -z_2|$,   (mod. p).

where e is any quadratic remainder (mod. p).

The groups of the third type are of order $24\,p^2(p-1)$. The form of their substitutions is different, according as $p \equiv 1$ or $p \equiv 3$ (mod. 4). In the former case the group contains beside the two substitutions

$|z_1, z_2 \quad z_1+a_1, z_2+a_2| \quad (a_1, a_2 \equiv 0, 1, 2, \ldots p-1),$
$|z_1, z_2 \quad az_1, az_2| \quad\quad\quad (a \equiv 1, 2, 3 \ldots p-1),$ (mod. p)

also the following four:

$|z_1, z_2 \quad iz_1, -iz_2|$, $|z_1, z_2 \quad iz_2, iz_1|$,
$|z_1, z_2 \quad z_1-iz_2, z_1+iz_2|$, $|z_1, z_2 \quad z_1+z_2, z_1-z_2|$,

where i is a root of the congruence $i^2 \equiv -1$ (mod. p). If $p \equiv 3$ (mod. 4), the group contains the first two substitutions above, together with the following four:

$|z_1, z_2 \quad z_2, -z_1|$, $|z_1, z_2 \quad sz_1+tz_2, tz_1-sz_2|$,
$|z_1, z_2 \quad -(1+st)z_1+(s-t^2)z_2, (t+s^2)z_1+(st-s-t)z_2|$,
$|z_1, z_2 \quad sz_1+(1+t)z_2, (t-1)z_1-sz_2|$,

where s and t satisfy the congruence $s^2+t^2 \equiv -1$ (mod. p).

For $p=3$ the first and second types, and for $p=5$ the second type are not general. These types are then included as special cases in the third type, which is always general.

§ 247. We return from the preceding special cases to the more general theory.

The same method which we have employed above in the case of p^2 can be applied in general to determine the substitutions of the group L which precedes M in the principal series of composition. L is obtained by adding to the substitutions

$$t \equiv |z_1, z_2, \ldots z_\kappa \quad z_1+a_1, z_2+a_2, \ldots z_\kappa+a_\kappa) \quad (\text{mod. } p)$$

of M a further substitution

$$s \equiv |z_1, z_2, \ldots z_\kappa \quad a_1z_1+b_1z_2+\ldots+c_1z_\kappa, a_2z_1+b_2z_2+\ldots+c_2z_\kappa, \ldots|$$
(mod. p),

where the first power of s to occur among the t's has a prime exponent. Since all the powers of s are of the same form as s itself, any power of s which occurs among the t's must be equal to 1. Consequently s must be of prime order. It is further necessary the group $L = \{t, s\}$ should not become non-primitive.

§ 248. From the form to which the substitutions of G are restricted, we have at once

Theorem XIV. *All the substitutions, except identity, which belong to the group*

$$M \equiv |z_1, z_2, \ldots z_\kappa \quad z_1 + a_1, z_2 + a_2, \ldots z_\kappa + a_\kappa|$$

affect all the elements.

The converse proposition, which was true for $\varkappa = 1$, does not hold in the general case. For the element $x_{z_1, z_2, \ldots z_\kappa}$ is unaffected by

$$s = |z_1, z_2, \ldots z \quad a_1 z_1 + b_1 z_1 + \ldots + c_1 z_\kappa + a_1, \; a_2 z_1 + b_2 z_2 + \ldots + c_2 z_2 + a_2, \ldots|$$

only in case the $\varkappa$ congruences

$$S) \quad \begin{aligned} (a_1 - 1) z_1 + & \quad b_1 z_2 + \ldots + & c_1 z_\kappa + a_1 \equiv 0, \\ a_2 z_1 + (b_2 - 1) z_2 + \ldots + & c_2 z_\kappa + a_2 \equiv 0, \\ \cdot \quad \cdot \quad \cdot \quad \cdot \quad \cdot \quad \cdot \quad \cdot \quad \cdot \quad \cdot \\ a_\kappa z_1 + & \quad b_\kappa z_2 + \ldots + (c_\kappa - 1) z_\kappa + a_\kappa \equiv 0 \end{aligned} \quad (\text{mod. } p)$$

are satisfied. Consequently, as soon as the determinant

$$D = \begin{vmatrix} a_1 - 1 & b_1 & \ldots c_1 \\ a_2 & b_2 - 1 & \ldots c_2 \\ \cdot & \cdot & \cdot \\ a_\kappa & b_\kappa & \ldots c_\kappa - 1 \end{vmatrix} \equiv 0 \quad (\text{mod. } p),$$

the $a_1, a_2, \ldots a_\kappa$ can be so chosen that the congruences $S)$ are not satisfied by any system $z_1, z_2, \ldots z_\kappa$.

We consider now all the substitutions of the group G which leave one element unchanged. Since the distinction between the elements is merely a matter of notation, we may regard $x_{0,0,\ldots 0}$ as the fixed element. Then the substitutions which leave this element unchanged are

$$\Gamma \equiv |z_1, z_2, \ldots z_\kappa \quad a_1 z_1 + b_1 z_2 + \ldots + c_1 z_\kappa, \; a_2 z_1 + b_2 z_2 + \ldots + c_2 z_\kappa \ldots|.$$

If we adjoin $x_{0,0\ldots 0}$ to the equation, the group G reduces to Γ.

Since all the substitutions of G are obtained by appending to those of $\varGamma$ the constants $a_1, a_2, \ldots$ and since the a's can be chosen in $p^\varkappa$ ways, it follows that the adjunction of a single root reduces G to its $(p^\varkappa)^{\text{th}}$ part.

§ 249. We will now consider the possibility that a substitution of G leaves $\varkappa+1$ elements $x_{s_1, s_2, \ldots s_\varkappa}$ unchanged. Then the congruences S) of the preceding Section are satisfied by $\varkappa+1$ systems of values $z_1, z_2, \ldots z_\varkappa$

$$z_1 = \zeta_1^{(\lambda)},\ z_2 = \zeta_2^{(\lambda)},\ \ldots\ z_\varkappa = \zeta_\varkappa^{(\lambda)}, \quad (\lambda = 0, 1, 2, \ldots \varkappa).$$

We will however regard not the coefficients $a, b, \ldots c$; a of the substitution but the values $\zeta_1^{(\lambda)}, \zeta_2^{(\lambda)} \ldots \zeta_\varkappa^{(\lambda)}$ as known, and attempt to determine the substitution from these data. If now the determinant

$$E = \begin{vmatrix} \zeta_1 & \zeta_2 & \ldots \zeta_\varkappa \\ \zeta'_1 & \zeta'_2 & \ldots \zeta'_\varkappa \\ \cdot & \cdot & \cdot \\ \zeta_1^{(\varkappa)} & \zeta_2^{(\varkappa)} & \ldots \zeta_\varkappa^{(\varkappa)} \end{vmatrix}$$

is not $\equiv 0$ (mod. p), then the $\varkappa$ systems $T_1), T_2), \ldots T_\varkappa)$ each of $\varkappa+1$ congruences with the unknown quantities $a, b, \ldots c$; a

$T_1)\ (a_1-1)\zeta_1^{(\lambda)} + \quad b_1\zeta_2^{(\lambda)} + \ldots + c_1\zeta_\varkappa^{(\lambda)} + a_1 \equiv 0,$

$T_2)\ a_2\zeta_1^{(\lambda)} \quad + (b_2-1)\zeta_2^{(\lambda)} + \ldots + c_2\zeta_\varkappa^{(\lambda)} + a_2 \equiv 0,$ $\quad (\lambda = 0, 1, \ldots \varkappa)$

$\ldots\ldots\ldots\ldots\ldots\ldots\ldots\ldots\ldots\ldots\ldots$

$T_\varkappa)\ a_\varkappa\zeta_1^{(\lambda)} \quad + b_\varkappa\zeta_2^{(\lambda)} \quad + \ldots + (c_\varkappa-1)\zeta_\varkappa^{(\lambda)} + a_\varkappa \equiv 0,$

have only one solution each, viz:

$L_1)\qquad a_1 = 1,\ b_1 = 0,\ \ldots c_1 = 0;\ a_1 = 0,$

$L_2)\qquad a_2 = 0,\ b_2 = 1,\ \ldots c_2 = 0;\ a_2 = 0,$

$\ldots\ldots\ldots\ldots\ldots\ldots\ldots\ldots$

$L_\varkappa)\qquad a_\varkappa = 0,\ b_\varkappa = 0,\ \ldots c_\varkappa = 1;\ a_\varkappa = 0,$

and these solutions furnish together the identical substitution 1.

We designate now a system of $\varkappa+1$ roots of an equation for which $E \equiv 0$ (mod. p) as a system of conjugate roots.

We have then

Theorem XV. *If a substitution of a primitive solvable group of degree $p^\varkappa$ leaves unchanged $\varkappa+1$ roots which do not form a conjugate system, the substitution reduces to identity.*

If therefore we adjoin $x+1$ such roots to the equation, the group G reduces to those substitutions which leave $x+1$ roots unchanged, *i. e.*, to the identical substitution. The equation is then solved.

Theorem XVI. *All the roots of a solvable primitive equation of degree p^κ can be rationally expressed in terms of any $x+1$ among them, provided these do not form a conjugate system.*

If we choose the notation so that one of the $x+1$ roots is $x_{0,0,\ldots 0}$, the determinant becomes

$$\pm E = \begin{vmatrix} \zeta'_1 & \zeta'_2 & \ldots & \zeta'_\kappa \\ \zeta''_1 & \zeta''_2 & \ldots & \zeta''_\kappa \\ \cdot & \cdot & & \cdot \\ \zeta^{(\kappa)}_1 & \zeta^{(\kappa)}_2 & \ldots & \zeta^{(\kappa)}_\kappa \end{vmatrix}.$$

If the roots are not to form a conjugate system, then $E \not\equiv 0$ (mod. p). The number r of systems of roots which satisfy this condition is determined in § 146. We found

$$r = (p^\kappa - 1)(p^\kappa - p)(p^\kappa - p^2) \ldots (p^\kappa - p^{\kappa-1}).$$

Theorem XVII. *For every root $x_{z_1, z_2, \ldots z_\kappa}$ we can determine*

$$\frac{(p^\kappa - 1)(p^\kappa - p) \ldots (p^\kappa - p^{\kappa-1})}{1, 2, \ldots \kappa}$$

systems of x roots each such that these $x+1$ roots do not form a conjugate system, so that all the other roots can be rationally expressed in terms of them. The system composed of the $x+1$ roots

$$x_{0,0,0,\ldots}, \quad x_{1,0,0\ldots 0}, \quad x_{0,1,0\ldots 0}, \ldots x_{0,0,0,\ldots 1}$$

is appropriate for the expression of all the roots.

These results throw a new light on our earlier investigations in regard to triad equations, in particular on the solution of the Hessian equation of the ninth degree (*cf.* §§ 203–6). It is plain that we can construct in the same way quadruple equations of degree p^3, and so on.

www.ingramcontent.com/pod-product-compliance
Lightning Source LLC
Chambersburg PA
CBHW022045230426
43672CB00008B/1077